A TEXTBOOK OF PROBABILITY AND STATISTICS

AF798014

A TEXTBOOK OF
PROBABILITY AND STATISTICS

(As Per JNTU Syllabus)

By

THIRUPATHI RAO PILLALAMARRY
M.Sc., M.Phil
Associate Professor, Head Department of Mathematics
Sai Spurthi Institute of Technology,
B. Gangaram, Sathupally, Khammam
(A.P.)

UNIVERSITY SCIENCE PRESS
(An Imprint of Laxmi Publications Pvt. Ltd.)
An ISO 9001:2015 Company

BENGALURU • CHENNAI • GUWAHATI • HYDERABAD
KOCHI • KOLKATA • LUCKNOW • RANCHI
NEW DELHI

Copyright © by Laxmi Publications (P) Ltd.

All rights reserved including those of translation into other languages. In accordance with the Copyright (Amendment) Act, 2012, no part of this publication may be reproduced, stored in a retrieval system, translated into any other language or transmitted in any form or by any means, electronic, mechanical, photocopying, recording or otherwise. Any such act or scanning, uploading, and or electronic sharing of any part of this book without the permission of the publisher constitutes unlawful piracy and theft of the copyright holder's intellectual property. If you would like to use material from the book (other than for review purposes), prior written permission must be obtained from the publishers.

Printed and bound in India

First Edition : 2011, *Reprint* : 2013, *Edition* : 2025

ISBN : 978-93-80856-96-4

Limits of Liability/Disclaimer of Warranty: The publisher and the author make no representation or warranties with respect to the accuracy or completeness of the contents of this work and specifically disclaim all warranties. The advice, strategies, and activities contained herein may not be suitable for every situation. In performing activities adult supervision must be sought. Likewise, common sense and care are essential to the conduct of any and all activities, whether described in this book or otherwise. Neither the publisher nor the author shall be liable or assumes any responsibility for any injuries or damages arising from here. The fact that an organization or Website if referred to in this work as a citation and/or a potential source of further information does not mean that the author or the publisher endorses the information the organization or Website may provide or recommendations it may make. Further, readers must be aware that the Internet Websites listed in this work may have changed or disappeared between when this work was written and when it is read.

All trademarks, logos or any other mark such as USP, Amanda, Golden Bells, Firewall Media, A-Cube, Triunity, Mathstick, Engpot, Laxmi appearing in this work are trademarks and intellectual property owned by or licensed to Laxmi Publications, its subsidiaries or affiliates. Notwithstanding this disclaimer, all other names and marks mentioned in this work are the trade names, trademarks or service marks of their respective owners.

Branches

✆	Bengaluru	080-26 75 69 30
✆	Chennai	044-24 34 47 26
✆	Guwahati	0361-254 36 69
✆	Hyderabad	040-27 55 53 83
✆	Kochi	0484-405 13 03
✆	Kolkata	033-40 04 77 79
✆	Lucknow	0522-430 36 13
✆	Ranchi	0651-224 24 64

Published in India by

UNIVERSITY SCIENCE PRESS

(An Imprint of Laxmi Publications Pvt. Ltd.)

An ISO 9001:2015 Company

113, GOLDEN HOUSE, GURUDWARA ROAD, DARYAGANJ,

NEW DELHI - 110002, INDIA

Corporate Office : B-33, Sector-5, Noida,

Gautambuddha Nagar (U.P.) - 201 301, INDIA

Telephone : +91-120-44 46 900

www.laxmipublications.com

info@laxmipublications.com

CONTENTS

PREFACE

The majority of engineering and technical students are obliged to study probability and statistics easily available and hand-outs prepared by their teacher. Also the teachers are faced with real problems while selecting suitable problems with technical bent. The author has all long been conscious of this want while of this want while teaching the subject to a large number of students in Computer Science, Mechanical, Civil, Chemical Engineering and Information Technology. This book has therefore, been written with the sole aim of providing a suitable textbook covering all topics to the Engineering courses.

To achieve this, more than the usual number of solved examples together with carefully designed diagrams and properly graded problems have been given. Many of the examples and problems have been selected from recent papers of university, so questions usually set there in. As far as possible it has been planned to make each chapter an independent unit so that the various topics could be studied in order.

The author is indebted to many authors, the faculty members, family members and other friends for the advice and help in preparing this book.

Suggestions for the improvement of the textbook as well as intimations of misprints will be most gratefully received.

—AUTHOR

ACKNOWLEDGEMENT

I would like to sincerely thank Dr. Bandi Pardhasarathi Reddy, Chairman SSIT, Sri M.V. Krishna Reddy Secretary, SSIT, Sri Bandi Rama Chandra Reddy, Principal, Govt. Junior College, Penuballi, Prof. T. Venu Gopal, Principal, SSIT for their whole-hearted exhort and encouragement in writing this book.

I would be failing in my duty if I miss to place my indebtedness to my colleagues and friends Mr. S.K. Meera Saheb, Mr. B. Rama Krishna, B. Sathyanarayna Reddy, Miss G. Jhansi and Mrs. M. Anitha, on record, for their consistent support and help.

My special thanks to Mr. K.N.P.S. Prasad SSIT but for whose unrelented support the emergence of this book would have been a mirage.

And my loving thanks to my children Harshini and Prasanth for their sensible understanding in not disturbing me while I slogged on the task.

My infinite gratitude to my wife Lalitha whose patience, support and service can never be ignored in giving shape to this book.

—AUTHOR

SAI SPURTHI INSTITUTE OF TECHNOLOGY

B. Gangaram (Po), Sathupally (Mandal), Khammam Dist.–507 303 (A.P.)

Ph. No. 08761-288278, 288179, 288180, Fax: 288249 www.saispurthi.ac.in

Mr. M. V. Krishna Reddy, Secretary

FOREWORD

This book **"A TEXTBOOK OF PROBABILITY AND STATISTICS"** is written to cater to the needs of Engineering students in learning the technical subtleties of Probability and Statistics in tune with the revised curriculum of all professional institutes and universities.

Emphasis on Probability and Statistics and of its great variety of applications in diverse fields of science and technology has been scholarly manifested by the author, keeping in view the intricacies involved in Quiz Questions thus making the book useful for other competitive exams like Gate also. The Exercises and Quiz questions are given at the end of each Unit for practice. This book is revised according to the JNTU new syllabus for the academic year 2008-09.

The author has evinced keen interest in explaining the concept in detail which leaves me come out with a strong opinion that this book is exceptionally useful to all the B.Tech students and teachers as well.

I whole-heartedly congratulate Mr. Thirupathirao Pillalamarry, the Associate Professor of Mathematics, SSIT, Sathupalli, Khammam, the author of this book , for his painstaking efforts in bringing out his versatility and mathematical prowess through this book for the benefit of student community.

(M.V. KRISHNA REDDY)

Dedicated to

My beloved Parents

Sri. Kailasam Pillalamarry
Smt. Subadra Devi
&
My Loving Wife

Lalitha Pillalamarry

Chapter 1

PROBABILITY

1.0 SAMPLE SPACE AND EVENTS

Trial and Events

Let an experiment be repeated under essentially identical conditions, does not give unique results, but may be any one of the several possible outcomes. Then, the experiment is called a trial and the possible outcomes are known as events or cases.

Examples:

1. Tossing of a coin is a trial and getting head or tail is an event.
2. Throwing a die is a trial and getting 1 or 2 or 3 or 4 or 5 or 6 is event.
3. Drawing a card from a pack of 52 cards is a trial and getting a king or a queen or an ace is an event.

Mutually Exclusive Events

Events are said to be mutually exclusive if the happening of anyone of them is ruled out the happening of all the others, *i.e.* if no two or more of them can happen simultaneously in the same trial. If two events A and B are mutually exclusive then, $A \cap B = \phi$ *i.e.,* A and B are disjoint.

Examples:

1. In tossing a coin, the events head and tail are mutually exclusive.
2. In throwing a die, all the six faces are mutually exclusive.

Equally Likely Events

Out comes of a trial are said to be equally likely if there is no reason to expect anyone in preference to any other.

Examples:

1. In tossing a coin, head and tail are equally likely.
2. In throwing a die, all the 6 faces are equally likely.

3. When a card is drawn from a pack, any card may appear in the draw so that the 52 different events are equally likely.

Exhaustive Events

Total number of all possible outcomes of a random experiment is known as exhaustive events.

Examples:

1. In tossing of a coin, there are two exhaustive events.
2. In throwing a die, there are six exhaustive events.
3. In drawing two cards from a pack of 52 cards, the exhaustive number of events is $^{52}C_2$. Since 2 cards can be drawn from 52 cards in $^{52}C_2$ ways.
4. In throwing two dice, the exhaustive number of events is $6^2 = 36$.

Note:

1. In tossing a coin 'n' times or n coins are tossed at a time, the possible outcomes are 2^n.
2. A die is thrown 'n' times or n dice are thrown at a time, the possible outcomes are 6^n.

Favourable Events

The number of cases favourable to an event in a trial is the number of outcomes which entail happening of the event.

Examples:

1. In tossing a coin two times, the favourable events of getting two heads is 1.
2. In a throw of two dice, the number of favourable events to getting same numbers on two dies is 6, *i.e.* (1, 1) (2, 2) (3, 3) (4, 4) (5, 5) (6, 6).
3. In drawing a card from a pack of 52 cards the number of events favourable to drawing a king is 4, for drawing a red card is 26.

Sample Space

The set of all possible outcomes in a random experiment is known as sample space and is denoted by S. Each element in a sample space is called sample point.

Example:

Two coins are tossed; the possible outcomes are HH, HT, TH, TT

$\therefore$ Sample space S = {HH, HT, TH, TT}

Complementary Events

Two events of a sample space whose intersection is ϕ and whose union is entire sample space S are called complementary events. Thus if E is an event of sample space S its complement is $\bar{E}$ or E^c and $E \cap \bar{E} = \phi$, $E \cup \bar{E} = S$.

1.1 MATHEMATICAL OR CLASSICAL PROBABILITY

If a random experiment results in 'n' exhaustive, mutually exclusive and equally likely events, out of which m are favourable to the happening of an event E, then the probability 'p' of happening of E, usually denoted by P(E), is given by:

$$p = P(E) = \frac{\text{Number of favourable events}}{\text{Total number of events}} = \frac{m}{n} \qquad ...(1)$$

Note:

1. The non-happening of the event E called the complementary event of E and is denoted by $\bar{E}$ or E^c. The number of events favourable to $\bar{E}$, *i.e.*, non-happening of E is $(n - m)$. Then the probability q that E will not happen is given by

$$q = P(\bar{E}) = \frac{n-m}{n}$$

$$= 1 - \frac{m}{n}$$

$$= 1 - p$$

$$\Rightarrow \quad p + q = 1 \qquad ...(2)$$

$$\therefore \quad q = P(\bar{E}) = 1 - P(E)$$

$$\Rightarrow \quad P(E) + P(\bar{E}) = 1 \qquad ...(3)$$

2. Some times (1)says that "the odds in favour of E are $m : (n - m)$ or the odds against E are $(n - m) : m$"
3. Probability 'p' of the happening of an event is also known as the probability of success and the probability 'q' of the non-happening of the event as the probability of failure.
4. If P(E) = 1, E is called certain event and if P(E) = 0, E is called an impossible event.
5. $P(\phi) = 0$.

1.2 STATISTICAL OR EMPIRICAL PROBABILITY

If an experiment is repeated under identical conditions, then the limiting value of the ratio of the number of times the event happen to the total number of trials, as the number of trials become infinitely large, is called the probability of happening of the event.

Symbolically, if in 'n' trials, an event E happens m times, then the probability of happening of E is given by

$$p = P(E) = \underset{n\to\infty}{\text{Lt}} \frac{m}{n}.$$

1.3 SOLVED PROBLEMS

Problem 1: A bag contains 3 red, 6 white and 7 blue balls. What is the probability that two balls drawn are white and blue?

Solution:

Total number of balls = 16

Out of 16 balls 2 balls can be drawn in $^{16}C_2$ ways

$\therefore$ Exhaustive number of events = 120.

Out of 6 white balls 1 can be drawn in $^{6}C_1$ ways.

Out of 7 blue balls 1 can be drawn in $^{7}C_1$ ways.

$\therefore$ Favourable number of events = $^{6}C_1 \times {}^{7}C_1 = 42$.

$$\therefore \text{ Required probability} = \frac{42}{120} = \frac{7}{20}$$

Problem 2: Two cards are drawn at random from well shuffled pack of 52 cards. Show that the chance of drawing two aces is 1/221.

Solution:

Out of 52 cards 2 can be drawn in $^{52}C_2$ ways.

$\therefore$ Total number of events = $^{52}C_2 = 1326$.

In a pack there are 4 aces and 2 can be drawn in $^{4}C_2 = 6$ ways.

$$\therefore \text{ Required probability} = \frac{6}{1326} = \frac{1}{221}.$$

Problem 3: Among the digits 1, 2, 3, 4, 5, at first one is chosen and then a second solution is made among the remaining four digits. Assuming that all twenty possible out comes have equal probabilities, find the probabilities that an odd digit will be selected

(*i*) The first time (*ii*) The second time (*iii*) Both the times.

Solution:

Total number of events = 4 × 5 = 20

(*i*) There are 12 events in which the first digit drawn is odd.

i.e., (1, 2), (1, 3), (1, 4), (1, 5), (3, 1), (3, 2), (3, 4), (3, 5), (5, 1), (5, 2), (5, 3), (5, 4)

$$\therefore \text{ The probability that the first digit drawn is odd} = \frac{12}{20} = \frac{3}{5}$$

(*ii*) Also there are 12 events in which the second digit drawn is odd.

i.e., (2, 1), (3, 1), (4, 1), (5, 1), (1, 3), (2, 3), (4, 3), (5, 3), (1, 5), (2, 5), (3, 5), (4, 5).

$$\therefore \text{ The probability that the second digit drawn is odd} = \frac{12}{20} = \frac{3}{5}$$

(*iii*) There are 6 events in which both the digits drawn are odd.

i.e., (1, 3), (1, 5), (3, 1), (3, 5), (5, 1), (5, 3).

$$\therefore \text{ The probability that both the digits drawn are odd} = \frac{6}{20} = \frac{3}{10}$$

Problem 4: If a test consists of a 12 true, false questions in how many different ways can a student make the test paper with one answer to each question. **(April 2003)**

Solution:

Number of question = 12

Each question can be answered in two ways (*i.e.*, True/ False)

2 questions can be answered in $2 \times 2 = 2^2$ ways

3 questions can be answer in $2 \times 2 \times 2 = 2^3$ ways .

$\therefore$ 12 questions can be answered in 2^{12} ways.

Problem 5: In how many different ways can one make a first second, third or fourth choice among 12 firm leasing construction equipments? **(April 2003)**

Solution:

The first choice can be made in 12 ways.

The second choice can be made in 11 ways.

The third choice can be made in 10 ways.

The fourth choice can be made in 9 ways.

$\therefore$ The total number of different ways $= 12 \times 11 \times 10 \times 9$

$= 11880$ ways.

Problem 6: Out of $(2n + 1)$ tickets numbered, consecutively three tickets are drawn at random. Find the probability that the numbers on them are in Arithmetical progression. **(Nov 2003]**

Solution:

Since out of $(2n + 1)$ tickets 3 can be drawn in ${}^{(2n+1)}C_3$ ways.

$$\therefore \quad \text{Total number of events} = \frac{(2n+1)\,2n\,(2n-1)}{3!}$$

$$= \frac{n(4n^2-1)}{3}$$

To find the number of favourable events we have to enumerate all the events in which the numbers on the drawn tickets are in Arithmetic progression with common difference.

If $d = 1$, the possible events are as follows:

1, 2, 3

2, 3, 4

....

$2n - 1$, $2n$, $2n + 1$ *i.e.*, $(2n - 1)$ events

If $d = 2$, the possible events are as follows:

1, 3, 5

2, 4, 6

....

$2n-3$, $2n-1$, $2n+1$ *i.e.*, $(2n-3)$ events.

....

....

If $d = n-1$, the possible events are

1,	n,	$2n-1$	
2,	$n+1$,	$2n$	
3,	$n+2$,	$2n+1$	*i.e.*, 3 events

If $d = n$, there is only one event

1, $n+1$, $2n+1$

Hence total number of favourable events = $(2n-1)+(2n-3)+\ldots\ldots\ldots.+5+3+1$

$= 1+3+5+\ldots\ldots\ldots(2n-3)+(2n-1)$

which is in arithmetic progression with $a = 1$, $d = 2$

$$\therefore \text{Number of favourable events} = \frac{n}{2}(1+2n-1)$$

$$= n^2$$

$$\therefore \text{Required probability} = \frac{3n^2}{n(4n^2-1)}$$

$$= \frac{3n}{4n^2-1}$$

Problem 7: From 25 tickets marked with the first 25 numerals, one is drawn at random. Find the chance that (*i*) It is multiple of 5 or 7, (*ii*) It is multiple of 3 or 7

Solution:

Total number of events = 25.

(*i*) Numbers which are multiple of 5 are 5, 10, 15, 20, 25, *i.e.*, 5

Numbers which are multiple of 7 are 7, 14, 21, *i.e.* 3

Hence number of favourable events = 5 + 3 = 8

$$\therefore \text{Required probability} = \frac{8}{25}$$

(*ii*) Numbers which are multiple of 3 are 3, 6, 9, 12, 15, 18, 21, 24 *i.e.*, 8

Hence favourable number of events = 8 + 3 − 1 = 10

$$\therefore \text{Required probability} = \frac{10}{25} = \frac{2}{5}$$

Problem 8: In a sample of 446 cars, stopped at a road block, only 67 of the drivers had their seat belts fastened. Estimate the probability that a driver stopped on the road, will have his or her seat belt fastened. **(Nov 2003)**

Solution:

Total number of events = 446

Favourable number of events = 67

Required probability = $\frac{67}{446}$ = 0.15

∴ 0.15 will have his or her seat belt fastened.

Problem 9: What is the chance that a leap year selected at random will contain 53 Sundays?

Solution:

In a leap year there are 366 days, *i.e.*, 52 complete weeks and 2 consecutive extra days.

The following are the possible combinations of two extra days.

(*i*) Sunday, Monday
(*ii*) Monday, Tuesday
(*iii*) Tuesday, Wednesday
(*iv*) Wednesday, Thursday
(*v*) Thursday, Friday
(*vi*) Friday, Saturday
(*vii*) Saturday, Sunday

Out of 7 possibilities 2 are favourable to this event.

∴ Required probability = $\frac{2}{7}$.

Problem 10: Two cards are selected at random from 10 cards numbered 1 to 10. Find the probability that the sum is odd if

(*i*) Two cards are drawn.
(*ii*) Two cards are drawn one after the other without replacement.
(*iii*) Two cards are drawn one after the other with replacement. **(May 2003)**

Solution:

(*i*) Out of 10 cards 2 can be drawn in $^{10}C_2$ ways.

Total number of events = 45

There are 25 events which the sum is odd.

i.e., (1, 2), (1, 4), (1, 6), (1, 8), (1, 10)
(3, 2), (3, 4), (3, 6), (3, 8), (3, 10)
....
....
(9, 2), (9, 4), (9, 6), (9, 10)

∴ Required probability = $\frac{25}{40} = \frac{5}{9}$.

(*ii*) The number of ways of selecting two cards one after the other without replacement is
$10 \times 9 = 90$.

There are 50 events which is the sum is odd.

i.e., (1, 2), (1, 4), (1, 6), (1, 8), (1, 10)
(2, 1), (2, 3), (2, 5), (2, 7), (2, 9)

$\therefore$ Required probability = $\frac{50}{90} = \frac{5}{9}$

(*iii*) The number of ways selecting 2 cards one after the other with replacement is

$$10 \times 10 = 100.$$

There are 50 events which the sum is odd.

$\therefore$ Required probability = $\frac{50}{100} = \frac{1}{2}$

Problem 11: A box contains 'n' tickets marked 1 through n. Two tickets are drawn without replacement. Determine the probability that the numbers on the tickets are consecutive integers.

(Nov. 2004, April 2007)

Solution:

A box contain 'n' tickets is numbered 1 to n from 'n' tickets 2 can be drawn in ${}^nC_2$ ways.

$\therefore$ Total number of events = $\frac{n(n-1)}{2}$

Consecutive integers without replacement are (1, 2), (2, 3), (3, 4), (4, 5) ......(n – 1, n).

$\therefore$ Favourable events $= n - 1$

$\therefore$ Required probability $= \frac{n-1}{n(n-1)} \times 2 = \frac{2}{n}$

Problem 12: A ten digit number is formed using the digits from 0–9. Every digit being used only once. Find the probability that the number is divisible by 4. **(April 2004)**

Solution:

The total number of ways in which the ten digits 0–9 can be arranged among themselves is 10!

Out of these number of arrangements which begin with 0 is 9!

Hence the total number of ten digits numbers that can be formed from the digits 0–9 is 10! – 9! = 3265920.

The number formed will be divisible by 4 if the number formed by two digits on extreme right is divisible by 4.

Such number are

04, 08, 12, 16, 20, 24, 28, 32, 36, 40, 48, 52, 56, 60, 64, 68, 72, 76, 80, 84, 92, 96.

The numbers of ten digit numbers ending with 04 are 8!

08, 20, 40, 60, 80 → total 6 numbers.

And other ten digit number ending with 12 are 8! – 7!, total 10 numbers.

Hence total number of 10 digits numbers divisible by 4 are $= 6 \times 8! + 16\,(8! - 7!)$

$$= 806400$$

Hence required probability $= \frac{806400}{3265920}$

Problem 13: Find the probability of getting a sum of 10 if we throw two dice.

(May 2007, Nov. 2007)

Solution:

When two dice are thrown the total number of events = 36

The sum is ten with two dice are thrown as follows (4, 6), (5, 5), (6, 4)

∴ Favourable number of evenıs = 3

∴ Probability of getting sum ten with two dice $= \frac{3}{36} = \frac{1}{12}$

Problem 14: A bag contains 5 red balls, 8 balls and 11 white balls. Three balls are drawn together from the box. Find the probability that (*i*) one is red, one is blue and one is white (*ii*) Two whites and one red (*iii*) Three white.

Solution:

Number of exhaustive events ${}^{24}C_3 = 2024$

(*i*) Number of possible events ${}^{5}C_1 \times {}^{8}C_1 \times {}^{11}C_1 = 440$

∴ Required Probability $= \frac{\text{No. of possible events}}{\text{No. of exhaustive events}}$

$= \frac{440}{2024} = \frac{55}{253}$

(*ii*) Number of possible events $= {}^{11}C_2 \times {}^{5}C_1 = 275$

∴ Required probability $= \frac{275}{2024} = \frac{25}{184}$

(*iii*) Number of possible events $= {}^{11}C_3 = 165$

∴ Required probability $= \frac{165}{2024} = \frac{15}{184}$

1.4 AXIOMS OF PROBABILITY

The probability is a function defined on a sample space, satisfying the following axioms.

1. For any event E is associated a real number between 0 to 1 called the probability of that event and is denoted by P(E) thus $0 \leq P(E) \leq 1$.
2. The sum of the all probabilities of all simple events constituting the sample space is 1. Thus P(S) = 1.
3. If A and B are mutually exclusive events in S then $P(A \cup B) = P(A) + P(B)$.

   i.e., Probability of sum is the sum of the probabilities.

1.5 SOME ELEMENTARY THEOREMS

Theorem 1: If A is any event in S, then $P(\bar{A}) = 1 - P(A)$.

Proof:

A and $\bar{A}$ are disjoint events

$$A \cup \bar{A} = S$$
$$P(A \cup \bar{A}) = P(S)$$
$$P(A) + P(\bar{A}) = 1$$
$$P(\bar{A}) = 1 - P(A)$$

Theorem 2: For any two events A and B, then

(*i*) $P(\bar{A} \cap B) = P(B) - P(A \cap B)$

(*ii*) $P(A \cap \bar{B}) = P(A) - P(A \cap B)$

Proof:

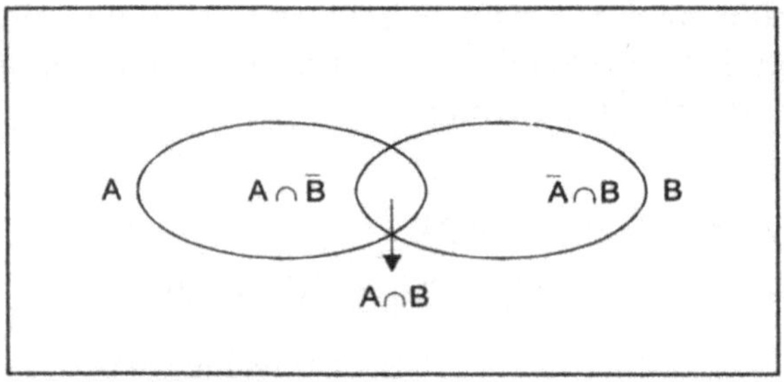

(*i*) $\bar{A} \cap B$ and $A \cap B$ are disjoint events and

$$(\bar{A} \cap B) \cup (A \cap B) = B$$
$$P[(\bar{A} \cap B) \cup (A \cap B)] = P(B)$$
$$P(\bar{A} \cap B) + P(A \cap B) = P(B)$$
$$P(\bar{A} \cap B) = P(B) - P(A \cap B)$$

(*ii*) $A \cap B$ and $A \cap B$ are disjoint events and

$$(A \cap \bar{B}) \cup (A \cap B) = A$$
$$P[(A \cap \bar{B}) \cup (A \cap B)] = P(A)$$
$$P(A \cap \bar{B}) + P(A \cap B) = P(A)$$
$$P(A \cap \bar{B}) = P(A) - P(A \cap B)$$

Theorem 3: If $B \subset A$, then (*i*) $P(A \cap \bar{B}) = P(A) - P(B)$ (*ii*) $P(B) \leq P(A)$

Proof:

When BIA

(*i*) B and $A \cap \bar{B}$ are disjoint events and their union is A

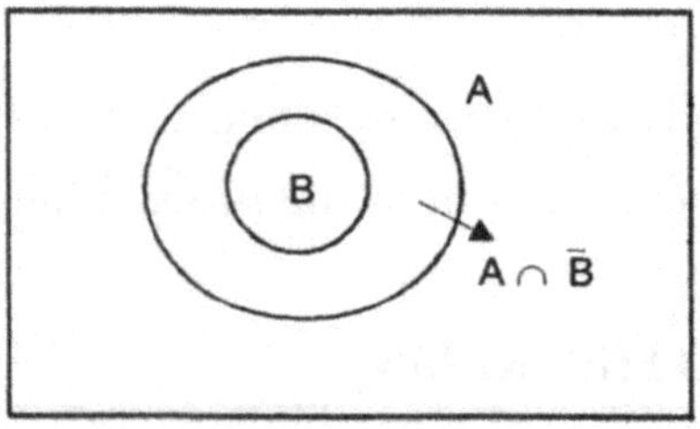

$$B \cup (A \cap \bar{B}) = A$$
$$P[(B \cup (A \cap \bar{B})] = P(A)$$

$$P(B) + P(A \cap \bar{B}) = P(A)$$

$$P(A \cap \bar{B}) = P(A) - P(B).$$

(*ii*) $P(A \cap \bar{B}) \geq 0 \Rightarrow P(A) - P(B) \geq 0$

$$\Rightarrow P(A) \geq P(B)$$

$$\Rightarrow P(B) \leq P(A)$$

Addition Theorem: If A and B are any two events of a sample space S and are not disjoint then

$P(A \cup B) = P(A) + P(B) - P(A \cap B)$ **(Jan 2003, 1999/S, May 2006, April 2007)**

Proof:

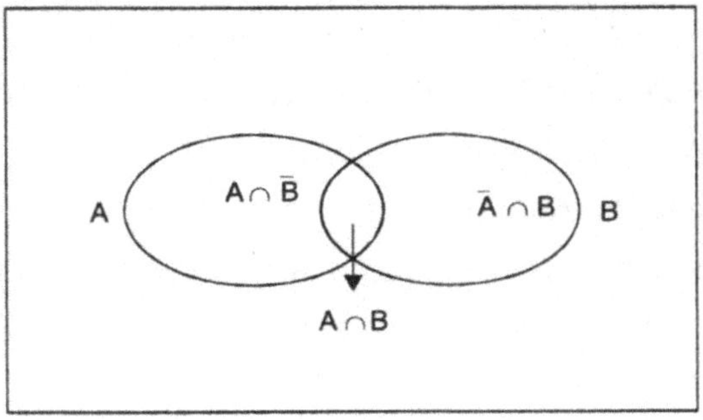

We have
$$A \cup B = A \cup (\bar{A} \cap B)$$

$$\begin{aligned} P(A \cup B) &= P[A \cup (\bar{A} \cap B)] \\ &= P(A) + P(\bar{A} \cap B) \\ &= P(A) + [P(\bar{A} \cap B) + P(A \cap B)] - P(A \cap B) \\ &= P(A) + P[(\bar{A} \cap B) \cup (A \cap B)] - P(A \cap B) \\ &= P(A) + P(B) - P(A \cap B) \end{aligned}$$

Theorem 4: If A, B, C are any three events, show that

$P(A \cup B \cup C) = P(A) + P(B) + P(C) - P(A \cap B) - P(B \cap C) - P(C \cap A) + P(A \cap B \cap C)$.

Proof:

$$\begin{aligned} P(A \cup B \cup C) &= P(D \cup C), \text{ where } D = A \cup B \\ &= P(D) + P(C) - P(D \cap C) \\ &= P(A \cup B) + P(C) - P[(A \cup B) \cap C] \\ &= P(A) + P(B) - P(A \cap B) + P(C) - P[(A \cap C) \cup (B \cap C)] \\ &= P(A) + P(B) + P(C) - P(A \cap B) - \{P(A \cap C) + P(B \cap C) \\ &\qquad - P(A \cap B \cap C)\} \\ &= P(A) + P(B) + P(C) - P(A \cap B) - P(B \cap C) - P(C \cap A) + \\ &\qquad P(A \cap B \cap C) \end{aligned}$$

Theorem 5: If A is an event in the finite sample space then prove that P(A) equals the sum of the probabilities of the individual outcomes comprising A.

Proof:

Let $A_1, A_2, A_3, \ldots\ldots\ldots\ldots, A_n$ be 'n' outcomes comprising A.

$\therefore$ $A = A_1 \cup A_2 \cup A_3 \cup .... \cup A_n$

Since A_i ($i = 1, 2, 3, ..., n$) are individual outcomes and they are mutually exclusive.

$\therefore$ $A_i \cap A_j = \phi, \qquad i \neq j$

$\therefore$ $P(A) = P(A_1 \cup A_2 \cup A_3 \cup .... \cup A_n)$

$= P(A_1) + P(A_2) + ... + P(A_n)$

***Theorem 6:* Boole's Inequality:** For n events $A_1, A_2, A_3, ..., A_n$, we have

$$(a)\ P\left(\bigcap_{i=1}^{n} A_i\right) \geq \sum_{i=1}^{n} P(A_i) - (n-1) \quad (b)\ P\left(\bigcup_{i=1}^{n} A_i\right) \leq \sum_{i=1}^{n} P(A_i)$$

Proof:

(*a*) $P(A_1 \cup A_2) = P(A_1) + P(A_2) - P(A_1 \cap A_2) \leq 1$

$\Rightarrow P(A_1 \cap A_2) \geq P(A_1) + P(A_2) - 1$

Hence (*a*) is true for $n = 2$

Let us now suppose that (*a*) is true for $n = r$ (say), such that

$$P\left(\bigcap_{i=1}^{r} A_i\right) \geq \sum_{i=1}^{r} P(A_i) - (r-1)$$

Then $$P\left(\bigcap_{i=1}^{r+1} A_i\right) = P\left(\bigcap_{i=1}^{r} A_i \cap A_{r+1}\right)$$

$$= P\left(\bigcap_{i=1}^{r} A_i\right) + P(A_{r+1}) - 1$$

$$\geq \sum_{i=1}^{r} P(A_i) - (r-1) + P(A_{r+1}) - 1$$

$$\Rightarrow P\left(\bigcap_{i=1}^{r+1} A_i\right) \geq \sum_{i=1}^{r} P(A_i) - r \Rightarrow (a) \text{ is true for } n = r + 1 \text{ also.}$$

The result now follows by the principle of mathematical induction.

(*b*)Applying the inequality (*a*) to the event $\bar{A}_1, \bar{A}_2, \bar{A}_3, ..., \bar{A}_n$, we get

$P(\bar{A}_1 \cap \bar{A}_2 \cap \bar{A}_3 \cap .... \cap \bar{A}_n) \geq \{P(\bar{A}_1) + P(\bar{A}_2) + P(\bar{A}_3) + ... + P(\bar{A}_n)\} - (n-1)$

$= \{1 - P(A_1)\} + \{1 - P(A_2)\} + ... + \{1 - P(A_n)\} - (n-1)$

$= 1 - P(A_1) - P(A_2) - .... - P(A_n)$

$\Rightarrow P(A_1) + P(A_2) + ... + P(A_n) \geq 1 - P(\bar{A}_1) \cap \bar{A}_2 \cap \bar{A}_3 \cap .... \cap \bar{A}_n$

$= 1 - P\left(\overline{A_1 \cup A_2 \cup ... A_n}\right)$

$= P(A_1 \cup A_2 \cup .... \cup A_n)$

$\Rightarrow P(A_1 \cup A_2 \cup .... \cup A_n) \leq P(A_1) + P(A_2) + ... + P(A_n)$ as desired.

Theorem 7: For any two events A and B show that $P(A \cap B) \leq P(A) \leq P(A \cup B) \leq P(A) + P(B)$.

Proof:

We have $A = (A \cap \bar{B}) \cup (A \cap B)$

Using axiom 3, we have $P(A) = P[(A \cap \bar{B}) \cup (A \cap B)]$

$P(A) = P(A \cap \bar{B}) + P(A \cap B)$ (from axiom 1)

Now $P(A \cap \bar{B}) \geq 0$ $\Rightarrow P(A) - P(A \cap B) \geq 0$

$\therefore$ $P(A) \geq P(A \cap B)$ ...(1)

Similarly, $P(B) \geq P(A \cap B)$

$\Rightarrow P(B) - P(A \cap B) \geq 0$

Now $P(A \cup B) = P(A) + [P(B) - P(A \cap B)]$ ...(2)

$\therefore$ $P(A \cup B) \geq P(A)$

$\Rightarrow P(A) \leq P(A \cup B)$ ...(3)

Also $P(A \cup B) \leq P(A) + P(B)$ from (2)

Hence from (1), (2) and (3), we get $P(A \cap B) \leq P(A) \leq P(A \cup B) \leq P(A) + P(B)$

1.6 SOLVED PROBLEMS

Problem 1: If $P(A) = \frac{1}{5}$, $P(B) = \frac{2}{3}$, $P(A \cap B) = \frac{1}{15}$.

Find (*i*) $P(A \cup B)$ (*ii*) $P(\bar{A} \cap B)$ (*iii*) $P(A \cap \bar{B})$ (*iv*) $P(\bar{A} \cap \bar{B})$ (*v*) $P(\bar{A} \cup \bar{B})$

Solution:

(*i*)
$$P(A \cup B) = P(A) + P(B) - P(A \cap B)$$
$$= \frac{1}{5} + \frac{2}{3} - \frac{1}{15}$$
$$= \frac{12}{15} = \frac{4}{5}$$

(*ii*)
$$P(\bar{A} \cap B) = P(B) - P(A \cap B)$$
$$= \frac{2}{3} - \frac{1}{15}$$
$$= \frac{9}{15} = \frac{3}{5}$$

(*iii*)
$$P(A \cap \bar{B}) = P(A) - P(A \cap B)$$
$$= \frac{1}{5} - \frac{1}{5} = \frac{2}{15}$$

(*iv*)
$$P(\bar{A} \cap \bar{B}) = P\left(\overline{A \cup B}\right)$$
$$= 1 - P(A \cup B)$$

$$= 1 - \frac{4}{5} = \frac{1}{5}$$

(v) $$P(\bar{A} \cup \bar{B}) = P(\overline{A \cap B})$$

$$= 1 - P(A \cap B)$$

$$= 1 - \frac{1}{15} = \frac{14}{15}.$$

Problem 2: If $P(A \cup B) = \frac{4}{5}$, $P(\bar{B}) = \frac{1}{3}$ and $P(A \cap B) = \frac{1}{5}$.

Find (*i*) P(B) (*ii*) P(A) (*iii*) $P(\bar{A} \cap B)$

Solution:

(*i*) $$P(\bar{B}) = 1 - P(B)$$

$$P(B) = 1 - P(\bar{B})$$

$$= 1 - \frac{1}{3} = \frac{2}{3}$$

(*ii*) $$P(A \cup B) = P(A) + P(B) - P(A \cap B)$$

$\therefore$ $$\frac{4}{5} = P(A) + \frac{2}{3} - \frac{1}{5}$$

$$P(A) = \frac{4}{5} - \frac{2}{3} + \frac{1}{5}$$

$$= 1 - \frac{2}{3} = \frac{1}{3}$$

(*iii*) $$P(\bar{A} \cap B) = P(B) - P(A \cap B)$$

$$= \frac{2}{3} - \frac{1}{5} = \frac{7}{15}.$$

Problem 3: Three students A, B, C are in a running race. A and B have the same probability of winning and each is twice as likely to win as C. Find the probability that B or C wins.

(April/May 2006, Nov 2007)

Solution:

Assume that the probability of winning A is $2x$.

The probability of winning B is $2x$

Then, the probability of winning C is x

We have to find $P(B \cup C)$.

We know that, $$P(B \cup C) = P(B) + P(C) - P(B \cap C)$$

$$P(B \cap C) = P(B).P(C)$$

$\therefore$ B and C are independent $= 2x^2$

$$P(A) + P(B) + P(C) = 1$$

$$2x + 2x + x = 1$$

$$\Rightarrow \qquad x = \frac{1}{5}$$

$$\therefore \qquad P(B) = \frac{2}{5},$$

$$P(C) = \frac{1}{5}.$$

Problem 4: Among 150 students 80 are studying Maths, 40 are studying Physics and 30 are studying Maths and Physics. If a student is chosen at random, find the probability that the student.

(*i*) Studying Maths or Physics (*ii*) Studying neither Maths nor Physics.

Solution:

Let A be the event that the student studying Maths.

Let B the event that the student studying Physics.

$A \cap B$ be the event that the student studying Maths and Physics

$$\therefore \qquad P(A) = \frac{80}{150} = \frac{8}{15}$$

$$P(B) = \frac{40}{150} = \frac{4}{15}$$

$$P(A \cap B) = \frac{30}{150} = \frac{3}{15}$$

(*i*) Probability that a student studying Maths or Physics is

$$P(A \cup B) = P(A) + P(B) - P(A \cap B)$$

$$= \frac{8}{15} + \frac{4}{15} - \frac{3}{15} = \frac{9}{15} = \frac{3}{5}$$

(*ii*) Probability that the student studying neither Maths nor Physics is

$$P(\bar{A} \cap \bar{B}) = P\left(\overline{A \cup B}\right)$$

$$= 1 - P(A \cup B)$$

$$= 1 - \frac{3}{5} = \frac{2}{5}.$$

Problem 5: An integer is chosen at random from the first 200 positive integers. What is the probability that the integer chosen is divisible by 6 or 8? **(Jan 2003, Jan 2002)**

Solution:

Let A and B be the events that the number is chosen is divisible by 6 or 8 respectively.

$A \cap B$ be the event that the number is divisible by 6 and 8.

$n(A) = \{6, 12, 18, \ldots\ldots, 198\} = 33$

$$\Rightarrow \qquad P(A) = \frac{33}{200}$$

$n(B) = \{8, 16, 24, \ldots\ldots, 200\} = 25$

$$\Rightarrow \qquad P(B) = \frac{25}{200}$$

$$n(A \cap B) = \{24, 48, \ldots\ldots\ldots, 192\} = 8$$

$$\Rightarrow \qquad P(A \cap B) = \frac{8}{200}$$

∴ Probability that the integer chosen is divisible by 6 or 8 is

$$P(A \cup B) = P(A) + P(B) - P(A \cap B)$$

$$= \frac{33}{200} + \frac{25}{200} - \frac{8}{200}$$

$$= \frac{50}{200} = \frac{1}{4}.$$

Problem 6: A card is drawn from a well shuffled pack of 52 cards. What is the probability that it is either a spade or an ace? **(Nov 2003)**

Solution:

Let A be the event that drawing a spade cards

Let B be the event that drawing an ace.

$\therefore\ P(A) = \frac{13}{52}$ $\qquad P(B) = \frac{4}{52}$

$A \cap B$ be the event that drawing a spade and ace.

$$\therefore \qquad P(A \cap B) = \frac{1}{52}$$

∴ Probability that the event drawing a spade or an ace is

$$P(A \cup B) = P(A) + P(B) - P(A \cap B)$$

$$= \frac{13}{52} + \frac{4}{52} - \frac{1}{52} = \frac{16}{52} = \frac{4}{13}.$$

Problem 7: If a card is drawn from a deck of playing cards, what is the probability that it will be either the King of Diamonds or the Queen of Hearts?

Solution:

Let A = the event of drawing the King of Diamonds

And B = the event of drawing the Queen of Hearts.

The two events are mutually exclusive.

$$P(A) = \frac{1}{52}$$

$$P(B) = \frac{1}{52}$$

$$P(A \cup B) = P(A) + P(B)$$

$$= \frac{1}{52} + \frac{1}{52} = \frac{1}{26}.$$

Problem 8: A bag contains 6 white, 5 black and 4 yellow balls. Find the chance of getting either a white or a black ball in a single draw.

Solution:

Let A = the event of getting a white ball

and B = the event of getting a black ball .

Total number of balls = 6 + 5 + 4 = 15

$$P(A) = \frac{6}{15}, \qquad P(B) = \frac{5}{15}$$

Since the two events are mutually exclusive

$$P(A \cup B) = P(A) + P(B)$$

$$= \frac{6}{15} + \frac{5}{15} = \frac{11}{15}.$$

1.7 CONDITIONAL PROBABILITY

If A and B are any two events then the probability of the happening of the event B given that A has already happened is denoted by P(B/A) is defined as $P(B/A) = \frac{P(A \cap B)}{P(A)}$.

Similarly, $P(A/B) = \frac{P(A \cap B)}{P(B)}$.

Multiplication Law of Probability

For any two events A and B, then

$$P(A \cap B) = P(A).P(B/A)$$

$$= P(B).P(A/B),$$

where P(B/A) is the conditional probability of B when A has already happened and P(A/B) is the conditional probability of A when B has already happened.

Proof:

Suppose a trial results in 'n' exhaustive mutually exclusive and equally likely events. 'm' of them being favourable to the happening of the event A.

$\therefore$ Probability of happening of event A is $P(A) = \frac{m}{n}$. ...(1)

Out of m outcomes m_1 are favourable to the happening of event B.

$\therefore$ Conditional probability of B given that A has happened is $P(B/A) = \frac{m_1}{m}$ ...(2)

Out of 'n' outcomes m_1 are favourable to the happening of A and B.

$\therefore$ Probability of simultaneous occurrence A and B is $P(A \cap B) = \frac{m_1}{n} = \frac{m}{n} \cdot \frac{m_1}{m}$

$= P(A).P(B/A)$

Hence $P(A \cap B) = P(A).P(B/A)$

Similarly, $P(A \cap B) = P(B).P(A/B)$

Note:

1. If P is the chance that an event will happen in one trial then that chance that it will happen in a succession of r trials is P.P.P................P, r times = P^r.
2. If $P_1, P_2, P_3, ...,P_n$ are the probability that certain events happen, then the probabilities of non-happening are $1-P_1, 1-P_2, ..., 1-P_n$.

   ∴ Probability of all these failing is $(1-P_1)(1-P_2)(1-P_3)............(1-P_n)$

   Hence the chance in which at least one of these events much happen is

   $1-[(1-P_1)(1-P_2)(1-P_3)............(1-P_n)]$.

Independent Events

Two events are said to be independent if the happening of an event is not effected by the happening of the other event. In other words happening of an event does not depend upon the other event. (OR)

An event A is said to be independent of another event B, if the conditional probability of A given B, *i.e.*, P(A/B) is equal to the unconditional probability of B, *i.e.*, if P(A/B) = P(A).

Pair-wise Independent Events

The events $A_1, A_2, A_3, ..., A_n$ are said to be pair-wise independent if and only if $P(A_i \cap A_j) = P(A_i).P(A_j)$ for $i \neq j = 1, 2, 3, ..., n$.

Mutually Independent Events

The n events $A_1, A_2, A_3, ..., A_n$ in a sample space S are said to be mutually independent if $P(A_1 \cap A_2 \cap A_3 \cap ....A_n) = P(A_1).P(A_2).P(A_3)............P(A_n)$.

Theorem 1: If A and B are two events with positive probabilities $[P(A) \neq 0, P(B) \neq 0]$, then A and B are independent if and only if $P(A \cap B) = P(A).P(B)$.

Proof: We have $P(A \cap B) = P(A).P(B/A) = P(B).P(A/B)$ ...(1)

If A and B are independent, *i.e.*, A is independent of B and B is independent of A, then we have

$P(A/B) = P(A)$ and $P(B/A) = P(B)$ ...(2)

From (1) and (2) $P(A \cap B) = P(A).P(B)$

Conversely, if $P(A \cap B) = P(A).P(B)$, then we get $\dfrac{P(A \cap B)}{P(B)} = P(A) \Rightarrow P(A/B) = P(A)$

$$\frac{P(A \cap B)}{P(A)} = P(B) \Rightarrow P(B/A) = P(B) \quad ...(3)$$

(3) ⇒ A and B are independent events.

Theorem 2: If A and B are independent events.

Then

(*i*) $\bar{A}$ and $\bar{B}$ are independent

(*ii*) $\bar{A}$ and B are independent

(*iii*) A and $\bar{B}$ are independent

Proof:

Given that A, B are independent, *i.e.*, $P(A \cap B) = P(A).P(B)$

(*i*)
$$\begin{aligned} P(\bar{A} \cap \bar{B}) &= P(\overline{A \cup B}) \\ &= 1 - P(A \cup B) \\ &= 1 - \{P(A) + P(B) - P(A \cap B)\} \\ &= 1 - P(A) - P(B) + P(A).P(B) \\ &= [1 - P(A)] - P(B)[1 - P(A)] \\ &= [1 - P(A)][1 - P(B)] \\ &= P(\bar{A}).P(B) \end{aligned}$$

$\therefore$ $\bar{A}$ and $\bar{B}$ are independent.

(*ii*)
$$\begin{aligned} P(\bar{A} \cap B) &= P(B) - P(A \cap B) \\ &= P(B) - P(A).P(B) \\ &= [1 - P(A)]\, P(B) \\ &= P(\bar{A}).P(B) \end{aligned}$$

$\therefore$ $\bar{A}$ and B are independent.

(*iii*)
$$\begin{aligned} P(A \cap \bar{B}) &= P(A) - P(A \cap B) \\ &= P(A) - P(A).P(B) \\ &= P(A)[1 - P(B)] \\ &= P(A).P(\bar{B}) \end{aligned}$$

$\therefore$ A and $\bar{B}$ are independent.

Theorem 3: If A, B and C are mutually independent events, then $A \cup B$ and C are also independent.

Proof:

A, B and C are mutually independent.

Given
$$\begin{aligned} P(A \cap B \cap C) &= P(A).P(B).P(C) \\ P[(A \cup B) \cap C] &= P[(A \cap C) \cup (B \cap C)] \\ &= P(A \cap C) + P(B \cap C) - P(A \cap B \cap C) \\ &= P(A)P(C) + P(B)\, P(C) - P(A)\, P(B)\, P(C) \\ &= [P(A) + P(B) - P(A)\,.\, P(B)]\, P(C) \end{aligned}$$

$$= [P(A) + P(B) - P(A \cap B)]\, P(C)$$
$$= P(A \cup B) \,.\, P(C)$$

$\therefore$ $A \cup B$ and C are independent.

Theorem 4: If A, B and C are pair-wise independent events and A is independent of $B \cup C$, then A, B and C are mutually independent.

Proof:

Given that A, B and C are pair-wise independent, *i.e.*, $P(A \cap B) = P(A)P(B)$

$$P(B \cap C) = P(B).P(C),\ P(A \cap C) = P(A)\,P(C)$$

Also
$$P\{A \cap (B \cup C)\} = P(A).P(B \cup C)$$
$$P[A \cap (B \cup C)] = P[(A \cap B) \cup (A \cap C)]$$
$$= P(A \cap B) + P(A \cap C) - P(A \cap B \cap C)$$
$$= P(A)\,P(B) + P(A)\,P\,(C) - P(A \cap B \cap C) \qquad ...(1)$$
$$P[A \cap (B \cup C)] = P(A).P(B \cup C)$$
$$= P(A)\,\{P(B) + P(C) - P(B \cap C)\}$$
$$= P(A)\,P\,(B) + P(A)\,P(C) - P(A)\,P(B)\,P(C) \qquad ...(2)$$

From (1) and (2),

$$P(A \cap B \cap C) = P(A)\,P(B)\,P(C)$$

$\therefore$ A, B, C are mutually independent.

1.8 SOLVED PROBLEMS

Problem 1: If A and B are events with P(A) = 1/3, P(B) = 1/4 and $P(A \cup B) = 1/2$, find (*i*) P(A/B) (*ii*) P(B/A) (*iii*) $P(A/\overline{B})$ (*iv*) $P(A \cap \overline{B})$.

(Nov 2004, May 2006, Nov 2007)

Solution:

(*i*)
$$P(A/B) = \frac{P(A \cap B)}{P(B)}$$
$$P(A \cup B) = P(A) + P(B) - P(A \cap B)$$
$$\frac{1}{2} = \frac{1}{3} + \frac{1}{4} - P(A \cap B)$$

$\therefore$
$$P(A \cap B) = \frac{1}{12}$$
$$P(A/B) = \frac{1/12}{1/4} = \frac{1}{3}.$$

(*ii*)
$$P(B/A) = \frac{P(A \cap B)}{P(A)}$$
$$= \frac{1/12}{1/3} = \frac{1}{4}$$

(*iii*)
$$P(A \cap \bar{B}) = P(A) - P(A \cap B)$$
$$\frac{1}{3} - \frac{1}{12} = \frac{1}{4}.$$

(*iv*)
$$P(A/\bar{B}) = \frac{P(A \cap \bar{B})}{P(\bar{B})} = \frac{P(A \cap \bar{B})}{1 - P(B)}$$
$$= \frac{1/4}{1 - 1/4}$$
$$= \frac{1/4}{3/4} = \frac{1}{3}.$$

Problem 2: If A and B are events such that $P(A) = \frac{1}{3}$, $P(B) = \frac{1}{4}$, $P(A \cap B) = \frac{1}{5}$, find

(*i*) $P(A/B)$ (*ii*) $P(B/A)$ (*iii*) $P(A \cup B)$ (*iv*) $P(\bar{B}/\bar{A})$.

Solution:

(*i*)
$$P(A/B) = \frac{P(A \cap B)}{P(B)}$$
$$= \frac{1/5}{1/4} = \frac{4}{5}.$$

(*ii*)
$$P(B/A) = \frac{P(A \cap B)}{P(A)}$$
$$= \frac{1/5}{1/3} = \frac{3}{5}.$$

(*iii*)
$$P(A \cup B) = P(A) + P(B) - P(A \cap B)$$
$$= \frac{1}{3} + \frac{1}{4} - \frac{1}{5}$$
$$= \frac{23}{60}.$$

(*iv*)
$$P(\bar{B}/\bar{A}) = \frac{P(\bar{A} \cap \bar{B})}{P(\bar{A})} = \frac{P(\overline{A \cup B})}{P(\bar{A})}$$
$$= \frac{1 - P(A \cup B)}{1 - P(A)}$$
$$= \frac{1 - 23/60}{1 - 1/3} = \frac{37}{40}$$

Problem 3: If A and B are two events such that P(A) = 0.4, P(B) = 0.8 and P(B/A) = 0.6, find P(A/B) and P(A ∪ B).

Solution:

$$P(B/A) = \frac{P(A \cap B)}{P(A)}$$

$$P(A \cap B) = P(A).P(B/A)$$

$$= 0.4 \times 0.6 = 0.24.$$

(*i*)

$$P(A/B) = \frac{P(A \cap B)}{P(B)}$$

$$= \frac{0.24}{0.8}$$

$$= 0.3.$$

(*ii*)

$$P(A \cup B) = P(A) + P(B) - P(A \cap B)$$

$$= 0.4 + 0.8 - 0.24$$

$$= 0.96.$$

Problem 4: A problem in statistics is given to the three students A, B, C whose chance of solving it are ½ , ¾ and ¼ respectively. What is the probability that the problem is solved?

(Nov 2003, Apr 2007, Nov 2007)

Solution:

Let P(A) denote the probability that A, to solve the problem $= \frac{1}{2}$

$$\Rightarrow \quad P(\bar{A}) = 1 - P(A) = \frac{1}{2}$$

Let P(B) denote the probability that B, to solve the problem $= \frac{3}{4}$

$$\Rightarrow \quad P(\bar{B}) = 1 - P(B) = \frac{1}{4}$$

Let P(C) denote the probability that C, to solve the problem $= \frac{1}{4}$

$$\Rightarrow \quad P(\bar{C}) = 1 - P(C) = \frac{3}{4}$$

∴ The probability that the problem is

$$= P(A \cup B \cup C)$$

$$= 1 - P\left(\overline{A \cup B \cup C}\right).$$

$$= 1 - P(\bar{A} \cap \bar{B} \cap \bar{C})$$

$$= 1 - P(\bar{A}).P(\bar{B}).P(\bar{C}) \text{ (Since A, B; C are independent)}$$

$$= 1 - \frac{1}{2}.\frac{1}{4}.\frac{3}{4} = \frac{29}{32}$$

Problem 5: A can hit a target 3 times in 5 shots, B 2 times in 5 shots and C 3 times in 4 shots. Find the probability of the target being hit, when all of them try? **(Nov 2003)**

Solution:

Let P(A) denote the probability that the target was hit by A.

Let P(B) denote the probability that the target was hit by B.

Let P(C) denote the probability that the target was hit by C.

Thus $$P(A) = \frac{3}{5}$$

$$\Rightarrow \quad P(\bar{A}) = 1 - P(A) = \frac{2}{5}$$

$$P(B) = \frac{2}{5}$$

$$\Rightarrow \quad (\bar{B}) = 1 - P(B) = \frac{3}{5}$$

$$P(C) = \frac{3}{4}$$

$$\Rightarrow \quad P(\bar{C}) = 1 - P(C) = \frac{1}{4}$$

$\therefore$ The probability that target was hit is = $P(A \cup B \cup C)$

$$= 1 - P\left(\overline{A \cup B \cup C}\right).$$

$$= 1 - P(\bar{A} \cap \bar{B} \cap \bar{C})$$

$$= 1 - P(\bar{A}).P(\bar{B}).P(\bar{C}) \text{ (Since A, B, C are independent)}$$

$$= 1 - \frac{2}{5} \times \frac{3}{5} \times \frac{1}{4} = 0.94$$

Problem 6: A and B throw alternately with a pair of ordinary dice. A wins if he throws 6 before B throws 7 and B wins if he throws 7 before A throws 6. If A begins, show that his chance of winning is 30/61. **(Apr 2003, 04)**

Solution:

When two dice are thrown, total number of events = 36.

The sum 6 can be obtained with two dice as follows. (1, 5), (2, 4), (3, 3), (4, 2), (5, 1)

$\therefore$ The probability of A's throwing 6 is

$$P(A) = \frac{5}{36}$$

The probability of A's not throwing 6 is

$$P(\bar{A}) = 1 - P(A) = \frac{31}{36}$$

Similarly,

The sum 7 can be obtained with two dice as follows (1, 6), (2, 5), (3, 4), (4, 3), (5, 2), (6, 1)

The probability of B's throwing 7 is $P(B) = \frac{6}{36}$

The probability of B's not throwing 6 is $P(\bar{B}) = 1 - P(B) = \frac{30}{36}$

If A begins, if A is to win, he should throw 6 in first or third or fifth ....throws.

$\therefore$ A's chance of winning is $= P(A) + P(\bar{A})P(\bar{B})P(A) + P(\bar{A})P(\bar{B})P(\bar{A})P(\bar{B})P(A) + \ldots\ldots\ldots$

$$= \frac{5}{36} + \frac{31}{36}\frac{30}{36}\frac{5}{36} + \frac{31}{36}\frac{30}{36}\frac{31}{36}\frac{30}{36}\frac{5}{36} + \ldots\ldots\ldots\ldots$$

$$= \frac{5}{36}\left[1 + \left(\frac{31}{36}\frac{30}{36}\right) + \left(\frac{31}{36}\frac{30}{36}\right)^2 \left(\frac{31}{36}\frac{30}{36}\right)^3 + \ldots\right]$$

$$= \frac{5}{36} \times \frac{1}{1 - \frac{31}{36}\frac{30}{36}} = \frac{30}{61}$$

Problem 7: The student in a class is selected at random, one after the other, for an examination. Find the probability that the boys and girls in the class alternate if (*i*) The class consists of 4 boys and 3 girls, (*ii*) The class consists of 3 boys and 3 girls. **(April 2006)**

Solution:

(*i*) Since there are 4 boys and 3 girls, they can sit alternately as follows

$$E = B\,G\,B\,G\,B\,G\,B$$

$$P(E) = P(B\,G\,B\,G\,B\,G\,B)$$

$$= \frac{4}{7}\cdot\frac{3}{6}\cdot\frac{3}{5}\cdot\frac{2}{4}\cdot\frac{2}{3}\cdot\frac{1}{2}$$

$$= \frac{1}{35}$$

(*ii*) Since there are 3 boys and 3 girls, hence we come across two mutually exclusive events.

They are, $E_1 = B\,G\,B\,G\,B\,G$ and $E_2 = G\,B\,G\,B\,G\,B$

$$P(E_1) = P(B\,G\,B\,G\,B\,G)$$

$$= \frac{3}{6}\cdot\frac{3}{5}\cdot\frac{2}{4}\cdot\frac{2}{3}\cdot\frac{1}{2} = \frac{1}{20}$$

$$P(E_2) = P(G\,B\,G\,B\,G\,B)$$

$$= \frac{3}{6}\cdot\frac{3}{5}\cdot\frac{2}{4}\cdot\frac{2}{3}\cdot\frac{1}{2} = \frac{1}{20}$$

$\therefore$ Probability that 3 boys and 3 girls in the class alternate is $P(E) = P(E_1 \cup E_2)$

$$= P(E_1) + P(E_2)$$

$$= \frac{1}{20} + \frac{1}{20}$$

$$= \frac{1}{10}.$$

Problem 8: Two bolts are drawn from a box containing 4 good and 6 bad bolts. Find the probability that the second bolt is good. If first one is found to be bad. **(Nov 2004)**

Solution:

Let A be the event that drawing a bad bolt in the first draw.

Let B be the event that drawing a good bolt in the second draw.

$$P(A) = \frac{{}^{6}C_1}{{}^{10}C_1} = \frac{6}{10} \text{ and } P(B) = \frac{{}^{4}C_1}{{}^{9}C_1} = \frac{4}{9}$$

$$P(A \cap B) = P(A).P(B) = \frac{6}{10} \times \frac{4}{9} = \frac{4}{15}$$

Probability of getting second bolt is good given that first bolt is bad = P(B/A)

$$= \frac{P(A \cap B)}{P(A)}$$

$$= \frac{4/15}{6/10} = \frac{4}{9}.$$

Problem 9: Two aeroplanes bomb a target in succession. The probability of each correctly scoring a hit is 0.3 and 0.2 respectively. The second will bomb only if the first misses the target. Find the probability that

(*i*) Target is hit, (*ii*) The target is hit by the second plane, (*iii*) Both fails to score hits. **(Nov 2004, Feb 2007)**

Solution:

Let A be the event that the first aeroplane hit.

Let B be the event that the second aeroplane hit.

$\therefore$ P(A) = 0.3, P(B) = 0.2

(*i*) Probability of the target is hit = P(A hit) + P(A fails and B hit)

$$= P(A) + P(\bar{A} \cap B)$$

$$= P(A) + P(\bar{A}) \,.\, P(B)$$

$$= 0.3 + (1 - 0.3) \,.\, (0.2)$$

$$= 0.3 + 0.14 = 0.44$$

(*ii*) Probability that the second airoplane hit the target = P(A fails and B hit)

$$= P(\bar{A} \cap B)$$

$$= P(\bar{A}) \,.\, P(B)$$

$$= (1 - 0.3).(0.2) = 0.14$$

(*iii*) Probability that both are fails the target = P(A fails and B fails)

$$= P(\bar{A} \cap \bar{B})$$

$$= P(\bar{A}) \,.\, P(\bar{B})$$

$$= (1 - 0.3) \cdot (1 - 0.2)$$
$$= 0.56$$

Problem 10: In a certain town 40% have brown hair, 25% have brown eyes and 15% have both brown hair and brown eyes. A person is selected at random from the town.

(May 2005, Nov 2006, April 2007)

(*i*) If he has brown hair, what is the probability that has brown eyes also?

(*ii*) If he has brown eyes, determine the probability that he does not have brown hair.

(*iii*) Determine the probability that he has neither brown hair nor brown eyes.

Solution:

Let A be the event that has brown eyes

$$\Rightarrow \quad P(A) = 0.25$$

Let B the event that has brown hair.

$$\Rightarrow \quad P(B) = 0.4,$$
$$P(A \cap B) = 0.15.$$

(*i*) Probability that a person has brown eyes given that he has brown hair is

$$P(A/B) = \frac{P(A \cap B)}{P(B)}$$
$$= \frac{0.15}{0.4} = \frac{3}{8}.$$

(*ii*) Probability that a person does not have brown hair given that he has brown eyes is

$$P(\bar{B}/A) = \frac{P(A \cap \bar{B})}{P(A)}$$
$$P(A \cap \bar{B}) = P(A) - P(A \cap B)$$
$$= 0.25 - 0.15$$
$$= 0.1$$
$$\therefore \quad P(\bar{B}/A) = \frac{0.1}{0.25} = \frac{2}{5}.$$

(*iii*) Probability that he has neither brown hair nor brown eyes is

$$P(\bar{A} \cap \bar{B}) = P\left(\overline{A \cup B}\right)$$
$$= 1 - P(A \cup B)$$
$$P(A \cup B) = P(A) + P(B) - P(A \cap B)$$
$$= 0.25 + 0.4 - 0.15$$
$$= 0.5.$$
$$\therefore \quad P(\bar{A} \cap \bar{B}) = 1 - 0.5 = 0.5.$$

Problem 11: Two marbles are drawn in succession from a box containing 10 red, 30 white, 20 blue and 15 orange marbles, with replacement being made after each drawing. Find the probability that

(*i*) Both are white (*ii*) First is red and second is white.

(*iii*) Neither is orange. **(May 2005, Dec 2005, May 2006, Apr 2007)**

Solution:

With replacement the events are independent.

Total number of marbles = 75.

(*i*) Let W_1 be the event that drawing a white marble in the first draw.

W_2 be the event that drawing a white marble in the second draw.

$\therefore \quad P(W_1) = \frac{30}{75} = \frac{2}{5}$ and $P(W_2) = \frac{30}{75} = \frac{2}{5}$

$\therefore$ Probability of drawing both are white marbles is $P(A \cap B) = P(A).P(B)$

$$= \frac{2}{5} \times \frac{2}{5} = \frac{4}{25}.$$

(*ii*) Let R be the event that drawing a red marble in the first draw.

W be the event that drawing a white marble in the second draw

$$\therefore \quad P(R) = \frac{10}{75} = \frac{2}{15}$$

$$P(W) = \frac{30}{75} = \frac{2}{5}$$

$\therefore$ Probability of drawing first marble is red and second is white is $P(R \cap W) = P(R).P(W)$

$$= \frac{2}{15} \times \frac{2}{5} = \frac{4}{75}.$$

(*iii*) Let A be the event that drawing an orange marble in the first draw.

B be the event that drawing an orange marble in the second draw.

$$\therefore \quad P(A) = \frac{15}{75}$$

$$\Rightarrow \quad (\bar{A}) = 1 - P(A)$$

$$= \frac{60}{75} = \frac{4}{5}$$

$$P(B) = \frac{15}{75}$$

$$\Rightarrow \quad P(\bar{B}) = 1 - P(B)$$

$$= \frac{60}{75} = \frac{4}{5}$$

$\therefore$ Probability of drawing neither orange marble is $P(\bar{A} \cap \bar{B}) = P(\bar{A}) \,.\, P(\bar{B})$

$$= \frac{4}{5} \cdot \frac{4}{5} = \frac{16}{25}.$$

Problem 12: If the probability that a research project will be well planned is 0.80 and the probability that it will be well planned and well executed is 0.72. The probability that the research will be well executed is 0.75. What is the probability that the project was well planned, given that it was well executed?

Solution:

Let A be the event that the project is well planned.

B be the event that the project is well executed.

Given

$$P(A) = 0.80,\ P(B) = 0.75,\ P(A \cap B) = 0.72$$

∴ Probability that the project was well planned given that it was well executed is

$$P(A/B) = \frac{P(A \cap B)}{P(B)}$$

$$= \frac{0.72}{0.75} = 0.96.$$

Problem 13: A class consists of 6 girls and 10 boys, if a committee of 3 is chosen at random from the class, find the probability that

(*i*) 3 boys are selected

(*ii*) Exactly two boys are selected.

(*iii*) At least one boy is selected

(*iv*) Exactly 2 girls are selected. **(May 2005)**

Solution:

(*i*) Total number of events = $^{16}C_3 = 560$

Out of 10 boys 3 are selected in $^{10}C_3$ ways.

∴ Favourable number of events = $^{10}C_3 = 120$

∴ Probability that 3 boys are selected is = $\dfrac{^{10}C_3}{^{16}C_3} = \dfrac{3}{14}$.

(*ii*) Out of 10 boys 2 are selected is $^{10}C_2 = 45$ ways .

Out of 6 girls one girl is selected in $^{6}C_1 = 6$ ways

∴ Favourable number of events = $45 \times 6 = 270$

∴ Probability of exactly two boys are selected = $\dfrac{270}{560} = \dfrac{27}{56}$.

(*iii*) Out of 6 girls 3 are selected in $^{6}C_3 = 20$ ways

∴ Probability of 3 girls are selected = $\dfrac{20}{560} = \dfrac{1}{28}$

∴ Probability of at least one boy is selected = $1 - \dfrac{1}{28} = \dfrac{27}{28}$.

(*iv*) Probability of 2 girls and 1 boy is selected = $\dfrac{^{10}C_1 \cdot {}^{6}C_2}{^{16}C_3} = \dfrac{15}{56}$.

Problem 14: In a certain college 25% of the students failed in mathematics, 15% failed in chemistry and 10% in mathematics and chemistry. A student is selected at random. (*i*) If he/she failed in chemistry, what is the probability that he/she failed in mathematics, (*ii*) If he/she failed in mathematics, what is the probability that he/she failed in chemistry? **(Nov 2006)**

Solution:

Here $n(S)=100$

Let A = students who failed in Maths, *i.e.*, $n(A) = 25$

$$P(A) = \frac{n(A)}{n(S)} = 25/100$$

Let B = students who failed in chemistry, *i.e.*, $n(B) = 15$

$$\therefore \quad P(B) = \frac{n(B)}{n(S)} = 15/100$$

Let $A \cap B$= students who failed in both subjects, *i.e.* $n(A \cap B) = 10$

$$\therefore \quad P(A \cap B) = \frac{n(A \cap B)}{n(S)} = 10/100$$

(*i*)
$$P(B/A) = \frac{P(A \cap B)}{P(A)} = 2/5$$

(*ii*)
$$P(A/B) = \frac{P(A \cap B)}{P(B)} = 2/3$$

1.9 TOTAL PROBABILITY

Statement: Let $B_1, B_2, \ldots, B_k$ constitute a partition of the sample space s with $P(B_i) \neq 0$, for $i = 1, 2, 3, 4, \ldots, k$. Then for any events A of S. $P(A) = \sum_{i=1}^{k} P(B_i \cap A) = \sum_{i=1}^{k} P(B_i)P(A/B_i)$.

(Nov 2006)

Proof: Since $B_1, B_2, \ldots, B_k$ constitute a partition of S.

$\therefore$ $S = B_1 \cup B_2 \cup B_3 \cup \ldots \cup B_k$ and $B_i \cap B_j = \phi$ for every i, j $B_1, B_2, \ldots, B_k$ are mutually disjoint.

$$\begin{aligned} A &= A \cap S \\ &= A \cap (B_1 \cup B_2 \cup B_3 \cup \ldots \cup B_k) \\ &= (A \cap B_1) \cup (A \cap B_2) \cup (A \cap B_3) \cup \ldots \cup (A \cap B_k) \\ P(A) &= P\{(A \cap B_1) \cup (A \cap B_2) \cup (A \cap B_3) \cup \ldots \cup (A \cap B_k)\} \\ &= P(A \cap B_1) + P(A \cap B_2) + P(A \cap B_3) + \ldots + P(A \cap B_k) \\ &= \sum_{i=1}^{k} P(B_i \cap A) \\ &= \sum_{i=1}^{k} P(B_i)\, P(A/B_i). \end{aligned}$$

1.10 BAYE'S THEOREM

Statement: Let the sample space S be partitioned into k sub events $B_1, B_2, \ldots, B_k$ with $P(B_i) \neq 0$ for I = 1, 2, 3, ..., k. For any event A in S with $P(A) \neq 0$.

$$P(B_i/A) = \frac{P(B_i)\,P(A/B_i)}{\sum_{i=1}^{k} P(B_i)\,P(A/B_i)}$$ **(Nov 2006, Apr 2007)**

Proof: Since $B_1, B_2, \ldots, B_k$ constitute a partition of S.

$\therefore S = B_1 \cup B_2 \cup B_3 \cup \ldots \cup B_k$ and $B_i \cap B_j = \phi \ \forall\ i, j$

$B_1, B_2, \ldots, B_k$ are mutually disjoint.

$$\begin{aligned} A &= A \cap S \\ &= A \cap (B_1 \cup B_2 \cup B_3 \cup \ldots \cup B_k) \\ &= (A \cap B_1) \cup (A \cap B_2) \cup (A \cap B_3) \cup \ldots \cup (A \cap B_k) \\ P(A) &= P\{(A \cap B_1) \cup (A \cap B_2) \cup (A \cap B_3) \cup \ldots \cup (A \cap B_k)\} \\ &= P(A \cap B_1) + P(A \cap B_2) + P(A \cap B_3) + \ldots + P(A \cap B_k) \\ &= \sum_{i=1}^{k} P(B_i \cap A) \\ &= \sum_{i=1}^{k} P(B_i)\,P(A/B_i) \qquad \ldots(1) \end{aligned}$$

Now $$P(A \cap B_i) = P(A).P(B_i/A)$$

$$P(B_i/A) = \frac{P(A \cap B_i)}{P(A)}$$

$$= \frac{P(B_i)P(A/B_i)}{\sum_{i=1}^{k} P(B_i)P(A/B_i)}$$ [From (1)]

1.11 SOLVED PROBLEMS

Problem 1: Box I contains 10 white and 3 black balls , while Box II contains 3 white and 5 black balls. Two balls are drawn at random from Box I and placed in the box II. What is the probability that it is a white ball?

Solution: The two balls drawn from the box I may be

(*i*) Both white (*ii*) Both black

(*iii*) One white and one black

Let E_1 be the event that drawing two white balls from box I.

Let E_2 be the event that drawing two black balls from box II

Let E_3 be the event that drawing one white and one black from boxes

$$P(E_1) = \frac{^{10}C_2}{^{13}C_2} = \frac{15}{26}.$$

$$P(E_2) = \frac{^{3}C_2}{^{13}C_2} = \frac{1}{26}.$$

$$P(E_3) = \frac{^{3}C_1\ ^{10}C_1}{^{13}C_2} = \frac{10}{26}.$$

When two balls are transformed from box I to box II, the second box will contains

(*i*) 5 white ; 5 black balls

(*ii*) 3 white ; 7 black balls

(*iii*) 4 white ; 6 black balls

Let A be the event that drawing a white ball from box II

$$P(A/E_1) = \frac{5}{10} \qquad P(A/E_2) = \frac{3}{10} \qquad P(A/E_3) = \frac{4}{10}$$

∴ Probability of drawing a white ball from box II

$$P(A) = \sum_{i=1}^{k} P(E_i)P(A/E_i) = 59/130.$$

Problem 2: A restaurant serves two special dishes A and B to its customers consisting of 60% men and 40% women, 80% of men order dish A and rest B. 70% of women order B and the rest A. In what ratio of A to B should be restaurant prepare the two dishes.

Solution:

Let E_1 be the event that the customer is male

Let E_2 be the event that the customer is female

$$\therefore \quad P(E_1) = 0.6 \quad P(E_2) = 0.4$$

Let A_1 be the event that the dish A is served.

Let B_1 be the event that the dish A is served.

$$\therefore \quad P(A_1/E_1) = 0.8$$

$$P(B_1/E_1) = 0.2$$

$$P(A_1/E_2) = 0.3$$

$$P(B_1/E_2) = 0.7$$

$$P(A) = P(E_1)\,P(A_1/E_1) + P(E_2)\,P(A_1/E_2)$$

$$= (0.6)(0.8) + (0.4)(0.3)$$

$$P(B) = P(E_1)\,P(B_1/E_1) + P(E_2)\,P(B_1/E_2)$$

$$= (0.6)(0.2) + (0.4)(0.7) = 0.40$$

$$\therefore \quad \frac{P(A)}{P(B)} = \frac{0.6}{0.4} = \frac{3}{2}.$$

Problem 3: Box I contains 3 red and 2 blue marbles while box II contains 2 red and 8 blue marbles. A fair coin is tossed. If the coin turns up heads a marble is chosen from box I. If it turns up tails a marble is chosen from box II. Find the probability that a red marble is chosen.

(Jan 2003)

Solution: Let E_1 be the event that box I is chosen.

Let E_2 be the event that box II is chosen.

$$\therefore \quad P(E_1) = P(E_2) = 1/2.$$

Let A be the event that a red marble is selected.

$$\therefore \quad P(A/E_1) = 3/5 \qquad P(A/E_2) = 2/10$$

Probability that a red marble is selected

$$P(A) = P(E_1)\, P(A/E_1) + P(E_2)\, P(A/E_2) = 2/5$$

Problem 4: There are three bags, first containing 1 white and 2 red, 3 green balls; second 2 white, 3 red, 1 green balls and third 3 white, 1 red and 2 green balls. Two balls are drawn from a bag chosen at random. These are found to be one white and one red. Find the probability that the balls are came from the second bag. **(Jan 2003)**

Solution:

Let E_1, E_2, E_3 be the events that the bags 1, 2, 3 are selected.

$$P(E_1) = P(E_2) = P(E_3) = 1/3$$

Let A be the event that two balls selected are white and red.

$$P(A/E_1) = \frac{{}^1C_1 \cdot {}^1C_1}{{}^6C_2} = \frac{2}{15}$$

$$P(A/E_2) = \frac{6}{15}$$

$$P(A/E_3) = \frac{3}{15}$$

Probability that drawing one white and one red ball from bag 2 is

$$P(E_2/A) = \frac{P(E_2)P(A/E_2)}{\sum_{i=1}^{3} P(E_i)P(A/E_i)} = \frac{6}{11}$$

Problem 5: In a bolt factory machines A, B and C manufacture respectively 25%, 35% and 40% of the total. Of their output 5, 4, 2 percent are defective bolts. A bolt is drawn at random from the product and is found to be defective. What are probabilities that is was manufactured by machines A, B and C? **(Apr 2003)**

Solution:

Let E_1, E_2, E_3 be the events that a ball is selected by the machines A, B and C.

$$\therefore \quad P(E_1) = 0.25,$$

$$P(E_2) = 0.35,$$

$$P(E_3) = 0.40$$

Let A be the event that drawing defective bolt.

∴ Probability of drawing defective bold by machine A is

$$P(A/E_1) = 0.05$$
$$P(A/E_2) = 0.04$$
$$P(A/E_3) = 0.02.$$

$$\therefore \quad P(E_1/A) = \frac{P(E_1)P(A/E_1)}{\sum_{i=1}^{3} P(E_i)P(A/E_i)} = \frac{25}{69}.$$

Similarly, $$P(E_2/A) = \frac{28}{69}$$
$$P(E_3/A) = \frac{16}{69}.$$

Problem 6: A business man goes to hotels X, Y, Z 20%, 50%, 30% of the time respectively. It is known that 5%, 4%, 8% of the rooms in X, Y, Z halts have faulty plumbing (*i*) Find the probability that the business man goes to the whole with faulty plumbing, (*ii*) What is the probability that business man's room having faulty plumbing is assigned to hotel Z?

(May 2006, Apr 2005, Apr 2007)

Solution:

Let E_1 be the event that the business man goes to hotel X

Let E_2 be the event that the business man goes to hotel Y

Let E_3 be the event that the business man goes to hotel Z

∴ $P(E_1) = 0.2$ $P(E_2) = 0.5$ $P(E_3) = 0.3$

Let A be the event of faulty plumbing

∴ $P(A/E_1) = 0.05$ $P(A/E_2) = 0.04$ $P(A/E_3) = 0.08$

Probability of faulty plumbing is $P(A) = \sum_{i=1}^{3} P(E_i)P(A/E_i) = 0.054.$

Problem 7: Of the three men the chances that a politician a business man and an educator will be appointed as a vice chancellor of a university are 0.5, 0.3 and 0.2 respectively. Probability that the research is promoted by these people, if they are appointed as a VC is 0.3, 0.79, 0.8 respectively,

(*i*) Find the probability that research is promoted in the university.

(*ii*) If research is promoted in the university, what is the probability that the V.C is an educator?

(*iii*) Businessman? **(Jan 2003, Apr 2007, Aug 2007, Nov 2007)**

Solution:

Let E_1 be the event that the politician is selected as V.C.

Let E_2 be the event that the business man is selected as V.C.

Let E_3 be the event that an educator is selected as V.C

$\therefore$ $P(E_1) = 0.5$ $P(E_2) = 0.3$ $P(E_3) = 0.2$

Let A be the event that the research is promoted

$$\therefore \quad P(A/E_1) = 0.3$$
$$P(A/E_2) = 0.79,$$
$$P(A/E_3) = 0.8$$

(*i*) Probability that the research is promoted

$$P(A) = \sum_{i=1}^{3} P(E_i)P(A/E_i) = 0.52$$

(*ii*) Probability that the research is promoted by educator is

$$P(E_3/A) = \frac{(0.2)(0.8)}{0.52} = 0.30769$$

(*iii*) Probability that the research is promoted by business man is

$$P(E_2/A) = 0.4038.$$

Problem 8: In a bolt factory, there are four machines, A, B, C and D manufacturing 20%, 15%, 25% and 40% of the total output respectively. Of their outputs 5%, 4%, 3% and 2% in the same order are defective bolts. A bolt is chosen at random from the factory's production and is found defective. What is the probability that the bolt was manufactured by machine A or machine D? **(April 2003)**

Solution:

Let E_1, E_2, E_3 and E_4 denote the events that s bolt selected at random is manufactured by the machines A, B, C and D respectively and let E denote the event of its being defective. Then

$$P(E_1) = \frac{20}{100} = \frac{2}{10} = 0.2 \qquad P(E_2) = \frac{15}{100} = 0.15$$
$$P(E_3) = \frac{25}{100} = 0.25 \qquad P(E_4) = \frac{40}{100} = 0.40$$

The probability that a defective bolt selected at random is manufactured by machine A is given by

$$P(E_1/E) = \frac{P(E_1)P(E/E_1)}{\sum_{i=1}^{4} P(E_i)P(E/E_i)}$$

The probability of drawing a defecting bolt manufactured by machine A is $P(E/E_1) = 0.05$

Similarly, $P(E/E_2) = 0.04$, $P(E/E_3) = 0.03$, $P(E/E_4) = 0.02$.

By Bayes theorem $$P(E_1/E) = \frac{P(E_1)P(E/E_1)}{\sum_{i=1}^{4} P(E_i)P(E/E_i)}$$

$$= \frac{0.2 \times 0.05}{0.2 \times 0.05 + 0.15 \times 0.04 + 0.25 \times 0.03 + 0.40 \times 0.2}$$

$$= 0.3175$$

The probability that a defective bolt selected at random is manufactured by machine D is

$$P(E_4/E) = \frac{P(E_4)P(E/E_4)}{\sum_{i=1}^{4} P(E_i)P(E/E_i)}$$

$$= \frac{0.40 \times 0.02}{0.2 \times 0.05 + 0.15 \times 0.04 + 0.25 \times 0.03 + 0.40 \times 0.2} = 0.254$$

Problem 9: There are three boxes. Box-I contains 10 light bulbs of which 4 are defective. Box II contains 6 light bulbs of which one is defective, Box III contains 8 light bulbs of which 3 are defective. A box is chosen and a bulb is drawn. Find the probability that the bulb is non defective. **(Feb 2007)**

Solution:

Let B_1 be the event where bulbs are of box-I

Let B_2 be the event where bulbs are of Box-II.

Let B_3 be the event where bulbs are of Box-III.

E be the event where bulbs are non-defective.

$P(B_1) = 1/10$ (probability if bulbs from Box-I

$P(B_2) = 1/6$ (probability of bulbs from Box-II)

$P(B_3) = 1/8$ (probability of bulbs from Box-III).

$P(E/B_1) = 6/10$ (probability of bulbs which are non defective of box-I)

$P(E/B_2) = 5/6$ (probability of bulbs which are non defective of Box-II)

$P(E/B_3) = 5/8$ (probability of bulbs which are non defective of Box-III)

$P(E) = P(B_1).\ P(E/B_1) + P(B_2)\ .P\ (E/B_2) + P(B_3)\ .P(E/B_3)$ [by theorem total probability]

$$P(E) = \frac{1}{10} \times \frac{6}{10} + \frac{1}{6} \times \frac{5}{6} + \frac{1}{8} \times \frac{5}{8}$$

$$= \frac{6}{100} + \frac{5}{36} + \frac{5}{64}$$

$$= \frac{864 + 2000 + 1125}{14400} = \frac{3989}{14400}$$

$$= 0.277.$$

Problem 10: In a bolt factory machines A, B, C manufacture 20%, 30% and 50% of the total of their output and 6%, 3 % and 2% are defective. A bolt is drawn at random and found to be defective. Find the probabilities that it is manufactured from (*i*) Machine A, (*ii*) Machine B, (*iii*) Machine C. **(May 2007, Feb 2007)**

Solution:

Let E_1 be the event that machine A manufactures.

Let E_2 be the event that machine B manufactures.

Let E_3 be the event that machine C manufactures

Let D be the event of defective output of machines.

$$P(E_1) = \frac{20}{100} \quad P(E_2) = \frac{30}{100} \quad P(E_3) = \frac{50}{100}$$

$$P(D/E_1) = \text{defective output from machine A} = \frac{6}{100}$$

$$P(D/E_2) = \text{defective output from machine B} = \frac{3}{100}$$

$$P(D/E_3) = \text{defective output from machine C} = \frac{2}{100}$$

$P(D) = P(E_1) .P(D/E_1) + P(E_2) P(D/E_2) + P(E_3) P(D/E_3)$ [by theorem of total probability]

$$P(D) = \frac{20}{100} \times \frac{6}{100} + \frac{30}{100} \times \frac{3}{100} + \frac{50}{100} \times \frac{2}{100}$$

$$= \frac{12}{1000} + \frac{9}{1000} + \frac{10}{1000} = \frac{31}{1000}$$

(*i*)
$$P(E_1/D) = \frac{P(E_1)P(D/E_1)}{P(D)}$$

$$= \frac{\frac{20}{100} \times \frac{6}{100}}{\frac{31}{1000}}$$

$$= \frac{12}{1000} \times \frac{1000}{31} = \frac{12}{31}$$

(*ii*)
$$P(E_2/D) = \frac{P(E_2)P(D/E_2)}{P(D)} = \frac{\frac{30}{100} \times \frac{3}{100}}{\frac{31}{1000}} = \frac{9}{1000} \times \frac{1000}{31} = \frac{9}{31}$$

(*iii*)
$$P(E_3/D) = \frac{P(E_3)P(D/E_3)}{P(D)} \frac{\frac{50}{100} \times \frac{2}{100}}{\frac{31}{1000}} = \frac{10}{1000} \times \frac{1000}{31} = \frac{10}{31}.$$

Problem 11: The probabilities of A, B, C to become M.D's of a factory are 5/10, 3/10, 2/10. The probabilities that bonus scheme will be introduced if they become M.D.s are 0.02, 0.03 and 0.04. Find the probabilities A, B, C to be become M.D 's if bonus scheme is introduced.

(Nov 2006, May 2007, Nov 2007)

Solution:

Let E_1 be the event (mutually exclusive) for A to become M.D.

Let E_2 be the event for B to become M.D.

Let E_3 be the event for C to become M.D.

Let A be the event where bonus scheme is introduced.

$$P(E_1) = \frac{5}{10} \quad P(E_2) = \frac{3}{10} \quad P(E_3) = \frac{2}{10}$$

$P(A/E_1) = 0.02 = \frac{2}{100}$ (probability that bonus scheme will be introduced if A becomes M.D.)

$P(A/E_2) = 0.03 = \frac{3}{100}$ (probability that the bonus scheme will be introduced if B becomes M.D.)

$P(A/E_3) = 0.04 = \frac{4}{100}$ (probability that the bonus scheme will be introduced if C becomes M.D.)

Probability that bonus scheme is introduced by theorem of total probability, we have

$$\begin{aligned} P(A) &= \sum_{i=1}^{3} P(E_i)P(A/E_i) \\ &= P(E_1) \,.\, P(A/E_1) + P(E_2)\, P(A/E_2) + P(E_3)\, P(A/E_3) \\ &= \frac{5}{10} \times \frac{2}{100} + \frac{3}{10} \times \frac{3}{100} + \frac{2}{10} \times \frac{4}{100} = \frac{27}{1000} \end{aligned}$$

(*i*) Probability to become M.D if bonus scheme is introduced for A

$$P(E_1/A) = \frac{P(E_1)P(A/E_1)}{P(A)} = \frac{10}{27}$$

(*ii*) Probability to become M.D if bonus scheme is introduced for B

$$P(E_2/A) = \frac{P(E_2)P(A/E_2)}{P(A)} = \frac{9}{27}$$

(*iii*) Probability to become M.D if bonus scheme is introduced for C

$$P(E_3/A) = \frac{P(E_3)P(A/E_3)}{P(A)} = \frac{8}{27}$$

Problem 12: Three machines produce 70%, 20% and 10% of the total number of a factory. The percentage of defective output of these machines is 4%, 3% and 2% respectively. An item is selected at random and found defective. Find the probability that it is from (*i*) Machine-I, (*ii*) Machine-II, (*iii*) Machine-III. **(Aug 2007)**

Solution:

Let E_1 be the event where factory is produced by Machine-I

Let E_2 be the event where factory is produced by Machine-II.

Let E_3 be the event where factory is produced by Machine-III.

D be the event that machine is defective.

$$P(E_1) = \frac{70}{100} \qquad P(E_2) = \frac{20}{100} \qquad P(E_3) = \frac{10}{100}.$$

$P(D/E_1)$ = Probability that the output of machine-I is defective = $\frac{4}{100}$

$P(D/E_2)$ = Probability that the output of machine-II is defective= $\frac{3}{100}$

$P(D/E_3)$ = Probability that the output of machine-III is defective= $\frac{2}{100}$

$$P(D) = \frac{70}{100} \times \frac{4}{100} + \frac{20}{100} \times \frac{3}{100} + \frac{10}{100} \times \frac{2}{100}$$

$$= \frac{36}{1000}$$

Probability that item is defective from

(*i*) Probability that the defective item is selected by Machine-I

$$P(E_1/D) = \frac{P(E_1)P(D/E_1)}{P(D)} = \frac{7}{9}$$

(*ii*) Probability that the defective item is selected by Machine-II

$$P(E_2/D) = \frac{P(E_2)P(D/E_2)}{P(D)} = \frac{1}{6}$$

(*iii*) Probability that the defective item is selected by Machine-III

$$P(E_3/D) = \frac{P(E_3)P(D/E_3)}{P(D)} = \frac{1}{18}$$

Problem 13: A box I contain 11 cards numbered 1 to 11, box-II contains 7 cards numbered 1 to 7. A box is selected at random and a card is drawn. If the number is even, find the probability that the card is from box-I.

Solution:

$P(E_1)$ = Probability that box-I is chosen.

$P(E_2)$ = Probability that box-II is chosen

$P(A)$ = Probability that the card is even.

$P(E_1/A)$ = Probability that number is even it is selected from box-I

$$P(E_1/A) = \frac{P(E_1).P(A/E_1)}{P(E_1)P(A/E_1)+P(E_2)P(A/E_2)}$$

$P(E_1) = P(E_2) = ½$

$P(A/E_1) = 5/11$

$P(A/E_2) = 3/7$

Since there are 5 even in box -I and 3 even in box -II

$\therefore \quad P(E_1/A) = 35/68.$

1.12 EXERCISE

1. What is the probability that a card drawn at random from the pack of playing cards may be either a queen or an ace?
2. A bag contains 4 red, 5 white and 7 black balls, if two balls are drawn at random. What is the probability that both of them will be black or both of them will be white?
3. There are two packs one card is drawn from each pack. What is the probability that (*i*) both of them are black (*ii*) they are of different colours?
4. The probabilities that a husband and wife will be alive 20 years from now are given by 0.8 and 0.9 respectively. Find the probability that in 20 years (*i*) both (*ii*) neither (*iii*) at least one will be alive.
5. Three light bulbs are chosen at random from 12 bulbs of which 5 are defective. Find the probability that (*i*) All are defective (*ii*) one is defective (*iii*) two are defective.
6. Three cards are drawn from a pack of 52 cards. Find the probability that (*i*) three are spades (*ii*) two spades one diamond (*iii*) one spade, one diamond, one heart.
7. Two cards are selected at random from 10 cards numbered 1 to 10. Find the probability that the sum is even. If (*i*) The two cards are drawn together (*ii*) The two cards are drawn one after the other with replacement. **(Jan 2006, Apr 2007, Nov 2007)**
8. A box contains 2 red and 3 marbles. Find the probability that if 2 marbles are drawn at random. (without replacement) (*i*) both are blue (*ii*) both are red (*iii*) one is red and one is blue.
9. What is the probability of drawing an ace from a well shuffled pack of 52 playing cards?
10. The student in a class is selected at random one after the other for an examination. Find the probability that the boys and girls are alternate if there are (*i*) 5 boys and 4 girls (*ii*) 4 boys and 4 girls **(Nov 2007)**
11. Determine (*i*) $P(B/A)$ (*ii*) $P(A/\bar{B})$ if A and B are events with $P(A) = 1/3$, $P(B) = 1/4$, $P(A \cap B) = 1/2$.
12. The probability that a regularly scheduled flight departs on time is $P(D) = 0.83$, the probability that it arrives on time is $P(A) = 0.82$ and the probability that it departs and arrive on time is $P(D \cap A) = 0.78$. Find the probability a plain (*i*) arrives on time given that it departs on time, (*ii*) departs on time given that it has arrived on time. **(Nov 2005)**
13. If at least one child in a family with two children is a boy what is the probability that both children are boys?
14. If $P(A \cup B) = 4/5$, $P(\bar{B}) = 1/3$ and $P(A \cap B) = 1/5$. Find (*i*) $P(B)$ (*ii*) $P(A)$ (*iii*) $P(\bar{A} \cap B)$.
15. If $P(A) = 1/5$, $P(B) = 2/3$, $P(A \cap B) = 1/15$, Find (*i*) $P(A \cup B)$ (*ii*) $P(\bar{A} \cap B)$ (*iii*) $P(A \cap \bar{B})$ (*iv*) $P(\bar{A} \cap \bar{B})$ (*v*) $P(\bar{A} \cup \bar{B})$.

16. A class has 10 boys and 5 girls. three students are selected at random, one after the other. Find the probability that (*i*) First two are boys and third is girl, (*ii*) First and third boys and second is girl, (*iii*) First and third of same sex and the second is opposite sex.

(Nov 2004, Apr 2007)

17. A and B through alternately with a pair of dice. One who first throws a total of nine wins. What are their respective chances of winning, if A starts the game?

(Nov 2004)

18. Box A contains 5 red and 3 white marbles and box B contains 2 red and 6 white marbles. (*i*) If a marble is drawn from each box, what is the probability that they are both of the same colour? (*ii*) If 2 marbles are drawn from each box, what is the probability that all four marbles of the same colour? **(May 2005, Feb 2007)**

19. A can hit a target once in 5 shots. B can hit two targets in two shots. C can hit a target once in 4 shots. What is the probability that two shots hit the target?

20. If the probability that a communication system will have high fidelity is 0.81 and the probability that it will have high fidelity and high selectivity is 0.18. What is the probability that a system high fidelity will also have high selectivity?

21. A man alternatively tosses a coin and throws a die beginning with point. What is the probability that he will get a head before he gets a 5 or 6 on the die?

22. Two digits are selected at random from the digits 1 through 9. (*i*) If the sum is odd, what is the probability that 2 is one of the numbers selected? (*ii*) If 2 is the one of the digits selected what is the probability that the sum is odd? **(Nov 2005, May 2006)**

23. Suppose 5 men out of 100 and 25 women out of 10, 000 are colour blind. A colour blind person is chosen at random. What is the probability of being a male? (Assume male and female are equal probable). **(Jan 2003)**

24. In a college 25% of boys and 10% of girls are studying mathematics. The girls constitute 60% of the student body. (*i*) What is the probability that mathematics is being studied? (*ii*) If a student is selected at random and is found to be studying mathematics, find the probability that the student is a girl. (*iii*) a boy. **(Nov 2004)**

25. Urn I has 2 white and 3 black balls. Urn II has 4 white and 1 black and Urn III has 3 white and 4 black balls. An Urn is selected at random and a ball is drawn at random is found to be white. Find the probability that Urn I was selected. **(Jan 2003)**

26. One bag contains 4 white balls and 3 black balls and second bag contains 3 white and 5 black balls. One ball is drawn from the first bag and placed unseen in the second bag. What is the probability that a ball now drawn from the second bag is black?

(Nov 2005)

27. Companies B_1, B_2, B_3 produce 30%, 45%, 25% of the cars respectively. It is know that 2%, 3%, 2% of these cars produced from B_1, B_2, B_3 are defective. (*i*) What is the probability that a car purchased is defective, (*ii*) If a car purchased is found to be defective. What is the probability that this car is produced by company B′.

(Nov 2005)

28. A can hit a target once in 5 shots. B can hit two targets in 3 shots. C can hit one target in 4 shots. What is the probability that 2 shots hit the target?

29. A problem is given to 4 students A, B, C, D. Probabilities of solving then independently are $\frac{2}{3}, \frac{2}{5}, \frac{1}{4}, \frac{3}{4}$. If all of them try to solve the problem, what is the probability that the problem is solved?

30. A, B and C in order toss a coin. The first one to toss a head wins the game. What are their probabilities of wining, assuming that the game may continue indefinitely?

31. The probabilities of passing in subjects A, B, C, D are $\frac{3}{4}, \frac{2}{3}, \frac{4}{5}, \frac{4}{5}$ and $\frac{1}{2}$ respectively. To qualify in the examination, a student should pass in A and two subjects among the three. What is the probability of qualifying in that examination?

32. In a certain college, 4% of the men and 1% of the women are taller than 1.8 m. Further more, 60% of the students are women. Now, if a student is selected, at random and is taller than 1.8 m, what is the probability that the student is a woman? **(April 2003)**

33. The students in a class are selected at random one after the other for an examination. Find the probability that the boys and girls are alternate, if there are (*i*) 5 boys and 4 girls, (*ii*) 4 boys and 4 girls. **(May 2006)**

34. The probabilities of A, B, C to become M.D's of a factory are $\frac{5}{10}, \frac{3}{10}, \frac{2}{10}$. The probabilities that bonus scheme will be introduced if they become M.D's are 0.02, 0.03 and 0.04. Find the probabilities A, B, C to be become M.D's if bonus scheme introduced.

(Nov 2006 , May 2007, Nov 2007)

35. There are two boxes in box I, 11 cards are there numbered 1 to 11 and in box II , 5 cards numbered 1 to 5. A box is chosen and a card is drawn. If the card shows an even number then another card is drawn from the same box. If card shows an odd number another card is drawn from the other box. Find the probability that (*i*) both are even (*ii*) both are odd (*iii*) if both are even. What is the probability that they are from box I.

(May 2007, Aug 2007, Nov 2007)

36. A card from a pack of 52 cards is lost. From the remaining cards of pack, two cards are drawn and are found to be hearts. Find the probability of the missing card to be hearts. **(May 2007)**

37. A can hit a target 4 times in 5 shots; B 3 times in 4 shots; C twice in 3 shots. They fire a volley. What is the probability that at least two shots hits?

38. An urn contains 10 white and 3 black balls, while another urn contains 3 white and 5 black balls. Two balls are drawn from the first urn and put into the second urn and then a ball is drawn from the latter. What is the probability that it is a white ball?

39. A five-figured number is formed by the digits 0, 1, 2, 3, 4 (without repetition). Find the probability that the number formed is divisible by 4.

40. A problem in statistics is given to five students. Their chances of solving it are $\frac{1}{2}, \frac{1}{3}, \frac{1}{4}, \frac{1}{4}$ and $\frac{1}{5}$. What is the probability that the problem will be solved?

41. A can hit a target 5 times in 6 shots, B 4 times in 5 shots and C 3 times in 4 shots. They fire a volley. What is the probability that two shots, at least hit the target?

42. A bag contains 10 balls, two of which are red, three are blue and five are black. Three balls are drawn at random from the bag. What is the probability that (*i*) the three balls are of different colours, (*ii*) two balls are of the same colour, (*iii*) the balls are all of the same colour.

ANSWERS

1. 2/3	**16.** 15/91, 15/91, 5/21	**30.** 4/7, 2/7, 1/7
2. 31/120	**17.** 9/17, 8/17	**31.** 11/20
3. 1/4, ½	**18.** 7/16, 55/784	**32.** 3/11
4. 0.72, 0.02, 0.98	**19.** ¼	**33.** 1/26, 1/35
5. 1/22, 21/44, 7/22	**20.** 2/9	**34.** 10/27, 1/3, 8/27
6. 11/850, 39/850, 169/1700	**21.** ¾	**35.** 31/220, 18/55, 20/31
7. 4/9, ½	**22.** ¼, 5/8	**37.** 5/6
8. 3/10, 1/10, 3/5	**23.** 0.952	**38.** 59/130
9. 1/3	**24.** 4/25, 3/8, 5/8	**39.** $\frac{5}{16}$
11. ¼, 1/3	**25.** 14/57	**40.** $\frac{17}{20}$
13. 1/3	**27.** 49/200, 12/49	**41.** $\frac{107}{120}$
14. 2/3, 1/3, 7/15	**28.** ¼	**42.** $\frac{1}{4}, \frac{79}{120}, \frac{11}{120}$
15. 4/5, 3/5, 2/15, 1/5, 14/15	**29.** 77/80	

1.13 QUIZ

1. There are 6 balls in a box. The number of ways two balls can be drawn with replacement = ...

   (*a*) 36 (*b*) 24

   (*c*) 32 (*d*) 40

2. In how many ways a student can answer 3 questions out of 5 questions in an examination................

(a) 12 (b) 10

(c) 15 (d) 20

3. A coin is tossed 3 times. The number of elements in sample space = ...

(a) 8 (b) 6

(c) 10 (d) 6

4. The probability of getting a sum of 6 if two dice are thrown is.....................

(a) 10 (b) 9

(c) 12 (d) 15

5. If $P(A) = \frac{1}{4}$, $P(B) = \frac{1}{3}$ and $P(A \cap B) = \frac{1}{5}$ then $P(A \cup B)$ = ...

(a) $\frac{23}{60}$ (b) $\frac{17}{60}$

(c) $\frac{1}{10}$ (d) $\frac{1}{4}$

6. If A and B are mutually exclusive events then $P(A \cup B)$ = ...

(a) P(A) + P(B) (b) P(A). P(B)

(c) P(A) – P(B) (d) P(A/B)

7. Probability of getting one Ace and one 10 when we draw 2 cards with replacement from a pack of 52 cards = ...

(a) $\frac{1}{13}$ (b) $\frac{1}{169}$

(c) $\frac{2}{169}$ (d) $\frac{3}{169}$

8. If a die is thrown, the probability of getting 5 or 6 is ....

(a) $\frac{1}{2}$ (b) $\frac{1}{3}$

(c) $\frac{2}{3}$ (d) $\frac{1}{4}$

9. Box I contains 3 Red balls 2 White balls, box II contains 2 Red balls, 2 White balls. One box is selected and one ball is selected. The probability that it is Red = ...

(a) $\frac{11}{20}$ (b) $\frac{13}{20}$

(c) $\frac{15}{20}$ (d) $\frac{17}{20}$

10. If two dice are thrown the probability that at least one of the faces is 2 is....................

(*a*) $\frac{11}{36}$ (*b*) $\frac{13}{36}$

(c) $\frac{1}{4}$ (*d*) $\frac{1}{6}$

11. If a coin weighed such that probability of getting head is $\frac{2}{3}$. Then probability of getting one head when two coins of same type are tossed is..................

(*a*) $\frac{1}{9}$ (*b*) $\frac{2}{9}$

(c) $\frac{3}{9}$ (*d*) $\frac{4}{9}$

12. If $P(A) = \frac{1}{3}$, $P(B) = \frac{1}{6}$, $P(A \cup B) = \frac{1}{4}$ then $P(B/A)$ =.........................

(*a*) $\frac{1}{4}$ (*b*) $\frac{1}{2}$

(c) $\frac{1}{3}$ (*d*) $\frac{3}{4}$

13. If A and B are independent events $P(A) = \frac{1}{4}$ and $P(A \cup B) = \frac{2}{3}$ then P(B)=...................

(*a*) $\frac{1}{9}$ (*b*) $\frac{3}{9}$

(c) $\frac{5}{9}$ (*d*) $\frac{7}{9}$

14. $P(A/B) = \frac{1}{10}$, $P(B) = \frac{1}{5}$ then $P(\bar{A} \cup \bar{B})$=....................

(*a*) $\frac{47}{50}$ (*b*) $\frac{49}{50}$

(c) $\frac{41}{50}$ (*d*) $\frac{39}{50}$

15. The probability of getting a multiple of 2 is $\frac{1}{2}$ and probability of getting a multiple of 3 is $\frac{1}{3}$ in throwing a die, then $P(A \cup B)$=.....................

(*a*) $\frac{1}{3}$ (*b*) $\frac{2}{3}$

(c) $\frac{1}{4}$ (*d*) $\frac{2}{5}$

16. Two persons are playing a game. Player A toss a coin. If head turns up he get one rupee if tail turns up he loses 1 rupee his expected gain is.......................

(*a*) 0 (*b*) 1

(c) 2 (*d*) 3

17. Two dice are thrown. The probability of getting at least one six is................

(a) $\frac{7}{36}$ (b) $\frac{11}{36}$

(c) $\frac{13}{36}$ (d) $\frac{5}{36}$

18. In a box there are Ten cards numbered 1 to 10. Two cards are drawn. The probability that both are even is.......................

(a) $\frac{1}{9}$ (b) $\frac{2}{9}$

(c) $\frac{4}{9}$ (d) $\frac{1}{3}$

19. In a box there are 50 cards numbered 1 to 50. If a card is drawn. The probability that it is a multiple of 3 and 5 is..................

(a) $\frac{3}{50}$ (b) $\frac{3}{25}$

(c) $\frac{7}{50}$ (d) $\frac{1}{50}$

20. The number of ways in which a true- false test consisting of 12 questions be answered if only one answer should be given to each equation..........................

(a) 12 (b) 24

(c) 2^{12} (d) 12^2

21. Getting heads or tails in a toss of a coin is called........................

(a) certain event (b) impossible event

(c) independent events (d) exhaustive events

22. In statistical definition of probability the value of $\frac{m}{n}$ is called ....

(a) frequency of E (b) relative frequency of E

(c) probability of E (d) none

23. The probability of drawing of king from a pack of cards is........................

(a) $\frac{1}{13}$ (b) $\frac{1}{26}$

(c) $\frac{1}{52}$ (d) none

24. The probability of drawing a queen of hearts from a pack of cards is.................

(a) $\frac{1}{13}$ (b) $\frac{1}{26}$

(c) $\frac{1}{52}$ (d) none

25. The probability that a leap year selected at random has 53 Sundays is.................

(*a*) $\frac{1}{7}$ (*b*) $\frac{2}{7}$

(*c*) $\frac{5}{7}$ (*d*) $\frac{6}{7}$

26. The probability that a non-leap year selected at random has 52 Sundays is..............

(*a*) $\frac{1}{7}$ (*b*) $\frac{2}{7}$

(*c*) $\frac{5}{7}$ (*d*) $\frac{6}{7}$

27. From 25 tickets marked 1 to 25, a ticket is drawn at random. What is the probability that it is divisible by 3 or 7 ....

(*a*) $\frac{8}{25}$ (*b*) $\frac{7}{25}$

(*c*) $\frac{6}{25}$ (*d*) $\frac{2}{5}$

28. If $A \cap B = \Phi$, P(A)=...............

(*a*) $= P(B)$ (*b*) $= P(\bar{B})$

(*c*) $\leq P(\bar{B})$ (*d*) $\leq P(B)$

29. $P(\bar{A} \cap B)$=.................

(*a*) $P(A) - P(B)$ (*b*) $P(A) - P(A \cap B)$

(*c*) $P(B) - P(A \cap B)$ (*d*) $P(B) - P(A)$

30. The probability that a problem will be solved of the probabilities that three student will solve independently are $\frac{1}{2}, \frac{1}{3}, \frac{1}{4}$ is......................

(*a*) $\frac{1}{4}$ (*b*) $\frac{3}{4}$

(*c*) $\frac{1}{2}$ (*d*) none

In a certain college 25% of boys and 10% girls are studying mathematics. The girls constitute 60% of the students.

31. If a student is selected at random what is the probability that the students studies mathematics..................

(*a*) $\frac{3}{8}$ (*b*) $\frac{5}{8}$

(*c*) $\frac{4}{25}$ (*d*) none

32. If a student is selected at random and is found to be Mathematics student, what is the probability that she is a girl...................

(a) $\frac{3}{8}$ (b) $\frac{5}{8}$

(c) $\frac{4}{25}$ (d) none

33. The probability that A happen is $\frac{1}{3}$. The odds against happening of A are...................

(a) 2 : 1 (b) 1 : 2

(c) 2 : 3 (d) 3 : 2

34. If P(A) = 0.35, P(B) = 0.73 and P(A $\cap$ B) = 0.14 then P(A $\cap$ $\bar{B}$)=....................

(a) 0.31 (b) 0.21

(c) 0.41 (d) 0.51

35. If A and B are independent, P(B) = 0.14 and P(A/B) = 0.24 then P(A)= ....

(a) 0.14 (b) 0.34

(c) 0.24 (d) 0.44

36. Maximum value of probability is......................

(a) 4 (b) 3

(c) 2 (d) 1

37. The probability of getting a total of 5 when a pair of dice is thrown is ....

(a) $\frac{1}{9}$ (b) $\frac{2}{9}$

(c) $\frac{3}{9}$ (d) $\frac{4}{9}$

38. If P(B) = 0.81 and P(A $\cap$ B) = 0.18 then P(A/B)=......................

(a) 0.1111 (b) 0.2222

(c) 0.3333 (d) 0.4444

39. If two unbiased dice are thrown simultaneously, the probability that the sum of the numbers on them is at least 10, is.........................

(a) $\frac{2}{9}$ (b) $\frac{3}{9}$

(c) $\frac{1}{6}$ (d) $\frac{4}{9}$

40. When 4 unbiased coins are tossed, the probability of getting 2 heads is..................

(a) 1/8 (b) 2/9

(c) 3/9 (d) 3/8

41. The chance of throwing 7 in a single throw with two dice is..................

(a) 1/6 (b) 2/9

(c) 3/9 (d) 3/8

42. In the role of an ideal die, the probability if getting a prime number is.....................

(a) 1/8 (b) 1/2

(c) 3/9 (d) 3/8

43. If A and B are mutually exclusive events, $P(A \cup B) = 0.6$, $P(B) = 0.4$, then P(A) =............

(a) 0.1 (b) 0.3
(c) 0.2 (d) 0.4

44. Two unbiased dice are thrown. Consider the event 'sum of points on the two dice is divisible by 3'. The number of favourable cases is................

(a) 10 (b) 12
(c) 14 (d) 8

45. From a bag containing 4 white and 5 black balls, 3 balls are drawn at random. The odds against all the balls being black is........................

(a) 35 : 7 (b) 5 : 37
(c) 37 : 5 (d) 7 : 35

46. If P(A)=1/2, P(B)=5/8 and $P(A \cap B)$=1/2, then P(AB) = ...

(a) 5/8 (b) 8/5
(c) 1/8 (d) 1/5

47. If A and B are two events such that P(A) is 1/3, P(B) is ¾ and $P(A \cup B)$=11/12, then P(A/B)=..................

(a) 1/9 (b) 3/9
(c) 4/9 (d) 2/9

48. 3 companies A, B and C produce 30%, 45%, 25% of computers respectively. It is known that 2%, 3% and 2% are defective. The probability that the computer purchased is defective is..........................

(a) 0.0245 (b) 0.0345
(c) 0.0124 (d) 0.045

49. Two biased dice are thrown, the number of favourable cases to get a total of more than 8 is.......................

(a) 8 (b) 14
(c) 12 (d) 10

50. 'A' speaks truth 4 out of 5 times. A die is thrown. He reports that there is a 6. The chance that actually there was actually 6 is..................

(a) 2/9 (b) 1/9
(c) 4/9 (d) 5/9

51. A has 3 shares in a lottery containing 3 prizes and 9 blanks . B has 2 shares in a lottery containing 2 prizes and 6 blanks. The ratio of their chances of success is...............

(a) 50/715 (b) 52/715
(c) 54/715 (d) 56/715

52. A purse contains 4 copper, 3 silver coins and a second purse contains 6 copper, 2 silver coins. If a coin is taken out of any purse, the probability that it is a copper coin is

(a) 37/56 (b) 36/56
(c) 35/56 (d) 34/56

53. Probability of getting 3 heads when 5 coins tossed is....................

(*a*) 1/16 (*b*) 3/16

(*c*) 5/16 (*d*) 7/16

54. Probability of null event is.....................

(*a*) 3 (*b*) 2

(*c*) 1 (*d*) 0

55. An unbiased coin is tossed 5 times then offs in favour of getting at least 1 tail is................

(*a*) 1 : 31 (*b*) 31 : 1

(*c*) 30 : 2 (*d*) 2 : 30

56. A box contains 4 white and 2 black marbles, another box containing 3 white and 5 black marbles. If 1 marbles is drawn from each box probability that both are white is......................

(*a*) 1/4 (*b*) 2/4

(*c*) 3/4 (*d*) 1

57. An unbiased coin is tossed until the first time 4 occurs. The expected number of tossed is.....................

(*a*) 1 (*b*) 12

(*c*) 3 (*d*) 4

58. If P(A)=0.4, P(A ∪ B)=0.7 and A, B are independent then P(B)=....................

(*a*) 1/2 (*b*) 1/3

(*c*) 1/2 (*d*) 1/5

59. 2 dice are thrown simultaneously, the probability of obtaining total score of 7 is.............

(*a*) 3/6 (*b*) 2/6

(*c*) 1/6 (*d*) none

60. A number is chosen at random of first 120 natural numbers, the probability that number divisible by 5 or 15 is........................

(*a*) 4/5 (*b*) 3/5

(*c*) 2/5 (*d*) 1/5

61. A speaks truth in 75% cases and B in 80%. The percentage of cases they are likely to contradict each other in starting the same fact is........................

(*a*) 30% (*b*) 35%

(*c*) 40% (*d*) 45%

62. There are 6 red and 5 black balls in a bag. A ball is drawn from bag at random and without utilizing it and another ball is drawn. The probability that both balls drawn is red is.......................

(*a*) 1/11 (*b*) 2/11

(*c*) 3/11 (*d*) 4/11

63. There are 4 letters and 4 addressed envelopes. If the letters are put at random in the envelopes then the chance that number of letter will occupy its correct place is..............

(*a*) 1/8 (*b*) 3/8

(*c*) 5/8 (*d*) 7/8

64. Two odds that a book will be received favourably by three independent critics are 3 to 2, 2 to 3 and 4 to 1 respectively. The probability that out of the 3 reviewers , majority will be favourable is........................

(*a*) 82/125 (*b*) 81/125

(*c*) 80/125 (*d*) 79/125

65. A card is taken out of a pack of 52 cards numbered 2 to 53. What is probability that the number on the card is prime number less than 20?

(*a*) $\frac{1}{3}$ (*b*) $\frac{2}{3}$

(*c*) 1 (*d*) $\frac{4}{3}$

66. The probability of getting a number between 1 to 100 which is divisible by 1 and itself only is....................

(*a*) 25/98 (*b*) 23/98

(*c*) 27/98 (*d*) none

67. When 2 dice are thrown the probability of setting the sum 10 or 11 is....................

(*a*) 4/36 (*b*) 5/36

(*c*) 6/36 (*d*) 7/36

68. The probability that an event happens in one trial of experiment is 0.4. If 3 independent trials performed, the probability that the event happens at least once is....................

(*a*) 0.874 (*b*) 0.487

(*c*) 0.784 (*d*) none

69. A die is cast and a coin is tossed the probability that a number less than 5 is thrown and H is tossed is..................

(*a*) 1/2 (*b*) 1/4

(*c*) 1/3 (*d*) 1/5

70. If 2 dice are thrown the probability that the sum show will be 8 or 12 is...............

(*a*) 1/3 (*b*) 1/4

(*c*) 1/5 (*d*) 1/6

71. 4 boys and 4 girls sit in row at random. What is probability that boys, girls sit alternately?

(*a*) 1/35 (*b*) 1/36

(*c*) 1/37 (*d*) 1/38

72. In drawing 2 cards from a pack of 52 cards, the exhaustive number of events is..........

(*a*) ${}^{52}C_1$ (*b*) ${}^{52}C_2$

(*c*) ${}^{52}C_4$ (*d*) ${}^{52}C_3$

73. If B is an event with P(B) > 0 then $P(\Phi/B)$ =..................

(*a*) 3 (*b*) 2

(*c*) 1 (*d*) 0

74. A and B throw with 3 dice. If A throws 16, B's chance of throwing a higher number is..................

(*a*) 1/52 (*b*) 1/53

(*c*) 1/54 (*d*) 1/55

75. If A and B are two events such that P(A)=1/3, P(B)=3/4 and P(A ∪ B)=11/12 then P(B/A) is..................

(*a*) 1/2 (*b*) 1/3

(*c*) 1/4 (*d*) 1/5

76. In a software company 25% of males and 10% of females are working in project design. The female employees constitute 60% of total work force. If a person is selected at random and is found to be project designer the probability is......................

(*a*) 1/8 (*b*) 2/8

(*c*) 3/8 (*d*) 4/8

77. A die is tossed if the number is prime then the probability that it is off is................

(*a*) 1/3 (*b*) 2/3

(*c*) 4/3 (*d*) 3/3

78. Let A and B are 2 events of an experiment such that P(A)=0.4, P(A ∪ B)=0.8, if A and B are mutually exclusive events the P(B) = ...

(*a*) 0.1 (*b*) 0.2

(*c*) 0.3 (*d*) 0.4

79. A bag cointains 4 green, 6 black and 7 white balls. A ball is drawn at random the probability that either a black or white ball is.......................

(*a*) 13/17 (*b*) 14/17

(*c*) 15/17 (*d*) 16/17

80. Two cards are drawn from a well shuffled ordinary deck if 52 cards. The probability that both are cards aces. If the 1st card is not replaced is....................

(*a*) 4/221 (*b*) 3/221

(*c*) 2/221 (*d*) 1/221

81. An unbiased coin is tossed until the 1st time head occurs. The expected number of tosses is..................

(*a*) 1 (*b*) 2

(*c*) 3 (*d*) 4

82. A wife and husband appear at an interview for 2 vacancies of the same post. The probability of husband selection is 1/7 and the wife selection is 1/5 and the probability that both of them will be selected is.......................

(*a*) 1/30 (*b*) 1/32

(*c*) 1/35 (*d*) 1/37

83. 2 digits are selected at random from the digits 1 to 9 if the sum is even the probability that both the numbers are even

(*a*) 1/15 (*b*) 1/14

(*c*) 1/13 (*d*) 1/16

84. A pair of fair dice is thrown the probability that sum is 10 or greater if a 5 appear on die is....................

(*a*) 1/3 (*b*) 1/4

(*c*) 1/13 (*d*) 1/16

85. 3 bulbs are chosen at random from 10 bulbs having 6 good bulbs to fit in a room. The chance that room lighted is..................

(*a*) 27/30 (*b*) 29/30

(*c*) 25/30 (*d*) None

86. A coin is tossed 6 times the probability of obtaining 4 or more bonds is.............

(*a*) 9/32 (*b*) 10/32

(*c*) 11/32 (*d*) 12/32

87. In a town 40% have motor cycle 25% have car and 15% have both. If a person is selected at random probability he has car also is................

(*a*) 1/8 (*b*) 2/8

(*c*) 3/8 (*d*) 4/8

88. 2 horses A and B run a race if the probability of A's win is twice that of B. The probability of B's win is

(*a*) 4/3 (*b*) 3/3

(*c*) 2/3 (*d*) 1/3

89. The probability of getting exactly 2 tails in tossing a fair coin thrice is.................

(*a*) 2/3 (*b*) 1/3

(*c*) 1 (*d*) none

90. 2 unbiased dice are thrown consider event sum of points on the 2 dice is divisible by 3 the no. of favourable events is......................

(*a*) 10 (*b*) 12

(*c*) 14 (*d*) 16

ANSWERS

1. (*a*)	**2.** (*b*)	**3.** (*a*)	**4.** (*a*)	**5.** (*a*)	**6.** (*a*)	**7.** (*b*)	**8.** (*b*)	**9.** (*a*)	**10.** (*a*)
11. (*b*)	**12.** (*d*)	**13.** (*c*)	**14.** (*b*)	**15.** (*b*)	**16.** (*a*)	**17.** (*b*)	**18.** (*b*)	**19.** (*a*)	**20.** (*d*)
21. (*a*)	**22.** (*b*)	**23.** (*a*)	**24.** (*c*)	**25.** (*b*)	**26.** (*d*)	**27.** (*d*)	**28.** (*c*)	**29.** (*c*)	**30.** (*b*)
31. (*c*)	**32.** (*a*)	**33.** (*a*)	**34.** (*b*)	**35.** (*c*)	**36.** (*d*)	**37.** (*a*)	**38.** (*b*)	**39.** (*c*)	**40.** (*d*)
41. (*a*)	**42.** (*b*)	**43.** (*c*)	**44.** (*b*)	**45.** (*c*)	**46.** (*a*)	**47.** (*d*)	**48.** (*a*)	**49.** (*d*)	**50.** (*c*)
51. (*b*)	**52.** (*a*)	**53.** (*c*)	**54.** (*d*)	**55.** (*b*)	**56.** (*a*)	**57.** (*b*)	**58.** (*a*)	**59.** (*c*)	**60.** (*d*)
61. (*b*)	**62.** (*c*)	**63.** (*b*)	**64.** (*a*)	**65.** (*b*)	**66.** (*a*)	**67.** (*b*)	**68.** (*c*)	**69.** (*c*)	**70.** (*d*)
71. (*a*)	**72.** (*b*)	**73.** (*d*)	**74.** (*c*)	**75.** (*a*)	**76.** (*c*)	**77.** (*b*)	**78.** (*d*)	**79.** (*a*)	**80.** (*d*)
81. (*b*)	**82.** (*c*)	**83.** (*d*)	**84.** (*a*)	**85.** (*b*)	**86.** (*c*)	**87.** (*c*)	**88.** (*d*)	**89.** (*a*)	**90.** (*b*)

Chapter 2

PROBABILITY DISTRIBUTIONS

2.0 INTRODUCTION

In the previous chapters, we have discussed that the concept of the random experiment leads to the notation of a sample space. The assignment and computation of probabilities of events were studied in detail. In many experiments, we are interested not in knowing which of the outcomes has occurred, but in the numbers associated with them.

Random Variables

Suppose S is the sample space of some experiment. We know that outcomes of the experiment are the elements of the sample space S and they need not be numbers. Some times we wish to assign a specific number to each outcome. For example the number of heads in tossing two or more coins, sum of the points on a pair of dice, when they are thrown. Such assignment is called a random variable.

Examples:

1. Suppose two coins are tossed simultaneously then the sample space is S={HH, HT, TH, TT}

   X denotes the number of heads. Then if X = 0 then the outcomes is {TT} and $P(X = 0) = \frac{1}{4}$.

   If X takes the value 1, then outcomes is {TH, HT} and $P(X = 1) = \frac{2}{4}$

   If X takes the value 2, then the outcome is {HH} and $P(X = 2) = \frac{1}{4}$

∴ The probability distribution of this random variable X is given by the following table

$X = x$	0	1	2	Total
$P(X = x)$	$\frac{1}{4}$	$\frac{2}{4}$	$\frac{1}{4}$	1

2. Out of 24 mangoes 6 are routine, two mangoes are drawn. Obtain the probability distribution of the number of routine mangoes that can be drawn.

   Let X denote the number routine mangoes drawn then X can take values 0, 1, 2

   $P(X = 0) = \frac{51}{92}$ $P(X = 1) = \frac{9}{23}$ $P(X = 2) = \frac{5}{92}$

   ∴ Probability distribution of X is

$X = x$	0	1	2	Total
$P(X = x)$	$\frac{51}{92}$	$\frac{9}{23}$	$\frac{5}{92}$	1

2.1 TYPES OF RANDOM VARIABLES

There are two types of random variables:

(*i*) Discrete random variables.

(*ii*) Continuous random variable.

Distribution Function

Let X be a one-dimensional random variable. The function F defined for all real x, by the equation $F(x) = P(X \leq x)$ is called the cumulative distribution function of X.

Note:

1. We write cumulative distribution function for cumulative distribution function. Sometimes, only distribution function is written instead of cumulative distributive function.
2. Suffix X in F is used to emphasize the fact that the distribution function is associated with the particular variate X. When the particular underlying variate is clear from the contexts, we shall simply write $F(x)$.

Properties of Distribution Function

1. If F is the distribution function of the random variable X and if $a < b$, then $P(a < X \leq b) = F(b) - F(a)$.
2. When $P(X = a) = 0$ and $P(X = b) = 0$, all four events $a \leq X \leq b$, $a < X < b$, $a \leq X < b$ and $a < X \leq b$ have the same probability $F(b) - F(a)$.
3. If F is distribution function of one-dimensional random variable X, then

   (*i*) $0 \leq F(x) \leq 1$.

   (*ii*) $F(x) \leq F(y)$ if $x < y$.

   In other words, all distribution functions are monotonically non-decreasing and lie between 0 and 1.

2.2 DISCRETE RANDOM VARIABLE

A variable which can assume only a finite number of real values and for which the value which the variable takes depends on chance, is called a discrete random variable. In other words, a real valued function defined on a discrete sample space is called a discrete random variable.

Examples of discrete random variables are marks obtained in a test, the number of children in the families of a colony, the number of rooms in the houses of a township.

Probability Mass Function

Let X be discrete random variable. The discrete probability function F(x) for X is given by F(x) = P(X = x) for all real x.

Since probabilities cannot be negative, a probability function F(x) cannot assume negative values.

Now, we can state the discrete probability distribution, probability function or probability mass function of a discrete random variable X as the function F(x) satisfying the following conditions:

(*i*) $P(x_i) \geq 0$

(*ii*) $\sum_{i=1}^{\infty} P(x_i) = 1$

(*iii*) $P(X = x) = F(x)$

2.3 SOLVED PROBLEMS

Problem 1: A random variable X has the following probability function:

x	0	1	2	3	4	5	6
$P(x)$	k	$3k$	$5k$	$7k$	$9k$	$11k$	$13k$

(*i*) Find k (*ii*) Evaluate $P(X < 4)$ (*iii*) $P(X \geq 5)$ (*iv*) $P(3 < X \leq 6)$

(*v*) What is the smallest value of x for which $P(X \leq x) > \frac{1}{2}$?

Solution:

(*i*) $\Sigma P(X = x_i) = 1$

$\Rightarrow P(X = 0) + P(X = 1) + P(X = 2) + \ldots\ldots\ldots\ldots + P(X = 6) = 1$

$\Rightarrow k + 3k + 5k + 7k + 9k + 11k + 13k = 1$

$\Rightarrow 49k = 1 \qquad \Rightarrow k = \frac{1}{49}$

(*ii*) $P(X < 4) = P(X = 0) + P(X = 1) + P(X = 2) + P(X = 3)$

$= \frac{1}{49} + \frac{3}{49} + \frac{5}{49} + \frac{7}{49} = \frac{16}{49}$

(*iii*) $P(X \geq 5) = P(X = 5) + P(X = 6)$

$= \frac{11}{49} + \frac{13}{49} = \frac{24}{49}$

(*iv*) $P(3 < X \leq 6) = P(X = 4) + P(X = 5) + P(X = 6)$

$$= \frac{9}{49} + \frac{11}{49} + \frac{13}{49} = \frac{33}{49}$$

(*v*) The smallest value of x is determined by trial method.

That is substituting various values of x

$$P(X \leq 6) = 1 > \frac{1}{2} \qquad P(X \leq 5) = \frac{36}{49} > \frac{1}{2}$$

$$P(X \leq 4) = \frac{25}{49} > \frac{1}{2} \quad P(X \leq 3) = \frac{16}{49} < \frac{1}{2}$$

∴ Smallest value of x for which $P(X < x) > 1/2$ is 4.

Problem 2: Two dice are thrown. Let X assign to each point (a, b) in S the maximum of its numbers *i.e.*, $X(a, b) = \max(a, b)$. Find the probability distribution X is a random variable with $X(S) = \{1, 2, 3, 4, 5, 6\}$. **(JNTU 2004)**

Solution:

$$P(1) = P(X = 1) = P(1, 1) = \frac{1}{36}$$

$$P(2) = P(X = 2) = P[(2, 1)(2, 2)(1, 2)] = \frac{3}{36}$$

$$P(3) = P(X = 3) = P[(1, 3)(3, 1)(2, 3)(3, 2)(3, 3)] = \frac{5}{36}$$

$$P(4) = P(X = 4) = P[(1, 4)(4, 1)(2, 4)(4, 2)(3, 4)(4, 3)(4, 4)] = \frac{7}{36}$$

$$P(5) = P(X = 5) = P[(1, 5)(5, 1)(2, 5)(5, 2)(3, 5)(5, 3)(4, 5)(5, 4)(5, 5)] = \frac{9}{36}$$

$$P(6) = P(X = 6) = P[(1, 6)(6, 1)(2, 6)(6, 2)(3, 6)(6, 3)(4, 6)(6, 4)(5, 6)(6, 5)(6, 6)] = \frac{11}{36}$$

x	1	2	3	4	5	6
$P(x)$	$\frac{1}{36}$	$\frac{3}{36}$	$\frac{5}{36}$	$\frac{7}{36}$	$\frac{9}{36}$	$\frac{11}{36}$

Problem 3: A random variable X has the following probability function:

x	0	1	3	4	5	6	7
$P(x)$	0	k	$2k$	$2k$	$3k$	k^2	$7k^2 + k$

(*i*) Find the value of k

(*ii*) Evaluate $P(X < 6)$, $P(X \geq 6)$

(*iii*) $P(0 < X < 5)$. **(Dec 2005)**

Solution:

(*i*) Given X is a discrete random variable

We know that $\sum_{i=1}^{n} p(x_i) = 1$

$$\Rightarrow \quad 0 + k + 2k + 2k + 3k + k^2 + 7k^2 + k = 1$$

$$\Rightarrow \quad 8k^2 + 9k - 1 = 0$$

$$k = \frac{-9 \pm \sqrt{113}}{16}$$

$$k = \frac{-9 \pm \sqrt{113}}{16} \quad \text{[Since } k \text{ must be +ve]}$$

(*ii*)

$$P(X < 6) = \sum_{i=1}^{5} p(x_i) = 0 + k + 2k + 2k + 3k = 8k$$

i.e., $P(X < 6) = 8k$

where $k = \dfrac{-9 \pm \sqrt{113}}{16}$

$$P(X > 6) = 1 - P(X < 6)$$

$$P(X > 6) = 1 - 8k$$

where $k = \dfrac{-9 \pm \sqrt{113}}{16} = \dfrac{-9 \pm 10.63}{16} = 0.10$

$$P(X > 6) = 1 - 8 \times 0.10 = 0.2$$

$$P(X > 6) = 0.2$$

(*iii*)

$$P(0 < X < 5) = \sum_{i=1}^{4} p(x_i) = k + 2k + 2k = 5k$$

$$P(0 < X < 5) = 5k$$

where $k = \dfrac{-9 + \sqrt{113}}{16} = 0.10$

$$\Rightarrow \quad P(0 < X < 5) = 5 \times 0.10 = 0.5.$$

2.4 CONTINUOUS RANDOM VARIABLES

A random variable X is said to be continuous if it can take all possible values between certain limits.

Probability Density Function (p.d.f.)

Consider the sma'l interval $(x, x + dx)$ of length dx around the point x. Let $f(x)$ be any continuous function of x, so that $f(x)\,dx$ represents the probability that X fails in the interval $(x, x + dx)$. Symbolically, $P(x \leq X \leq x + dx) = f(x)\,dx$.

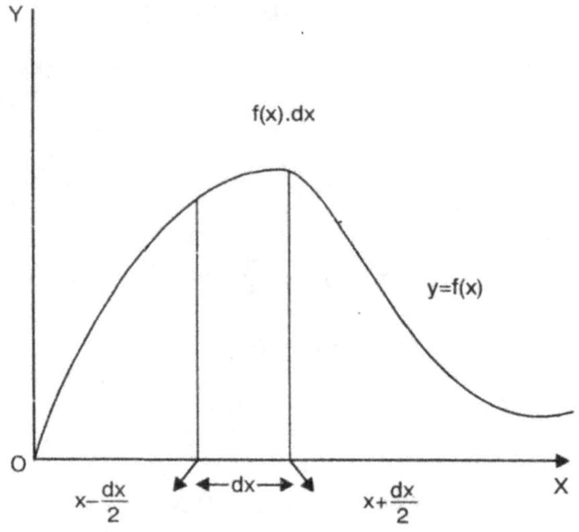

In the figure, $f(x)dx$ represents the area bounded by the curve $y = f(x)$, X-axis and the ordinates at the points x and $x + dx$. The function $f(x)$ so defined is known as probability density function or simply density function of random variable X.

Note:

1. The expression, $f(x)\ dx$ is known as the probability differential.
2. The curve $y = f(x)$ is known as the probability density curve or simply probability curve.

Definition

Probability density function of the random variable X is defined as, *i.e.*, derivative of cumulative distribution function gives the probability density function. The plot of probability density function of a discrete random variable is an impulse train.

The probability for a variate value to lie in the interval dx is $f(x)dx$ and hence, the probability for a variate value to fall in the finite interval $[a, b]$ is

$P(a \le x \le b) = \int_a^b f(x)dx$ which represents the area between the curve $y = f(x)$, X-axis and the ordinates at $x = a$ and $x = b$.

Since the total probability is unity $\int_a^b f(x)dx = 1$. The probability density function (p.d.f.) of a random variable X usually denoted by $f(x)$ has the following properties:

(*i*) $f(x) \ge 0$ (*ii*) $\int_{-\infty}^{\infty} f(x)dx = 1$.

Note:

1. For a continuous variate, point probabilities are zero.
2. Area under the probability curve $y = f(x)$ is unity; the fact $f(x) \ge 0$ implies that graph of $f(x)$ is above X-axis.

3. Area under the graph of probability curve $y = f(x)$ bounded by $x = a$, $x = b$ simply $P(a \le x \le b)$.
4. Relation between probability density function and cumulative distribution function: The density function $f(x)$ and cumulative distribution function $F(x)$ are always connected by

$$(i)\ F(x) = \int_{-\infty}^{x} f(x)dx \qquad (ii)\ \frac{d}{dx}[F(x)] = f(x)$$

5. $P(a \le X \le b) = \int_a^b f(x)dx$
6. $P(a < X < b) = P(a \le X < b) = P(a < X \le b) = P(a \le X \le b)$.

2.5 SOLVED PROBLEMS

Problem 1: A continuous random variable X has the probability density function given by $f(x) = kx^2$, $0 \le x \le 1$. Find the value of k. With this value of k, find $P(X < \frac{1}{2})$ and $P(X \ge \frac{3}{4})$.

Solution:

Since $f(x)$ is the p.d.f.

$$\therefore \quad \int_0^1 f(x)dx = 1, \int_0^1 kx^2\, dx = 1$$

$$\Rightarrow \quad \left[k\frac{x^3}{3}\right]_0^1 = 1$$

$$\Rightarrow \quad \frac{k}{3} = 1 \Rightarrow k = 3$$

$$\therefore \quad f(x) = 3x^2, 0 \le x \le 1$$

$$P(X < \frac{1}{2}) = P(0 < X < \frac{1}{2})$$

$$= \int_0^{1/2} f(x)dx = \int_0^{1/2} 3x^2\, dx = \frac{1}{8}$$

$$P(X \ge \frac{3}{4}) = P(\frac{3}{4} \le X \le 1)$$

$$= \int_{3/4}^{1} f(x)dx = \int_{3/4}^{1} 3x^2\, dx = \left[3\frac{x^3}{3}\right]_{3/4}^{1} = 1 - \frac{27}{64} = \frac{37}{64}$$

Problem 2: Is the function $f(x)$ defined as follows, a density function?

$$f(x) = 0, x < 2$$

$$= \frac{1}{18}(3 + 2x), 2 \le x \le 4$$

$$= 0, x > 4.$$

Solution:

$$\int_{-\infty}^{\infty} f(x)dx = \int_{-\infty}^{2} f(x)\,dx + \int_{4}^{\infty} f(x)dx + \int_{4}^{\infty} f(x)\,dx$$

$$= \int_{2}^{4} \frac{1}{18}[3+2x]\,dx$$

$$= \frac{1}{18}\left[3x + x^2\right]_2^4 = 1$$

Hence $f(x)$ is density function.

Problem 3: If the probability density of a random variable is given by $f(x) = k(1 - x^2)$; $0 < x < 1 = 0$; elsewhere.

Find the value of k and the probabilities that a random variable having this probability density will taken on a value (*i*) between 0.1 and 0.2 (*ii*) greater than 0.5

Solution:

$$\int_0^1 f(x)dx = 1$$

$$k\int_0^1 f(x)dx = 1$$

$$k\left[x - \frac{x^3}{3}\right]_0^1 = 1$$

$$\frac{2k}{3} = 1$$

$$k = \frac{3}{2}$$

(*i*)

$$P(0.1 < X < 0.2) = \int_{0.1}^{0.2} f(x)dx$$

$$= \frac{3}{2}\int_{0.1}^{0.2}\left(1 - x^2\right)dx$$

$$= \frac{3}{2}\int_{0.1}^{0.2}\left(1 - x^2\right)dx$$

$$= \frac{3}{2}\left[x - \frac{x^3}{3}\right]_{0.1}^{0.2} = 0.146$$

(*ii*)

$$P(X > 0.5) = \int_{0.5}^{1} f(x)\,dx$$

$$= \frac{3}{2}\int_{0.5}^{1}\left(1 - x^2\right)dx = 0.3125$$

Problem 4: A random process gives measurements X between 0 and 1 with a probability density function.

$$f(x) = 12x^3 - 21x^2 + 10x \quad 0 < x < 1$$
$$= 0 \quad \text{otherwise}$$

(*i*) Find $P(X \leq \frac{1}{2})$ and $P(X > \frac{1}{2})$ (*ii*) Find a number k such that $P(X \leq k) = \frac{1}{2}$.

Solution:

(*i*)
$$P(X \leq \frac{1}{2}) = \int_0^{1/2} f(x)dx$$
$$= \int_0^{1/2} \left(12x^3 - 21x^2 + 10x\right)dx$$
$$= \left[3x^4 - 7x^3 + 5x^2\right]_0^{1/2}$$
$$= \frac{3}{16} - \frac{7}{8} + \frac{5}{4}$$
$$= \frac{3 - 14 + 20}{16} = \frac{9}{16}$$
$$P(X > \frac{1}{2}) = 1 - P(X \leq \frac{1}{2})$$
$$= 1 - \frac{9}{16}$$
$$= \frac{7}{16}$$

(*ii*)
$$P(X < k) = \int_0^{k} f(x)dx = \frac{1}{2}$$

$$\Rightarrow \quad 3k^4 - 7k^3 + 5k^2 = \frac{1}{2}$$

$$\Rightarrow \quad 6k^4 - 14k^3 + 10k^2 - 1 = 0 \quad \Rightarrow \quad k = 0.452$$

Problem 5: Let $f(x) = 3x^2$, when $0 \leq x \leq 1$ be the probability density function of a continuous variable X. Determine a and b such that (*i*) $P(X \leq a) = P(X > a)$, (*ii*) $P(X > b) = 0.05$. **(Nov 2004)**

Solution:

(*i*)
$$P(X \leq a) = P(X > a)$$

$$\int_{-\infty}^{a} f(x)\,dx = \int_{a}^{\infty} f(x)\,dx$$

$$\int_0^a 3x^2\,dx = \int_a^1 3x^2\,dx$$

$$\left[x^3\right]_0^a = \left[x^3\right]_a^1$$

$$a^3 = 1 - a^3$$

$$2a^3 = 1 \Rightarrow a^3 = 0.5 \Rightarrow a = 0.79$$

(*ii*) $$P(X > b) = 0.05$$

$$= \int_b^\infty f(x)dx = 0.05$$

$$= \int_b^1 3x^2\,dx = 0.05$$

$$\left[x^3\right]_b^1 = 0.05$$

$$1 - b^3 = 0.05$$

$$\Rightarrow \quad b = (0.95)^{1/3} = 0.98.$$

Problem 6: A continuous random variable X has the distribution function

$$F(x) = 0\ ; \text{ if } x \le 1$$
$$= k\,(x-1)^4\ ; \quad \text{if } 1 < x \le 3$$
$$= 1\ ; \text{ if } x > 3$$

Find (*i*) The probability density function of X (*ii*) Find k (*iii*) Mean. **(Nov 2003)**

Solution:

(*i*) We have $f(x)$ = The probability density function of X = $\frac{d}{dx}[F(x)]$

i.e., $$f(x) = \frac{d}{dx}[F(x)]$$

$$f(x) = 0 \quad \text{if } x \le 1$$
$$= 4k\,(x-1)^3 \quad \text{if } i < x \le 3$$
$$= 0 \quad \text{if } x > 3$$

(*ii*) $$\int_{-\infty}^{\infty} f(x)dx = 1$$

$$\Rightarrow \quad \int_1^3 4k(x-1)^3 dx = 1$$

$$4k\left[\frac{(x-1)^4}{4}\right]_1^3 = 1$$

$$k(3-1)^4 = 1$$

$$k = \frac{1}{16}$$

(*iii*)

$$\text{Mean } \mu = \int_{-\infty}^{\infty} xf(x)\,dx$$

$$= \int_{-\infty}^{1} xf(x)\,dx + \int_{1}^{3} xf(x)dx + \int_{3}^{\infty} xf(x)dx$$

$$= \int_{1}^{3} xf(x)\,dx$$

$$= 4k\int_{1}^{3} (x-1)^3\,dx$$

$$= 4 \times \frac{1}{16}\left[\frac{(x-1)^4}{4}\right]_1^3 = 1 \quad \text{Since } k = \frac{1}{16}$$

Problem 7: Let X denote the number of heads in a single toss of 4 fair coins. Determine

(*i*) $P(X < 2)$ (*ii*) $P(1 < X \leq 3)$. **(May 2006)**

Solution:

X	0	1	2	3	4
P(X)	$\frac{1}{16}$	$\frac{1}{16}$	$\frac{1}{16}$	$\frac{1}{16}$	$\frac{1}{16}$

(*i*) $P(X < 2) = P(X = 0) + P(X = 1)$

$$= \frac{1}{16} + \frac{1}{16} = \frac{1}{8}$$

(*ii*) $P(1 < X \leq 3) = P(X = 2) + P(X = 3)$

$$= \frac{1}{16} + \frac{1}{16} = \frac{1}{8}.$$

2.6 MATHEMATICAL EXPECTATION

Let X be a discrete random variable which takes the values $x_1, x_2, x_3, \ldots\ldots, x_n$ with probabilities $P(x) = P(X = x_i)$, where $i = 1, 2, 3, \ldots\ldots, n$ then the expectation of X is denoted by E(X) and defined as

$$E(X) = \sum_{i=1}^{n} x_i \, P(X = x_i)$$

E(X) is also called the mean of random variable X and is denoted by μ.

$$\therefore \quad \mu = E(X) = \sum_{i=1}^{n} x_i \, P(X = x_i)$$

$$E(X^2) = \sum_{i=1}^{n} x^2 \, i P(X = x_i)$$

Similarly, if X is continuous random variable taking the values $-\infty < X < \infty$ with $f(x)$ as the probability density function, the $\mu = E(X) = \int_{-\infty}^{\infty} x f(x) dx$

$$E(X^2) = \int_{-\infty}^{\infty} x^2 f(x) dx$$

Properties of Expectation

Let X and Y are two random variables and a and b are constants then

(*a*) $E(a) = a$

(*b*) $E(aX) = aE(X)$

(*c*) $E(aX + b) = aE(X) + b$

(*d*) $E(aX \pm bY) = aE(X) \pm bE(Y)$

(*e*) $E(X + a) = E(X) + a$

(*f*) If X, Y, Z are independent random variables then $E(XYZ) = E(X)\,E(Y)\,E(Z)$

Variance

If X is a random variable then the variance of X is denoted by V(X) or σ_x^2 is defined as

$$\sigma_x^2 = V(X) = E\left[\{X - E(X)\}^2\right]$$

$$= E(X^2) - [E(X)]^2$$

$$= E(X^2) - \mu^2$$

If X is discrete random variable then the variance is given by

$$\sigma_x^2 = \sum_{i=1}^{n} (x_i - \mu)^2 P(X = x_i)$$

For a continuous random variable X, the variance is written as

$$\sigma_x^2 = \int_{-\infty}^{\infty} (x - \mu)^2 f(x) dx$$

The positive square root of variance is standard deviation and is denoted by σ.

Properties of Variance

Let X and Y are two random variables and a and b are two constants, then

1. $V(X + a) = V(X)$
2. $V(aX) = a^2 V(X)$
3. $V(a) = 0$
4. $V(aX + b) = a^2V(X)$
5. $V(aX \pm bY) = a^2V(X) + b^2V(Y)$
6. $V(X \pm Y) = V(X) + V(Y) \pm 2\,[E(XY) - E(X)\,E(Y)]$

The quantity $[E(XY) - E(X)\,E(Y)]$ is called covariance of X and Y and is written as Cov (X, Y).

2.7 SOLVED PROBLEMS

Problem 1: If X and Y are discrete random variables and k is a constant then prove that
(*i*) $E(X + k) = E(X) + k$ (*ii*) $E(X + Y) = E(X) + E(Y)$ **(April 2005, Nov 2005)**

Solution:

(*i*) Let X is random variable taking values x_i with probabilities p_i

$$E(X + k) = \sum_i (x_i + k)p_i = \sum_i x_i p_i + \sum_i kp_i$$

$$= \sum_i x_i p_i + k\sum_i p_i$$

$$= E(X) + k.1 \quad \text{(Since X is discrete random variable)}$$

$$= E(X) + k$$

(*ii*) Let X take value x_i with probabilities p_i

$$\therefore \quad E(X) = \sum_i x_i p_i$$

Let Y take value y_j with probabilities p_j

$$\therefore \quad E(Y) = \sum_j y_j p_j$$

then X + Y takes the values $x_i + y_j$ with probabilities p_{ij}

$$\therefore \quad E(X + Y) = \sum_i \sum_j (x_i + y_j)p_{ij}$$

$$= \sum_i \sum_j x_i p_{ij} + \sum_i \sum_j y_j p_{ij}$$

$$= \sum_i x_i \left\{\sum_j p_{ij}\right\} + \sum_j y_j \left\{\sum_i p_i\right\}$$

$$= \sum_i x_i p_i + \sum_j y_j p_j$$

$$= E(X) + E(Y)$$

Problem 2: Given the following table:

X	−3	−2	−1	0	1	2	3
P(X)	0.05	0.1	0.3	0	0.3	0.15	0.1

Compute

(*i*) E(X) (*ii*) E(2X ± 3) (*iii*) E(4X + 5) (*iv*) E(X^2) (*v*) V(X) (*vi*) V(2X ± 3)

Solution:

(*i*) $$E(X) = \sum_{i=0}^{6} x_i P(X = x_i) = 0.25$$

(*ii*) $$E(2X \pm 3) = 2E(X) \pm 3 = 3.5 \text{ or } -2.5$$

(*iii*) $$E(4X + 5) = 4E(X) + 5 = 6$$

(*iv*) $$E(X^2) = \sum_{i=0}^{6} x_i^2 P(X = x_i) = 2.95$$

(*v*) $$V(X) = E(X^2) - \mu^2 = 2.8875$$

(*vi*) $$V(2X \pm 3) = 4V(X) = 4 \times 2.8875 = 11.55$$

Problem 3: Find the mean and variance of the discrete random variable X whose probability distribution is given below:

X	1	2	3	4	5
P(X)	0.1	0.1	0.3	0.3	0.2

Solution:

$$\text{Mean } \mu = \sum_{i=0}^{5} x_i P(X = x_i)$$

$$= 1(0.1) + 2(0.1) + 3(0.3) + 4(0.3) + 5(0.2)$$

$$= 3.4$$

$$\text{Variance } \sigma^2 = \sum_{i=0}^{5} (x_i - \mu^2) P(X = x_i) = 1.44$$

Problem 4: Find the mean and variance of the uniform probability distribution given by

$$f(x) = \frac{1}{n} \text{ for } x = 1, 2, \ldots, n$$

Solution:

Probability distribution is

x:	1	2	3	4	....	n
$f(x)$:	$\frac{1}{n}$	$\frac{1}{n}$	$\frac{1}{n}$	$\frac{1}{n}$	....	$\frac{1}{n}$

$$\text{Mean } \mu = \sum_{i=1}^{n} x_i f(x_i) = 1.\frac{1}{n} + 2.\frac{1}{n} + 3.\frac{1}{n} + ...+ n.\frac{1}{n}$$

$$= (1 + 2 + 3 + 4......+ n)\frac{1}{n}$$

$$= \frac{n(n+1)}{2}\frac{1}{n} = \frac{n+1}{2}$$

$$\text{Variance } \sigma^2 = \sum_{i=1}^{n} x_i^2 f(x_i) - \mu^2$$

$$= 1^2.\frac{1}{n} + 2^2.\frac{1}{n} + 3^2.\frac{1}{n} + ...+ n^2.\frac{1}{n} - \frac{(n+1)^2}{4}$$

$$= (1^2 + 2^2 + 3^2 + ...+ n^2)\frac{1}{n} - \frac{(n+1)^2}{4}$$

$$= \frac{n(n+1)(2n+1)}{6}\frac{1}{n} - \frac{(n+1)^2}{4}$$

$$= (n+1)\left(\frac{(2n+1)}{6} - \frac{n+1}{4}\right) = \frac{(n^2-1)}{12}$$

Problem 5: The probability density function of a random variable is given by

$$f(x) = y_0 e^{-|x|}, \quad -\infty < x < \infty. \text{ Prove that } y_0 = \frac{1}{2}.$$

Find the mean and variance of the distribution. **(Feb 2007)**

Solution: Given function is density function then $\int_{-\infty}^{\infty} f(x)\,dx = 1$

$$\int_{-\infty}^{\infty} y_0 e^{-|x|}\,dx = 1$$

$$\int_{-\infty}^{0} y_0 e^{+x}\,dx + \int_{0}^{\infty} y_0 e^{-x}\,dx = 1$$

$$\left[y_0 e^{x}\right]_{-\infty}^{0} + \left[y_0 e^{-x}\right]_{0}^{\infty} = 1$$

$$y_0 + y_0 = 1$$

$$\Rightarrow \quad 2y_0 = 1$$

$$y_0 = \frac{1}{2}$$

$$\text{Mean } \mu = \int_{-\infty}^{\infty} x.f(x)dx$$

$$= \int_{-\infty}^{0} \frac{x}{2} e^{+x}\, dx + \int_{0}^{\infty} \frac{x}{2} e^{-x}\, dx$$

$$= \frac{1}{2}\left[xe^{x} - e^{x}\right]_{-\infty}^{0} + \frac{1}{2}\left[x\left(-e^{x}\right) - e^{x}\right]_{0}^{\infty}$$

$$= -\frac{1}{2} + \frac{1}{2} = 0$$

$$\text{Variance } \sigma^2 = \int_{-\infty}^{\infty} x^2 . f(x) dx - \mu^2$$

$$= \int_{-\infty}^{0} \frac{x^2}{2} e^{+x}\, dx + \int_{0}^{\infty} \frac{x^2}{2} e^{-x}\, dx = 2$$

Problem 6: Let $f(x)$ be the distribution function of a random variable x given by

$$F(x) = cx^3 \text{ when } 0 \le x \le 3$$
$$= 1 \quad x \ge 3$$
$$= 0 \quad x < 0$$

If $P(X = 3) = 0$ determine

(*i*) The density function (*ii*) C (*iii*) mean (*iv*) $P(X > 1)$ **(Nov 2004, Apr 2007)**

Solution:

(*i*) Let $F(x)$ be the distribution function

Density function $f(x) = \frac{d}{dx}[F(x)]$

$$= 3x^2C \quad 0 \le x < 3$$
$$= 0 \quad x > 3$$
$$= 0 \quad x < 0$$

(*ii*)
$$\int_0^3 f(x)dx = 1$$

$$\int_0^3 3Cx^2 .dx = 1 \Rightarrow C = \frac{1}{27}$$

(*iii*)
$$\text{Mean } \mu = \int_{-\infty}^{\infty} x.f(x)dx$$

$$= \int_0^3 \frac{x}{27} 3x^2 .dx$$

$$= \left[\frac{1}{9}.\frac{x^4}{4}\right]_0^3 = \frac{81}{4}$$

(*iv*) $$P(X > 1) = \int_1^{\infty} f(x)dx$$

$$= \int_0^3 \frac{3x^2}{27} dx = \left[\frac{1}{9}.\frac{x^3}{3}\right] = 1$$

Problem 7: A continuous random variable has the p.d.f. $f(x) = Kx\, e^{-\lambda x}$ if $x \geq 0, \lambda > 0 = 0$ otherwise. Determine the constant k and find mean, variance. **(Nov 2003)**

Solution:

Given $$f(x) = Kxe^{-\lambda x} \quad x \geq 0 \; \lambda > 0$$

$$= 0 \quad \text{otherwise}$$

Given function is probability density function then $\int_{-\infty}^{\infty} f(x)dx = 1$

$$\Rightarrow \int_{-\infty}^{0} f(x)dx + \int_0^{\infty} f(x)dx = 1$$

$$\int_{-\infty}^{\infty} Kxe^{-\lambda x}\, dx = 1$$

$$K\left[x\frac{e^{-\lambda x}}{-\lambda} - \frac{e^{-\lambda x}}{\lambda^2}\right]_0^{\infty} = 1$$

$$\frac{K}{\lambda^2} = 1$$

$$\Rightarrow \quad K = \lambda^2$$

$$\text{Mean } \mu = \int_0^{\infty} x.\, f(x)\, dx = K\int_0^{\infty} x^2 e^{-\lambda x}\, dx$$

$$= K\left[x^2\frac{e^{-\lambda x}}{-\lambda} - 2x.\frac{e^{-\lambda x}}{\lambda^2} + 2\frac{e^{-\lambda x}}{-\lambda^3}\right]_0^{\infty}$$

$$= \frac{2K}{\lambda^3} = \frac{2}{\lambda}$$

$$\text{Variance } \sigma^2 = E(X^2) - \mu^2$$

$$E(X^2) = \int_0^{\infty} x^2.f(x)\, dx = K\int_0^{\infty} x^3 e^{-\lambda x}\, dx = \frac{6}{\lambda^2}$$

$$\therefore \quad \text{Variance } \sigma^2 = \frac{6}{\lambda^2} - \frac{4}{\lambda^2} = \frac{2}{\lambda^2}$$

Problem 8: Given that $f(x) = \frac{K}{2^x}$ is a probability distribution for a random variable X that can taken on the values $x = 0, 1, 2, 3, 4$ (*i*) Find K (*ii*) Mean and variance of x

Solution:

(*i*)
$$f(x) = \frac{K}{2^x}$$

Sum of all probabilities $= 1$

$$\Rightarrow \quad \sum f(x) = 1$$

$$K\left\{\frac{1}{2^0}+\frac{1}{2}+\frac{1}{2^2}+\frac{1}{2^3}+\frac{1}{2^4}\right\} = 1$$

$$K = \frac{16}{31}$$

$$\therefore \quad f(x) = \frac{16}{31}\cdot\frac{1}{2^x}$$

(*ii*)
$$\text{Mean } \mu = \sum x.\, f(x)$$
$$= \frac{16}{31}\sum_{x=0}^{4}\frac{x}{2^x}$$
$$= \frac{16}{31}\left\{0+\frac{1}{2}+\frac{2}{2^2}+\frac{3}{2^3}+\frac{4}{2^4}\right\}$$
$$= \frac{26}{31} = 0.839$$

$$\text{Variance } \sigma^2 = E(X^2) - \mu^2$$
$$= \sum x^2 f(x) - \mu^2$$
$$= \frac{16}{31}\left\{0+\frac{1}{2}+\frac{4}{2^2}+\frac{9}{2^3}+\frac{16}{2^4}\right\} - \left(\frac{26}{31}\right)^2 = 1.168$$

Problem 9: If X is discrete random variable and K is constant then prove that

(*i*) $\text{Var}(X + K) = \text{Var}(X)$

(*ii*) $\text{Var}(KX) = K^2\text{Var}(X)$.

Solution:

(*i*)
$$\text{Var}(X) = E(X^2) - [E(X)]^2$$
$$\text{Var}(X + K) = \Sigma(x_i + K)^2 f(x_i) - [E(X + K)]^2$$
$$= \Sigma(x_i^2 + 2Kx_i + K^2) f(x_i) - [E(X) + K]^2$$

[Since $E(X + K) = E(X) + K$]

$$= \Sigma x_i^2 f_i + 2K\,\Sigma x_i f_i + K^2\,\Sigma f_i - [(E(X))^2 + 2KE(X) + K^2]$$
$$= \Sigma x_i^2 f_i + 2KE(X) + K^2 - [E(X)]^2 - 2KE(X) - K^2$$

$$= E(X^2) - [E(X)]^2$$
$$= Var(X)$$

(*ii*) $$Var(KX) = \Sigma(Kx_i)^2 f_i - [E(KX)]^2$$
$$= K^2 \Sigma x_i^2 f_i - [KE(X)]^2$$
$$= K^2[\Sigma x_i^2 f_i - [E(X)]^2]$$
$$= K^2 Var(X)$$

Problem 10: If 3 cars are drawn from a lot of 6 cars containing 2 defective cars, find the probability distribution of the number of defective cars. **(Dec 2004)**

Solution:

Let X be the random variable indicating the number of defective cars.

Probability that a defective car is selected.

$$P = \frac{2}{6} = \frac{1}{3}$$

$$q = (1 - P) = 1 - \frac{1}{3} = \frac{2}{3}$$

Now $$P(x) = {}^nC_x\, p^x\, q^{n-1}$$

$$P(0) = P(X = 0) = {}^3C_0 \left(\frac{1}{3}\right)^0 \left(\frac{2}{3}\right)^3 = \frac{8}{27}$$

$$P(1) = P(X = 1) = {}^3C_1 \left(\frac{1}{3}\right)^1 \left(\frac{2}{3}\right)^{3-1} = \frac{4}{9}$$

$$P(2) = P(X = 0) = {}^3C_2 \left(\frac{1}{3}\right)^2 \left(\frac{2}{3}\right)^{3-2} = \frac{2}{9}$$

Probability distribution

x	0	1	2
$P(x)$	$\frac{8}{27}$	$\frac{4}{9}$	$\frac{2}{9}$

Problem 11: A pair of fair die is tossed. Let X denotes the maximum of the number appearing, *i.e.*, $X(a, b) = Max(a, b)$ and let Y denotes the sum of the numbers appearing *i.e.*, $Y(a, b) = a + b$. find the variance and standard deviation of X and Y. **(May 2004, April 2007)**

Solution:

Given $$X(a, b) = Max(a, b)$$
$$Y(a, b) = a + b$$

Now $$X(a, b) = Max(a, b)$$

(1, 1) max is 1 $\Rightarrow P(X = 1) = \frac{1}{36}$

(1, 2), (2, 2), (2, 1) maximum is 2 $\Rightarrow P(X = 2) = \frac{3}{36}$

(1, 3), (2, 3), (3, 3), (3, 2), (3, 1) maximum is 3 $\Rightarrow P(X = 3) = \frac{5}{36}$

Similarly, $P(X = 4) = \frac{7}{36}$

$P(X = 5) = \frac{9}{36}$

$P(X = 6) = \frac{11}{36}$

$$\text{Mean } \mu = E(X) = \sum_{i=1}^{6} x_i P(X = x_i)$$

$$= 1.\frac{1}{36} + 2.\frac{3}{36} + 3.\frac{5}{36} + 4.\frac{7}{36} + 5.\frac{9}{36} + 6.\frac{11}{36}$$

$$= 4.472$$

$$E(X^2) = \sum_{i=1}^{6} x_i^2 P(X = x_i) = 21.972$$

$$\text{Variance } \sigma^2 = E(X^2) - \mu^2$$

$$= 21.972 - 19.998 = 1.974$$

$$\therefore \quad \text{Standard deviation} = \sqrt{\text{variance}} = \sqrt{1.974} = 1.4049$$

$$Y(a, b) = a + b$$

$$(1, 1) ; Y = 1 + 1 = 2 \Rightarrow P(Y = 2) = \frac{1}{36}$$

$$(1, 2), (2, 1) ; Y = 1 + 2 = 3 \Rightarrow F(Y = 3) = \frac{2}{36}$$

$$(1, 3), (2, 2), (3, 1); Y = 1 + 3 = 4 \Rightarrow P(Y = 4) = \frac{3}{36}$$

Similarly, $P(Y = 5) = \frac{4}{36}$

$P(Y = 6) = \frac{5}{36}$

$P(Y = 7) = \frac{6}{36}$

$P(Y = 8) = \frac{5}{36}$

$P(Y = 9) = \frac{4}{36}$

$$P(Y = 10) = \frac{3}{36}$$

$$P(Y = 11) = \frac{2}{36}$$

$$P(Y = 12) = \frac{1}{36}$$

$\therefore$ Mean $\mu = E(Y)$

$$= 2.\frac{1}{36} + 3.\frac{2}{36} + 4.\frac{3}{36} + 5.\frac{4}{36} + 6.\frac{5}{36} + 7.\frac{6}{36} + 8.\frac{5}{36}$$
$$+ 9.\frac{4}{36} + 10.\frac{3}{36} + 11.\frac{2}{36} + 12.\frac{1}{36}$$
$$= 7$$

$$E(Y^2) = \sum_{i=2}^{12} y_i^2 P(X = y_i) = 54.833$$

Variance $\sigma^2 = 5.833$

$\therefore$ Standard deviation $= \sqrt{\text{Variance}} = \sqrt{5.833} = 2.415$.

Problem 12 : For the discrete probability distribution

x	0	1	2	3	4	5	6	7
f	0	k	$2k$	$2k$	$3k$	k^2	$2k^2$	$7k^2 + k$

Determine (*i*) k, (*ii*) mean, (*iii*) variance, (*iv*) smallest value of x such that $P(X \le x) > 1/2$

Solution: (*i*) We have $\sum P(x) = 1$ *i.e.,* $\sum_{i=0}^{7} F_i$

$$0 + k + 2k + 2k + 3k + k^2 + 2k^2 + 7k^2 + k = 1$$
$$10k^2 + 9k - 1 = 0$$
$$10k^2 + 10k - k - 1 = 0$$
$$10k(k + 1) - 1(k + 1) = 0$$
$$k = -1, \frac{1}{10}$$

k cannot be negative

$$k = \frac{1}{10}, \text{ (since } P(X) \ge 0)$$

(*ii*) Mean $\mu = \sum xP(x)$

$$= 0(0) + 1(k) + 2(k) + 3(2k) + 4(3k) + 5(k^2) + 6(k^2) + 7(7k^2 + k)$$
$$= k + 4k + 6k + 12k + 5k^2 + 12k^2 + 49k^2 + 7k$$
$$= 66k^2 + 30k$$
$$= 66 \times \left(\frac{1}{10}\right)^2 + 30\left(\frac{1}{10}\right)$$

$$= \frac{66}{100} + 3$$
$$= 3.66$$

(*iii*) Variance $= E(X^2) - [E(X)]^2$

$$= \sum x^2 P(x) - (3.66)^2$$
$$= 0^2(0) + 1^2(k) + 2^2(2k) + 3^2(2k) + 4^2(3k) + 5^2(k^2) + 6^2(2k^2) + 7^2(7k^2 + k)$$
$$= k + 8k + 18k + 48k + 25k^2 + 72k^2 + 343k^2 + 49k$$
$$= 440k^2 + 124k$$
$$= 440\left(\frac{1}{10}\right)^2 + 124\left(\frac{1}{10}\right)$$
$$= 4.4 + 12.4 = 16.8$$

(*iv*) The smallest value of 'x' is determined by trial method.

$$P(X \le 0) = 0$$
$$P(X \le 1) = P(X = 0) + P(X = 1) = 0 + \frac{1}{10} = \frac{1}{10}$$
$$P(X \le 2) = P(X = 0) + P(X = 1) + P(X = 2) = 0 + \frac{1}{10} + \frac{2}{10} = \frac{3}{10}$$
$$P(X \le 3) = P(X = 0) + P(X = 1) + P(X = 2) + P(X = 3)$$
$$= 0 + \frac{1}{10} + \frac{2}{10} + \frac{2}{10} = \frac{5}{10}$$
$$P(X \le 4) = P(X = 0) + P(X = 1) + P(X = 2) + P(X = 3) + P(X = 4)$$
$$= 0 + \frac{1}{10} + \frac{2}{10} + \frac{2}{10} + \frac{3}{10} = \frac{8}{10} > \frac{1}{2}$$

Therefore, the smallest value of x for which $P(X \le x) > \frac{1}{2}$ is $x = 4$.

Problem 13: If X is a continuous random variable with distribution. $f(x) = \frac{x}{6} + k$ if $0 \le x \le 3$
$= 0$ elsewhere

Determine (*i*) The value of k, (*ii*) The mean, (*iii*) $P(1 \le x \le 2)$ **(Nov 2006)**

Solution:

(*i*) Given. $f(x) = \frac{x}{6} + k$ if $0 \le x \le 3$.

Given function is density function then $\int_{-\infty}^{\infty} f(x)dx = 1$

$$\therefore \int_0^3 f(x)dx = 1$$
$$\int_0^3 \left(\frac{x}{6} + k\right) dx = 1$$

$$\int_0^3 \frac{x}{6}\,dx + \int_0^3 k\,dx = 1$$

$$\frac{1}{6}\left[\frac{x^2}{2}\right]_0^3 + k\,[x]_0^3 = 1$$

$$k = \frac{1}{12}$$

(*ii*)

$$\text{Mean} = \int_0^3 x f(x)\,dx$$

$$= \int_0^3 x\left[\frac{x}{6} + k\right]dx$$

$$= \int_0^3 \left(\frac{x^2}{6} + kx\right)dx$$

$$= \frac{1}{6}\int_0^3 x^2\,dx + k\int_0^3 x\,dx$$

$$= \frac{1}{6}\left[\frac{x^3}{3}\right]_0^3 + k\left[\frac{x^2}{2}\right]_0^3$$

$$= \frac{1}{6}\left[\frac{27}{3}\right] + k\left[\frac{9}{2}\right]$$

$$= \frac{3}{2} + \frac{9}{2}k$$

$$= \frac{3}{2} + \frac{9}{2} \times \frac{1}{12}$$

$$= \frac{3}{2} + \frac{3}{8} = \frac{12+3}{8} = \frac{15}{8} = 1.875$$

$$\therefore \quad \text{Mean} = 2.25$$

(*iii*)

$$P(1 \le x \le 2) = \int_1^2 f(x)\,dx$$

$$= \int_1^2 \left(\frac{x}{6} + k\right)dx$$

$$= \frac{1}{6}\int_1^2 x\,dx + k\int_1^2 1\,dx$$

$$= \frac{1}{6}\left[\frac{x^2}{2}\right]_1^2 + k\left[\frac{x}{1}\right]_1^2$$

$$= \frac{1}{6}\left[\frac{4}{2}\right] - \left[\frac{1}{2}\right] + k[2-1]$$

$$= \frac{1}{6}\left[\frac{3}{2}\right] + k \qquad \left[\text{Since } k = \frac{1}{12}\right]$$

$$= \frac{1}{4} + \frac{1}{12} = \frac{3+1}{12} = \frac{4}{12} = \frac{1}{3}$$

Problem 14: For the continuous probability function $f(x) = kx^2 e^{-x}$ when $x \geq 0$,
Find (*i*) k (*ii*) mean (*iii*) variance.

Solution:

(*i*) We have
$$\int_{-\infty}^{\infty} f(x)\,dx = 1$$

$$\int_{-\infty}^{\infty} kx^2 e^{-x}\,dx = 1$$

$$\Rightarrow \quad k\left[x^2 \int e^{-x}\,dx - \int\left(\left(\frac{d}{dx}x^2\right)\int e^{-x}\,dx\right)dx\right]_0^{\infty} = 1$$

$$\Rightarrow \quad k\left[-x^2 e^{-x} - \int (2x)e^{-x}\,dx\right]_0^{\infty} = 1$$

$$\Rightarrow \quad k\left[-x^2 e^{-x} + 2\left\{-xe^{-x} + \int e^{-x}\,dx\right\}\right]_0^{\infty} = 1$$

$$\Rightarrow \quad k\left[-x^2 e^{-x} + 2\left\{-xe^{-x} - e^{-x}\right\}\right]_0^{\infty} = 1$$

$$\Rightarrow \quad k\left[-x^2 e^{-x} - 2xe^{-x} - 2e^{-x}\right]_0^{\infty} = 1$$

$$\Rightarrow \quad k\,[(-0-0-0) - (0-0-2e^{-0})] = 1$$

$$\Rightarrow \quad k = \frac{1}{2}$$

(*ii*) If x is a continuous random variable

Then mean $\mu = E[x]$

$$= \int_{-\infty}^{\infty} xf(x)\,dx$$

$$= \int_{-\infty}^{\infty} x\left(kx^2\right)e^{-x}\,dx$$

$$= k\int_{-\infty}^{\infty} x^3 e^{-x}\,dx$$

$$= k\left[-x^3 e^{-x} + \int 3x^2 e^{-x}\,dx\right]_0^{\infty}$$

$$= k\left[-x^3 e^{-x} + 3\left[-x^2 e^{-x} + \int 2x\, e^{-x}\, dx\right]\right]_0^\infty$$

$$= k\left[-x^3 e^{-x} + 3\left[-x^2 e^{-x} + 2\left[-xe^{-x} + \int e^{-x}\, dx\right]\right]\right]_0^\infty$$

$$= k\left[-x^3 e^{-x} + 3\left[-x^2\, e^{-x} - 2x e^{-x} - 2e^{-x}\right]\right]_0^\infty$$

$$= k\left[-x^2 e^{-x} - 3x^2 e^{-x} - 6xe^{-x} - 6e^{-x}\right]_0^\infty$$

$$= k[(-0-0-0-0)-(0-0-0-6)] = \frac{1}{2}\,[6]$$

$$\therefore \quad \text{Mean} = 3$$

(*iii*) Variance

$$\sigma_x^2 = \text{Var}(x) = \text{E}(x^2) - [\text{E}(x)]^2$$

Now consider

$$\text{E}(x^2) = \int_{-\infty}^{\infty} x^2 f(x)dx$$

$$= \int_{-\infty}^{\infty} x^2\left(kx^2\right)e^{-x}\, dx$$

$$= k\int_{-\infty}^{\infty} x^4\, e^{-x}\, dx$$

$$= k\left[-x^4 e^{-x} + \int 4x^3 e^{-x}\, dx\right]$$

$$= k\left[-x^4 e^{-x} + 4\left(-x^3 e^{-x} + \int 3x^3 e^{-x}\, dx\right)\right]$$

$$= k\left[-x^4 e^{-x} - 4x^3 e^{-x} + 12\left(-x^2 e^{-x} + \int 2x e^{-x}\, dx\right)\right]$$

$$= k\left[-x^4 e^{-x} - 4x^3 e^{-x} - 12x^2 e^{-x} + 24\left(-xe^{-x} + \int e^{-x}\, dx\right)\right]$$

$$= k\left[-x^4\, e^{-x} - 4x^3\, e^{-x} - 12x^2 e^{-x} - 24x e^{-4} - 24e^{-x}\right]_0^\infty$$

$$= \frac{1}{2}\,[24]$$

$$\text{E}(x^2) = 12$$

$$\text{Variance} = \text{E}(x^2) - [\text{E}(x)]^2 = 12 - 3^2$$

$$\text{Variance} = 3.$$

2.8 CHEBYCHEV'S INEQUALITY

Statement

If X is a random variable with mean μ and variance σ^2, then for any positive number k, we have

$$\left.\begin{aligned} P\{|X-\mu| \geq k\sigma\} &\leq \frac{1}{k^2} \\ P\{|X-\mu| \geq k\sigma\} &\leq 1-\frac{1}{k^2} \end{aligned}\right\} \quad ...(1)$$

Proof:

Case *i* : X is continuous random variable by definition

$$\sigma^2 = \sigma x^2 = E\{X - E(X)\}^2 = E(X-\mu)^2$$

$$= \int_{-\infty}^{\infty} (x-\mu)^2 f(x)dx \text{, where } f(x) \text{ is p.d.f. of X.}$$

$$= \int_{-\infty}^{\mu-k\sigma} (x-\mu)^2 f(x)dx + \int_{\mu-k\sigma}^{\mu+k\sigma} (x-\mu)^2 f(x)dx + \int_{\mu+k\sigma}^{\infty} (x-\mu)^2 f(x)dx$$

$$= \int_{-\infty}^{\mu-k\sigma} (x-\mu)^2 f(x)dx + \int_{\mu-k\sigma}^{\infty} (x-\mu)^2 f(x)dx \quad ...(2)$$

We know that $x \leq \mu - k\sigma$ and $x \geq \mu + k\sigma \Leftrightarrow |x-\mu| \geq k\sigma$ ...(3)

Substituting in (2), we get

$$\sigma^2 \geq k^2\sigma^2 \left[\int_{-\infty}^{\mu-k\sigma} f(x)dx + \int_{\mu+k\sigma}^{\infty} f(x)dx\right] \quad \text{[From (3)]}$$

$$= k^2\sigma^2 [P(X \leq \mu - k\sigma) + P(X \geq \mu + k\sigma)]$$

$$= k^2\sigma^2 P(|X-\mu| \geq k\sigma) \quad \text{[From (3)]}$$

$$\Rightarrow \quad P(|X-\mu| \geq k\sigma) \leq 1/k^2 \quad ...(4)$$

Also since $P(\{|X-\mu| \geq k\sigma\} + P\{|X-\mu| < k\sigma\} = 1$, we get

$$P\{|X-\mu| < k\sigma\} = 1 - P\{|X-\mu| \geq k\sigma\} \geq 1 - \{1/k^2\} \quad \text{[From (4)]}$$

Case *ii*. In case of discrete random variable, the proof follows exactly similarly, on replacing integration by summation.

2.9 SOLVED PROBLEMS

Problem 1: Use Chebychev's inequality to determine how many times a fair coin must be tossed that the probability will be at least 0.90 that the ratio of the observed number of heads to the number of tosses will lie between 0.4 to 0.6.

Solution:

As in the proof of Bernoulli's law of Large numbers, we get for any $\varepsilon > 0$

$$P\left\{\left|\frac{X}{n} - p\right| \geq \varepsilon\right\} \leq \frac{1}{4n\varepsilon^2} \Rightarrow P\left\{\left|\frac{X}{n} - p\right| \geq \varepsilon\right\} \geq 1 - \frac{1}{4n\varepsilon^2}$$

Since $p = 0.5$ and we want the proportion of successes X/n to lie between 0.4 and 0.6

We have $|(X/n) - p| \leq 0.1$

Thus choosing $\varepsilon = 0.1$ we have $P\left\{\left|\frac{X}{n} - p\right| \geq 0.1\right\} \geq 1 - \frac{1}{4n\,(0.1)^2} = 1 - \frac{1}{0.04n}$

Since we want this probability to be 0.9 we fix $1 - \dfrac{1}{0.04n} = 0.90$

$$\Rightarrow \qquad 0.10 = \frac{1}{0.04n} \Rightarrow n = \frac{1}{0.10 \times 0.04} = 250$$

Hence the required number of tosses is 250.

Problem 2: For geometric distribution $p(x) = 2^{-x}$; $x = 1, 2, 3, \ldots\ldots\ldots$, prove that Chebychev's inequality gives $P\{|X - 2| \leq 2\} > \dfrac{1}{2}$, while the actual probability is $\dfrac{15}{16}$.

Solution:

$$E(X) = \sum_{x=1}^{\infty} \frac{x}{2^x} = \frac{1}{2} + \frac{2}{2^2} + \frac{3}{2^3} + \frac{4}{2^4} \ldots$$

$$= \frac{1}{2}\,(1 + 2A + 3A^2 + 4A^3 \ldots\ldots\ldots\ldots)$$

$$= \frac{1}{2}\,(1 - A)^{-2} = 2, \qquad (A = \frac{1}{2})$$

$$E(X^2) = \sum_{x=1}^{\infty} \frac{x^2}{2^x} = \frac{1}{2^2} + \frac{4}{2^3} + \frac{9}{2^4} \ldots$$

$$= \frac{1}{4}\,\{1 + 4A + 9A^2 + \ldots\}$$

$$= \frac{1}{4}\,(1 + A)\,(1 - A)^{-3} = 6, \quad (A = \frac{1}{2})$$

$\therefore$ $Var(X) = E(X^2) - [E(X)]^2 = 6 - 4 = 2 \Rightarrow \sigma = \sqrt{2}$

Using Chebychev's inequality, we get $P\{|X - E(x)| \leq k\sigma\} > 1 - \dfrac{1}{k^2}$

With $k = \sqrt{2}$, we get $P\{|X - 2| \leq \sqrt{2}\cdot\sqrt{2}\} > 1 - \dfrac{1}{2} = \dfrac{1}{2}$ $\therefore$ $P(|X - 2| \leq 2\} > \dfrac{1}{2}$

And the actual probability is given by

$$P\{|X - 2| \leq 2\} = P\{0 \leq X \leq 4\} = P\{X = 1, 2, 3 \text{ or } 4\} = \frac{1}{2} + \left(\frac{1}{2}\right)^2 + \left(\frac{1}{2}\right)^3 + \left(\frac{1}{2}\right)^4 = \frac{15}{16}$$

Problem 3: Two unbiased dice are thrown. If X is the sum of the numbers showing up. Prove that $P(|X - 7| \geq 3) \leq \frac{35}{54}$. Compare this with the actual probability

Solution:

Probability distribution of the random variable X is

X	2	3	4	5	6	7	8	9	10	11	12
$P(X = x)$	$\frac{1}{36}$	$\frac{2}{36}$	$\frac{3}{36}$	$\frac{4}{36}$	$\frac{5}{36}$	$\frac{6}{36}$	$\frac{5}{36}$	$\frac{4}{36}$	$\frac{3}{36}$	$\frac{2}{36}$	$\frac{1}{36}$

$$\text{Mean } \mu = E(X) = \sum x_i P(X = x_i) = 7$$

$$E(X^2) = \sum x_i^2 P(X = x_i) = \frac{329}{6}$$

$$\text{Variance } \sigma^2 = E(X^2) - \mu^2$$

$$= \frac{329}{6} - 49 = \frac{35}{6}$$

By Chebychev's inequality $K\sigma = 3 \Rightarrow K = \frac{3}{!}$

$$P(|X - 7| \geq 3) \leq \frac{!^2}{9}$$

$$P(|X - 7| \geq 3) \leq \frac{35}{54}$$

Actual probability is $P(|X - 7| \geq 3) = 1 - P(|X - 7| < 3)$

$$= 1 - P(-3 < X - 7 < 3)$$

$$= 1 - P(4 < X < 10).$$

$$= 1 - \frac{1}{36}(4 + 5 + 6 + 5 + 4) = \frac{1}{3}.$$

Problem 4: A symmetric die is thrown 600 times. Find the lower bound for the probability of getting 80 to 120 sixes.

Solution:

Let S be total number of successes. Then $E(S) = np = 600 \times \frac{1}{6} = 100$ and

$V(S) = npq = 600 \times \frac{1}{6} \times \frac{5}{6} = \frac{500}{6}$

Using Chebychev's inequality, we get

$P\{|S - E(S)| < k\sigma \geq 1 - \frac{1}{k^2} \Rightarrow P\{|S - 100| < k\sqrt{500/6}\} \geq 1 - \frac{1}{k^2}$

$\therefore P\{100 - k\sqrt{500/6} < S < 100 + k\sqrt{500/6}\} \geq 1 - \frac{1}{k^2}$

Taking $k = \frac{20}{\sqrt{500/6}}$, we get $P(80 \le S \le 120) \ge 1 - \frac{1}{400 \times (6/500)} = \frac{19}{24}$.

Problem 5: If X is the number appearing on a die when it is thrown, show that the Chebychev's inequality gives $P(|X - \mu| > 2.5) < 0.47$. While the actual probability is zero.

Solution:

Let X is the number appearing on the die. The probability distribution of X is

X	1	2	3	4	5	6
P(X)	$\frac{1}{6}$	$\frac{1}{6}$	$\frac{1}{6}$	$\frac{1}{6}$	$\frac{1}{6}$	$\frac{1}{6}$

$$\text{Mean } \mu = E(X) = \sum x_i \, P(X = x_i) = \frac{7}{2}$$

$$E(X^2) = \sum x_i^2 \, P(X = x_i) = \frac{91}{6}$$

$$\text{Variance } \sigma^2 = E(X^2) - \mu^2 = \frac{91}{6} - \frac{49}{4} = 2.9167$$

$$\sigma = 1.708$$

By Chebychev's inequality, $k\sigma = 2.5 \Rightarrow k = 2.5/\sigma$

$$P(|X - \mu| \ge k\sigma) < \frac{1}{k^2}$$

$$P(|X - 3.5| \ge 2.5) < \frac{\sigma^2}{6.25}$$

$$P(|X - 3.5| \ge 2.5) < 0.47$$

Actual probability $= P(|X - 3.5| \ge 2.5)$

$= P(6 < X < 1) = 0$

$\therefore$ The random variable X lies outside the limits (1, 6) which is zero for die.

Problem 6: X is a random variable such that $E(X) = 3$ and $E(X^2) = 13$. Determine the lower bound for $P(-2 < X < 8)$. Using Chebychev's inequality.

Solution:

Given $E(X) = 3$

$E(X^2) = 13$

$$\text{Variance } \sigma^2 = E(X^2) - \mu^2 = 13 - 9 = 4 \Rightarrow \sigma = 2$$

Chebychev's inequality

$$P(|X - \mu| < k\sigma) \ge 1 - \frac{1}{k^2}$$

$$P(|X - 3| < 2k) \ge 1 - \frac{1}{k^2}$$

$$3 - 2k = -2$$

$$k = \frac{5}{2}$$

$$P(-2 < X < 8) \geq 1 - \frac{4}{25}$$

$$P(-2 < X < 8) = \frac{21}{25}$$

$\Rightarrow$ Lower bond for $P(-2 < X < 8)$ is $\frac{21}{25}$.

2.10 EXERCISE

1. If a random variable has the probability density function defined as

$$f(x) = 2e^{-2x} \quad x > 0$$
$$= 0 \quad x \leq 0$$

Find the probability that it will take on a value

(*i*) between 1 and 3 (*ii*) greater than 0.5

2. The diameter of an electric cable, say X is assumed to be a continuous random variable with p.d.f. $f(x) = 6x(1 - x)$ $0 \leq x \leq 1$

(*i*) Check that above is p.d.f.

(*ii*) Determine *a* number *b* such that $P(X < b) = P(X > b)$

3. A continuous random variable X is defined by

$$f(x) = (3 + x)^2/16 \quad -3 \leq x < -1$$
$$= (6 - 2x^2)/16 \quad -1 < x < 1$$
$$= (3 - x^2)/16 \quad -1 \leq x \leq 3$$

Verify that $f(x)$ is density and find also the Mean of X. **(Nov 2003)**

4. For the continuous random X whose probability density function is given by

$f(x) = cx(2 - x)$ if $0 \leq x \leq 2$

$= 0$ otherwise

Find *c*, mean and variance of X **(Nov 2003)**

5. If X is a continuous random variable with probability density function given by

$$f(x) = kx \quad \text{if } 0 \leq x < 2$$
$$= 2k \quad \text{if } 2 \leq x < 4$$
$$= -kx + 6k \quad \text{if } 4 \leq x < 6$$

Find *k* and mean value of X. **(May 2003)**

6. A game between the players is fair if each player has the same mathematical expectation. Of someone gives us Rs. 5 each time we roll a 1 or a 2 with a balanced die, how much must we pay that person each time we roll a 3, 4, 5 or 6 to make the game fair. **(Nov 2003)**

7. When the American league and national league champions are evenly matched, the probabilities that a world series will end in 4, 5, 6 or 7 games are respectively

$\frac{1}{8}, \frac{1}{4}, \frac{5}{16}$ and $\frac{5}{16}$. What is expected length of a world series when the two teams are evenly matched? **(Nov 2003)**

8. If $E(X) = 1$, $E(X^2) = 4$, find the mean and variance of $Y = 2X - 3$.

9. A random variable X has the following probability function value of X

X	0	1	2	3	4	5	6
$P(X)$	k	$3k$	$5k$	$7k$	$9k$	$11k$	$13k$

(*i*) Find k (*ii*) Evaluate $P(X < 4)$, $P(X \geq 5)$ and $P(X \leq x) > 1/2$?

10. Consider the distribution function for X defined by

$$F(X) = \begin{cases} 0, & x < 0 \\ 1 - \frac{1}{4}e^{-x}, & x \geq 0 \end{cases}$$

Determine $P(X = 0)$ and $P(X \geq 0)$.

11. Determine whether the following is a valid distribution function :

$$F(x) = \begin{cases} 1 - e^{-x/2}, & x \geq 0 \\ 0 & x < 0 \end{cases}$$

12. A continuous random variable X that can assume any value between $x = 2$ and $x = 5$ has a density function given by $f(x) = k(1 + x)$, find $P(X < 4)$.

13. A random variable X has the density function $f(x) = \begin{cases} \frac{1}{4}, & -2 < x < 2 \\ 0, & \text{elsewhere} \end{cases}$

Obtain (*i*) $P(X < 1)$ (*ii*) $P(|X| > 1)$ (*iii*) $P(2X + 3 > 5)$

14. If the probability density of a random variable is given by $f(x) = \begin{cases} k(1 - x^2), & 0 < x < 1 \\ 0, & \text{otherwise} \end{cases}$

Find (*i*) k (*ii*) the distribution function of the random variable.

15. A random variable X has the density function $f(x) = K.\frac{1}{1 + x^2}$, if $-\infty < x < \infty = 0$, otherwise

Determine K and the distribution function.

16. A random variable X takes the values –2, –1, 0 and 1 with probabilities $\frac{1}{8}, \frac{1}{8}, \frac{1}{4}$ and $\frac{1}{2}$ respectively. Find and draw the probability distribution and density function.

17. Find the expected value and dispersion of the discrete random variable X whose probability distribution is given below.

X	1	2	3	4	5
$P(X = x)$	0.1	0.1	0.3	0.3	0.2

18. Find the mean value of the random variable whose probability density function is given by $f(x) = \frac{3}{5} 10^{-5} \times (100 - x),\ 0 \le x \le 100$.

19. For a geometric distribution, $f(x) = \frac{1}{2^x}$; $x = 1, 2, 3, 4, \ldots\ldots$ Prove that Chebyshev's inequality gives $P[|X - 2| \le 2] > \frac{1}{2}$ while the actual probability is $\frac{15}{16}$.

20. A random variable X has mean of 12 and variance of 9. If the probability distribution is unknown, determine $P(6 < X < 18)$.

21. The diameter of an electric cable, say X is assumed to be a continuous random variable with probability density function $f(x) = 6x(1 - x),\ \ 0 \le x \le 1$. (*i*) Check that $f(x)$ is probability density function and (*ii*) Determine *a* number b such that $P(X < b) = P(X > b)$.

22. Five defective bulbs are accidentally mixed with twenty good ones. It is not possible to just look at a bulb and tell whether or not it is defective. Find the probability distribution of the number of defective bulbs, if four bulbs are drawn at random from this lot.

23. A die is tossed thrice. A success is getting 1 or 6 on a toss. Find the mean and variance of the number of successes.

ANSWERS

1. 0.367, 0.1326	**3.** 0.25	**4.** $c = 3/4,\ \mu = 1,\ \sigma^2 = 0.2$
5. $k = 1/8$, Mean = 6.166	**6.** 1.66	**7.** 15.8125
8. Mean = –1, Variance = 12	**17.** 3.4, 1.44	**18.** 50
20. $\ge \frac{3}{4}$	**21.** $b = 1/2$	**23.** 1 and 2/3

2.11 QUIZ

A continuous function X has the probability density function given by $f(x) = cx^2,\ 0 \le x \le 1$ then

1. The value of c is

(*a*) 1 (*b*) 2

(*c*) 3 (*d*) 4

with this value of c

2. $P(X < \frac{1}{2}) =$

(*a*) $\frac{1}{2}$ (*b*) $\frac{1}{4}$

(*c*) $\frac{1}{6}$ (*d*) $\frac{1}{8}$

3. $P(X \ge \frac{3}{4}) =$

(*a*) $\frac{37}{64}$ (*b*) $\frac{35}{64}$

(*c*) $\frac{1}{16}$ (*d*) $\frac{1}{8}$

4. A random variable X has the probability density function given by $f(x) = \frac{1}{8}, (3 + 2x)$ $2 \le x \le 4$ then $P(2 \le X \le 3) =$

(a) $\frac{1}{9}$ (b) $\frac{2}{9}$

(c) $\frac{4}{9}$ (d) $\frac{8}{9}$

5. The frequency function of a random variable X is given by $f(x) = cx(2 - x), 0 \le x \le 2$ then $c =$

(a) $\frac{1}{2}$ (b) $\frac{1}{4}$

(c) $\frac{3}{4}$ (d) none of these

6. In a continuous distribution if $P(X \le \frac{1}{2}) = \frac{9}{16}$ then $P(X < \frac{1}{2}) =$

(a) $\frac{7}{16}$ (b) $\frac{9}{16}$

(c) $\frac{5}{16}$ (d) none of these

7. Given that $f(x) = \frac{k}{2^x}$ is the probability distribution of a random X taking value x, $x = 0, 1, 2, 3, 4$.
Then $c =$

(a) $\frac{4}{31}$ (b) $\frac{3}{31}$

(c) $\frac{12}{31}$ (d) $\frac{16}{31}$

Two coins are tossed simultaneously. Let X denote the number of heads, then

8. The mean of X is

(a) 1 (b) $\frac{1}{2}$

(c) $\frac{3}{4}$ (d) $\frac{1}{\sqrt{2}}$

9. The variance of X is

(a) 1 (b) $\frac{1}{2}$

(c) $\frac{3}{4}$ (d) $\frac{1}{\sqrt{2}}$

10. The standard deviation of X is

(a) 1 (b) $\frac{1}{2}$

(c) $\frac{3}{4}$ (d) none of these

11. If it rains, a dealer rain coat earns Rs. 500 per day and if it is fair, he losses Rs. 50 per day. If the probability of a rainy day is 0.4, his average daily income is

(a) Rs. 130 (b) Rs. 140

(c) Rs. 170 (d) none of these

12. If a is a constant then $V(a) =$

(a) a (b) a^2

(c) $\sqrt{a}$ (d) none of these

13. The joint density function of the random variable X and Y is $f(x, y)$; $-\infty < x < \infty$, $-\infty < y < \infty$ then the marginal density of X is

(a) $\int_{-\infty}^{\infty}\int_{-\infty}^{\infty} f(x,y)dxdy$ (b) $\int_{-\infty}^{\infty} f(x,y)dy$

(c) $\int_{-\infty}^{\infty} f(x,y)dx$ (d) none of these

14. The mean of the probability of the numbers on face of die in throwing a die is..........

(a) $\frac{5}{2}$ (b) $\frac{3}{2}$

(c) $\frac{7}{2}$ (d) $\frac{1}{2}$

If the probability distribution of the random variable X is

X	0	1	2	3	4	5
P(X)	$\frac{1}{25}$	$\frac{2}{25}$	$\frac{3}{25}$	$\frac{1}{5}$	$\frac{1}{5}$	$\frac{9}{5}$

15. The mean of the probability distribution in the above table =

(a) $\frac{88}{25}$ (b) $\frac{66}{25}$

(c) $\frac{77}{25}$ (d) $\frac{55}{25}$

16. If a coin so weighed that probability of getting head is $\frac{2}{3}$. X is the random variable assignment to the no. of head in tossing a coin. Mean is....................

(a) $\frac{1}{3}$ (b) $\frac{2}{3}$

(c) $\frac{1}{2}$ (d) $\frac{3}{4}$

17. Variance of getting head in tossing a coin for (8)..............

(a) 0 (b) 1

(c) 2 (d) 3

18. If probability of getting 1 or 6 is a success. The probability of getting at least once 1 or 6 if we throw a dice four times is...................

(*a*) $\frac{61}{81}$ (*b*) $\frac{63}{81}$

(*c*) $\frac{65}{81}$ (*d*) $\frac{59}{81}$

19. Two dice are thrown; getting doubled is a success. The probability of getting one doublet in throwing two dice, three times is.................

(*a*) $\frac{7}{24}$ (*b*) $\frac{23}{72}$

(*c*) $\frac{25}{72}$ (*d*) $\frac{19}{72}$

20. Two items are selected from 5 items of which 3 are defective. The expected number of defected items is.....................

(*a*) $\frac{1}{5}$ (*b*) $\frac{2}{5}$

(*c*) $\frac{3}{5}$ (*d*) $\frac{6}{5}$

21. If X is the random variable assigned to no. of heads in tossing two coins. Then the mean is

(*a*) 0 (*b*) 1

(*c*) 2 (*d*) 1.5

22. If the distribution is given by

X	2	3	11
$f(x)$	1/3	1/2	1/6

Then E(x) = ...

(*a*) 0 (*b*) 1

(*c*) 2 (*d*) 4

23. If the distribution is

X	1	2	3	5	7
$f(x)$	0.2	0.3	0.2	0.25	0.05

Then E(x) = ...

(*a*) 1 (*b*) 2

(*c*) 3 (*d*) 5

24. If the distribution is

X	1	2	3	5	7
$f(x)$	0.2	0.3	0.2	0.25	0.05

Then the variance is

(*a*) 3.1 (*b*) 2.9

(*c*) 2.5 (*d*) 2.7

25. A fair die is thrown. Let X denote the twice the number appearing then the standard deviation of X is....................

(*a*) 5 (*b*) 0

(*c*) 1 (*d*) 7

26. If the probability density function of a random variable X is $f(x) = kx(x-1)$ in $1 \le x \le 4$ = 0 elsewhere

And $p(1 \le x \le 3) = \frac{28}{3}$ then the value of k is

(*a*) 0 (*b*) 1

(*c*) 2 (*d*) 3

27. If the probability density function of a random variable X is $f(x) = 2x(x - 1)$ in $1 \le x \le 4 = 0$ elsewhere, then the variance σ^2

(*a*) 251 (*b*) 262

(*c*) 281.6 (*d*) 246.4

28. The probability density function of a random variable X is $f(x) = \frac{1}{2} \sin x$ in $0 \le x \le \pi = 0$ elsewhere, then the mode of x is

(*a*) 0 (*b*) π

(*c*) $\pi/4$ (*d*) $\pi/2$

29. The probability density function of a random variable X is $f(x) = \frac{1}{2} \sin x$ in $0 \le x \le \pi$ = 0 elsewhere, then the variance σ^2 =

(*a*) $\frac{\pi^2}{4}$ (*b*) $\frac{\pi^2}{4} - 2$

(*c*) $\frac{\pi^2}{4} + 2$ (*d*) $\frac{\pi^2}{4} + 4$

30. If X is a continuous random variable and $y = ax + b$ then the expected valued of y =

(*a*) aE(X) (*b*) aE(X) + E(b)

(*c*) aE(X) + b (*d*) E(a) L (X)

31. If a continuous random variable has the probability density $f(x) = 2e^{-2x}$, $x > 0 = 0$, $x \le 0$ then $P(1 \le x \le 3)$ =

(*a*) 0.2 (*b*) 0.312

(*c*) 0.425 (*d*) 0.133

32. Two dice are thrown let X assign to each point (a, b) in S the maximum of its numbers $X(a, b) = \max(a, b)$ then $P(X = 3) =$

(a) $\frac{1}{36}$ (b) $\frac{3}{36}$

(c) $\frac{5}{36}$ (d) $\frac{7}{36}$

33. Two dice are thrown let X assign to each point (a, b) in S the maximum of its numbers $X(a, b) = \max(a, b)$ then $p(X = 6) =$

(a) $\frac{5}{36}$ (b) $\frac{7}{36}$

(c) $\frac{1}{4}$ (d) $\frac{11}{36}$

34. If a random variable x has the distribution

X	1	2	3	4	5
P(X)	$\frac{1}{2}$	$\frac{1}{4}$	$\frac{1}{8}$	$\frac{1}{16}$	$\frac{1}{16}$

Then $E(x) =$

(a) 1 (b) 0

(c) 2.2 (d) 1.9

35. The mean of uniform probability distribution $f(x) = \frac{1}{n}$ for $x = 1, 2, 3, \ldots, n$ is

(a) n (b) n^2

(c) $\frac{n-1}{2}$ (d) $\frac{n+1}{2}$

36. If a continuous random variable X has the probability density function $f(x) = k(1 - x^2)$, $0 < x < 1 = 0$ elsewhere then the value of $k =$ ....

(a) 1 (b) $\frac{3}{2}$

(c) 2 (d) $\frac{5}{2}$

37. If a continuous random variable X has the probability density function $f(x) = \frac{3}{2}(1 - x^2)$, $0 < x < 1 = 0$ elsewhere then the mean of X is

(a) $\frac{1}{8}$ (b) $\frac{1}{4}$

(c) $\frac{3}{8}$ (d) 1

38. If X is continuous random variable then $P(X = c)\ \forall\ c$ is..........

(a) 0 (b) 1

(c) 2 (d) 3

39. The distribution of a random variable X is as follows

X	1	2	3	4
$P(X = x)$	c	$2c$	$3c$	$4c$

Then the value of c is ....

(*a*) 1/10 (*b*) 2/10

(*c*) 3/10 (*d*) 4/10

40. If the mean of the binomial distribution is 25, the standard deviation lies in the interval ....

(*a*) 40 (*b*) 45

(*c*) 15 (*d*) none

41. If $f(x)$ is a distribution function of a random variable x then $\underset{x \to 0}{\text{Lt}} f(x)$ is ....

(*a*) 0 (*b*) 1

(*c*) 2 (*d*) 3

42. If a random variable x has the following probability distribution

X	0	1	2	3
$P(x)$	$2k^2$	$3k^2$	$5k^2$	$6k^2$

Then the value of k is ....

(*a*) 1/5 (*b*) 1/6

(*c*) $\pm \frac{1}{4}$ (*d*) $\pm$ 1/5

43. The probability density function of a random variable x is given by $f(x) = 2e^{-2x} \quad x \geq 0 = 0$. The value of $p(x > 2)$ is ....

(*a*) e^{-1} (*b*) e^{-2}

(*c*) e^{-3} (*d*) e^{-4}

44. If x is random variable with the following probability distribution is

$X = x$	-3	6	9
$P(X = x)$	1/6	$\frac{1}{2}$	1/3

Then variance of x is ....

(*a*) 65/4 (*b*) 4/65

(*c*) $-65/4$ (*d*) $-4/65$

45. A continuous random variable x has a distributive function

$$f(x) = \begin{cases} 0 & x < 2 \\ \lambda (x-2)^4 & 2 \leq x < 3 \\ 1 & x \geq 3 \end{cases}$$

Then the value of λ is ....

(*a*) 4 (*b*) 3

(*c*) 2 (*d*) 1

46. If a continuous random variable has the density function $f(x) = 0.003\, x^2$, $0 < x \le 10$ then $P(X < 3)$ is ....

(*a*) 0.27 (*b*) 0.027

(*c*) 0.0027 (*d*) 0.127

47. If a random variable x has probability density function $f(x) = c/(1 + x^2)$, $-\infty < x < \infty$, then $c =$ ....

(*a*) $4/\pi$ (*b*) $3/\pi$

(*c*) $1/\pi$ (*d*) $2/\pi$

48. The distribution function of a random variable x is $f(x) = 1/16(x - 1)^4$, $1 < x < 3$ then the probability density function of x is ....

(*a*) $(1/4)(x-1)$ (*b*) $(1/4)(x-1)^2$

(*c*) $(1/4)(x-1)^4$ (*d*) $(1/4)(x-1)^3$

49. If the probability density function of a random variable x is given by $f(x) = x^2/9$, $0 < x < 3$ then the cumulative distribution function $f(x)$ is ....

(*a*) $x^3/27$ (*b*) $x^3/9$

(*c*) $x^3/25$ (*d*) $x^3/18$

50. If $V(x) = \sigma^2$, then $V(ax + b) =$ ....

(*a*) $aV(x) + V(b)$ (*b*) $a^2V(x)$

(*c*) $a^2V(x) + b$ (*d*) none

ANSWERS

1. (*c*)	**2.** (*d*)	**3.** (*a*)	**4.** (*c*)	**5.** (*c*)	**6.** (*a*)	**7.** (*d*)	**8.** (*a*)	**9.** (*b*)	**10.** (*d*)
11. (*c*)	**12.** (*d*)	**13.** (*b*)	**14.** (*c*)	**15.** (*a*)	**16.** (*b*)	**17.** (*a*)	**18.** (*c*)	**19.** (*c*)	**20.** (*d*)
21. (*b*)	**22.** (*d*)	**23.** (*c*)	**24.** (*b*)	**25.** (*d*)	**26.** (*c*)	**27.** (*c*)	**28.** (*d*)	**29.** (*c*)	**30.** (*c*)
31. (*d*)	**32.** (*c*)	**33.** (*d*)	**34.** (*d*)	**35.** (*d*)	**36.** (*b*)	**37.** (*c*)	**38.** (*a*)	**39.** (*a*)	**40.** (*b*)
41. (*b*)	**42.** (*c*)	**43.** (*d*)	**44.** (*a*)	**45.** (*d*)	**46.** (*b*)	**47.** (*c*)	**48.** (*d*)	**49.** (*a*)	**50.** (*b*)

Chapter 3

DISTRIBUTIONS

3.0 BINOMIAL DISTRIBUTION

Binomial distribution was discovered by James Bernoulli (1654–1705) in the year 1700 and was first published posthumously in 1713, eight years after his death. Let a random experiment be performed repeatedly, each repetition being called a trial and let the occurrence of an event in a trial be called a success and its non-occurrence is called a failure. Consider a set of 'n' independent Bernoulli trials (n being finite) in which the probability 'p' of success and the probability 'q' of failure in any trial, such that $p + q = 1$.

The probability of getting x successes is p^x.

The probability of getting $n - x$ failures is q^{n-x}.

$\therefore$ The probability of getting x successes and $n - x$ failures is $p^x . q^{n-x}$.

The number of combinations of x successes selected among 'n' trials is ${}^nC_x$.

Hence the probability of x successes in 'n' trials in any order is ${}^nC_x\, p^x\, q^{n-x}$.

Definition

A random variable X is said to follow binomial distribution if it assumes only non-negative values and its probability mass function is given by

$$P(X = x) = p(x) = {}^nC_x\, p^x\, q^{n-x}\ ;\ x = 0, 1, 2, 3, ..., n = 0 \text{ otherwise} \qquad ...(1)$$

Note:

1. The assignment of probabilities is permissible because

$$\sum_{x=0}^{n} p(x) = \sum_{x=0}^{n} {}^nC_x\, p^x\, q^{n-x} = (q + p)^n = 1$$

2. Let us suppose that n trials constitute an experiment. Then, if this experiment is repeated N times, frequency function of the Binomial distribution is given by

$$f(x) = Np(x) = N \sum_{x=0}^{n} {}^nC_x\, p^x\, q^{n-x} \quad ; x = 0, 1, 2, ..., n$$

and the expected frequencies of 0, 1, 2 , ..., n successes are the successive terms of the Binomial expansion, $N(q + p)^n$, $q + p = 1$.

3. Each trial results in two exhaustive and mutually disjoint outcomes, termed as success and failure.
4. The number of trials 'n' is finite.
5. The trials are independent of each other.
6. The probability of success 'p' is constant for each trial.

3.1 MEAN AND VARIANCE OF THE BINOMIAL DISTRIBUTION

(May 2007, Nov 2006)

For the Binomial distribution $P(X = x) = {}^nC_x\, p^x q^{n-x}$

$$\text{Mean} \quad \mu = \sum_{x=0}^{n} x.P(X = x)$$

$$= \sum_{x=0}^{n} x.{}^nC_x\, p^x q^{n-x}$$

$$= {}^nC_1\, p.q^{n-1} + 2\,{}^nC_2\, p^2 q^{n-2} + 3.{}^nC_3\, p^3 q^{n-3} + \ldots + n.{}^nC_n p^n$$

$$= {}^nC_1\, p.q^{n-1} + \frac{2n(n-1)}{1.2} p^2 q^{n-2} + 3\frac{n(n-1)(n-2)}{1.2.3} p^3 q^{n-3} + \ldots\ldots + \frac{n.n(n-1)\ldots.1}{1.2.3\ldots n-1} p^n$$

$$= np\,[q^{n-1} + (n-1)\,p\,q^{n-2} + \frac{(n-1)(n-2)}{1.2} p^2 q^{n-3} + \ldots\ldots + \frac{(n-1)\ldots 1}{1.2.3.\ldots n-1}\; p^{n-1}]$$

$$= np(q + p)\; n - 1 = np \qquad [\text{Since } p + q = 1]$$

$\therefore$ The mean of Binomial Distribution $= np$

Variance of Binomial Distribution:

$$\text{Variance} = \sigma^2 = \sum_{x=0}^{n} x^2 P(X=x) - \mu^2$$

$$= \sum_{x=0}^{n} {}^nC_x\, p^x q^{n-x} x^2 - \mu^2$$

$$= \sum_{x=0}^{n} {}^nC_x\, p^x q^{n-x} \left[x(x-1) + x\right] - \mu^2$$

$$= 1.2\ {}^nC_2\, p^2 q^{n-2} + 3.2,\ {}^nC_3\, p^3 q^{n-3} + \ldots {}^nC_n\, n\,(n-1)\, p^n + \sum {}^nC_x\,.\,x p^x q^{n-x} - \mu^2$$

$$= n(n-1)\,p^2 \sum_{x=2}^{n} {}^{n-2}C_{x-2}\, p^{x-2} q^{n-x} + np - n^2p^2 \quad [\text{Since} \sum x.{}^nC_x\, p^x q^{n-x} = \mu = np]$$

$= n(n-1)p^2(p+q)^{n-2} + np - n^2p^2 = n(n-1)p^2 + np - n^2p^2$ (Since $p + q = 1$)

$= n^2p^2 - np^2 + np - n^2p^2$

$= np(1-p) = npq$

∴ Variance of Binomial Distribution is npq.

Mode of Binomial Distribution

Mode is the value of x at which $P(x)$ has maximum value.

The probability of getting x successes among n trials is ${}^nC_x\, p^x q^{n-x}$.

Let x be the mode of Binomial distribution.

∴ $P(x) \geq P(x+1)\ .\ p(x) \geq p(x-1)$

$P(x) \geq P(x+1)$

∴ $$\frac{P(x)}{P(x+1)} \geq 1$$

i.e., $$\frac{{}^nC_x\, p^x q^{n-1}}{{}^nC_{x+1} p^{x+1} q^{n-x-1}}$$

$$P(x+1) = {}^nC_{x+1} p^{x+1} q^{n-x-1} = \frac{n(n-1)(......n-x)}{(x+1)!} . p^{x+1} q^{n-x-1}$$

$$\frac{P(x)}{P(x+1)} = \frac{n(n-1)(.......n-x)}{x!} . p^{x+1} q^{n-x} \times \frac{(x+1)!}{n(n-1).........(n-x)p^{x+1} q^{n-x-1}}$$

$$= \frac{(x+1)}{n-x} \frac{q}{p} \geq 1$$

$(x+1)q \geq (n-x)p$

$qx + q \geq np - px$

$px + qx \geq np - q$

$(p+q)x \geq np - (1-p)$

$x \geq np - 1 + p$

$x \geq (n+1)p - 1$ ...(1)

Taking $P(x) \geq P(x-1)$ we will get $x \leq (n+1)p$ ...(2)

From (1) and (2) $(n+1)p - 1 \leq x \leq (n+1)p$

Case *i*. If $(n+1)p$ is not an integer, Mode is the integral part of $(n+1)p$. In this case the distribution is called unimodal.

Case *ii*. If $(n+1)p$ is an integer both $(n+1)p$ and $(n+1)p - 1$ will represent modes and the distribution is bimodal.

3.2 SOLVED PROBLEMS

Problem 1: Ten coins are tosses simultaneously. Find the probability of getting at least seven heads. **(May 2004, May 2007, Nov 2007)**

Solution:

p = Probability of getting a head = 1/2

q = Probability of not getting a head = 1/2

The probability of getting x heads in a random throw of 10 coins is ;

$$P(x) = {}^{10}C_x \left(\frac{1}{2}\right)^x \left(\frac{1}{2}\right)^{10-x} = {}^{10}C_x \left(\frac{1}{2}\right)^{10} ; x = 0, 1, 2, \ldots, 10$$

$\therefore$ Probability of getting at least seven heads is given by ;

$$P(X \geq 7) = P(7) + P(8) + P(9) + P(10)$$

$$\left(\frac{1}{2}\right)^{10} \left\{{}^{10}C_7 + {}^{10}C_8 + {}^{10}C_9 + {}^{10}C_{10}\right\} = \frac{120 + 45 + 10 + 1}{1024} = \frac{176}{1024}.$$

Problem 2: A and B play a game in which their chances of winning are in the ratio 3 : 2. Find A's chance of winning at least three games out of the five games played.

Solution:

Let p be the probability that 'A' wins the game.

Then we are given: $n = 5, p = \frac{3}{5}$

$\Rightarrow$ $q = 1 - p = \frac{2}{5}$.

Hence, by Binomial Probability law, the probability that out of 5 games played, A wins 'x' games is given by:

$$P(X = x) = P(x) = {}^5C_x \left(\frac{3}{5}\right)^x \left(\frac{2}{5}\right)^{5-x} ; x = 0, 1, 2, \ldots, 5$$

The required probability that 'A' wins at least three games is given by:

$$P(X \geq 3) = \sum_{r=3}^{5} {}^5C_r \frac{3^r \cdot 2^{5-r}}{5^5} = \frac{3^3}{5^5}\left[{}^nC_3\, 2^2 + {}^5C_4\, 3 \times 2 + 1.3^2 \times 1\right]$$

$$= \frac{27 \times (40 + 30 + 9)}{3125} = 0.68.$$

Problem 3: An irregular six faced die is thrown and the expectation that in 10 throws it will give five even numbers is twice the expectation that it will give four even numbers. How many items in 10,000 sets of 10 throws each, would you expect it to give no even number?

Solution:

Let p be the probability of getting an even number in a throw of a die. Then the probability of getting x even numbers in ten throws of a die is given by:

$$P(X = x) = {}^{10}C_x\, p^x\, p^{10-x}; x = 0, 1, 2, \ldots, 10$$

We are given that: $P(X = 5) = 2P(X = 4) \Rightarrow {}^{10}C_5\, p^5 q^5 = {}^{210}C_4\, p^4 q^6$ ...(1)

$$\Rightarrow \quad \frac{10!p}{5!5!} = 2\,\frac{10!q}{4!6!}$$

$$\Rightarrow \quad \frac{p}{5} = \frac{2q}{6} = \frac{q}{3}$$

$$\Rightarrow \quad \frac{p}{5} = \frac{q}{3}$$

$$\Rightarrow \quad 3p = 5q$$

$$\Rightarrow \quad 3p = 5(1-p)$$

$$\Rightarrow \quad p = \frac{5}{8} \text{ and } q = \frac{3}{8}.$$

Thus $$P(X = x) = {}^{10}C_x \left(\frac{5}{8}\right)^x \left(\frac{3}{8}\right)^{10-x}$$ ...[From (1)]

Hence the required number of times that in 10,000 sets of 10 throws each we get no even number $10,000 \times P(X = 0) = 10000 \times {}^{10}C_0 \left(\frac{5}{8}\right)^0 \left(\frac{3}{8}\right)^{10-0}$

$$= 10000 \times \left(\frac{3}{8}\right)^{10} = 1 \text{ (Approx.)}$$

Problem 4: With usual notations, find p for a Binomial variate X, if $n = 6$ and $9P(X = 4) = P(X = 2)$.

Solution:

For the Binomial random variable X with parameters $n = 6$ and p, the probability function is

$$P(X = x) = {}^6C_x\, p^x q^{6-x}\ ;\ x = 0, 1, 2, \ldots\ldots, 6$$

We are given: $$9P(X = 4) = P(X = 2)$$

$$\Rightarrow \quad 9 \times {}^6C_4\, p^4 q^2 = {}^6C_2\, p^2 q^4$$

$$\Rightarrow \quad 9p^2 = q^2 \Rightarrow 9p^2 = (1-p)^2 = 1 + p^2 - 2p$$

$$\Rightarrow \quad 8p^2 + 2p - 1 = 0$$

$$\Rightarrow \quad p = \frac{-2 \pm \sqrt{4+32}}{2 \times 8} = \frac{-2 \pm 6}{16} = -\frac{1}{2}, \frac{1}{4}$$

Since probability cannot be negative, $p = -\frac{1}{2}$ is rejected. Hence $p = \frac{1}{4}$

Problem 5: The probability of a man hitting a target is 1/3.

(*i*) If he fires 5 times. What is the probability of his hitting the target at least twice?

(*ii*) How many times must he fire so that the probability of his hitting the target at least once is more than 90%? **(May 2006)**

Solution:

(*i*) Given

$$p = 1/3,\ n = 5$$

$$P(X = x) = {}^nC_x\, p^x\, q^{n-x}$$

$$P(X \geq 2) = 1 - P(x < 2)$$

$$= 1 - P(X = 0) - P(X = 1)$$

$$= 1 - {}^5C_0\left(\frac{1}{3}\right)^0\left(\frac{2}{3}\right)^{5-0} - {}^5C_1\left(\frac{1}{3}\right)^1\left(\frac{2}{3}\right)^{5-1}$$

$$= 1 - \frac{32}{243} - \frac{80}{243} = \frac{131}{243}$$

(*ii*) Given

$$P(\text{hit a target}) = \frac{1}{3}$$

The number of times he should fire, so that the probability of hitting the target at least once is more than 90% is calculated as follows:

$$1 - P(\text{no hit}) = 1 - \left(1 - \frac{1}{3}\right)^n > 0.9$$

$$= 1 - \left(\frac{2}{3}\right)^n > 0.9$$

For $n = 6$, the inequality is satisfied because,

$$1 - \left(\frac{2}{3}\right)^6 = 0.912 > 0.9$$

Problem 6: Determine the probability of getting 9 exactly twice in 3 throws with a pair of fair dice. **(April 2006)**

$$n = 3,\ x = 2$$

Solution: Probability of getting a sum 9 with two dice $= p = \frac{4}{36} = \frac{1}{9}$

$$\Rightarrow \quad q = 1 - \frac{1}{9} = \frac{8}{9}$$

Probability of getting 9 exactly two dice $= P(X = 2) = {}^nC_2\, p^2\, q^{n-2}$

$$= {}^3C_2\, p^2\, q^{3-2} = {}^3C_2\left(\frac{1}{9}\right)^2\left(\frac{8}{9}\right)^{3-2} = \frac{8}{243}$$

Problem 7: The mean and variance of a Binomial Distribution are 4 and 4/3 respectively. Find $P(X \geq 1)$. **(April 2004)**

Solution:

Given mean $= np = 4$, Variance $= npq = \frac{4}{3}$

$$\frac{\text{variance}}{\text{mean}} = \frac{npq}{np} = \frac{4/3}{4} = \frac{4}{3} \times \frac{1}{4} = \frac{1}{3}$$

$$P = 1 - \frac{1}{3} = \frac{2}{3}$$

$$np = 4,\ n = \frac{4}{p} = \frac{4}{2/3} \Rightarrow n = 6$$

$$P(X \geq 1) = 1 - P(X < 1)$$
$$= 1 - P(X = 0)$$
$$= 1 - {}^6C_n\, p^0\, q^{6-0}$$
$$= 1 - 1 \times \left(\frac{2}{3}\right)^0 \left(\frac{1}{3}\right)^0$$
$$= 1 - \left(\frac{1}{3}\right)^0 = 0.9986.$$

Problem 8: The mean and variance of a Binomial variable X with parameters n and p are 16 and 8. Find $P(X \geq 1)$ and $P(X > 2)$. **(April 2004)**

Solution:

$$\mu = 16,\ \lambda^2 = 8$$
$$\mu = np = 16$$
$$npq = 8 \Rightarrow 16\,q = 8$$
$$q = \frac{1}{2} \Rightarrow p = \frac{1}{2} \qquad np = 16 \Rightarrow n = 32$$
$$P(X \geq 1) = 1 - P(X < 1) = 1 - P(X = 0)$$
$$= 1 - {}^nC_x\, p^x\, q^{n-x}$$
$$= 1 - {}^{32}C_0\, p^0\, q^{32}$$
$$= 1 - {}^{32}C_0 \left(\frac{1}{2}\right)^0 \left(\frac{1}{2}\right)^{32}$$
$$= 1 - 1\left(\frac{1}{2}\right)^{32}$$
$$P(X \geq 1) = 0.99999$$
$$P(X > 2) = 1 - P(X \leq 2)$$
$$= 1 - [P(X = 0) + P(X = 1) + P(X = 2)]$$
$$= 1 - \left[{}^{32}C_0\left(\frac{1}{2}\right)^{32} + {}^{32}C_1\left(\frac{1}{2}\right)^{32} + {}^{32}C_2\left(\frac{1}{2}\right)^{32}\right]$$
$$= 1 - \left(\frac{1}{2}\right)^{32}\left[1 + 32 + 496\right]$$
$$= 1 - \left(\frac{1}{2}\right)^{32}[529]$$
$$= 0.9999.$$

Problem 9: In eight throws of a die 5 or 6 is considered a success. Find the mean number of success and the standard deviation. **(April 2004)**

Solution:

When a die is thrown,

The probability of getting 5 or 6 is $\frac{1}{6} + \frac{1}{6}$

i.e., $p = \frac{1}{3}$ and $q = \frac{2}{3}$

Given that the number of trials (n) = 8.

From Binomial distribution

$B(x, n, p) = {}^nC_x\, p^x q^{n-x} = {}^nC_x\, p^x (1-p)^{n-x}$ for $x = 0, 1, 2, \ldots, n$

Mean of the Binomial distribution $= np = 8x\, \frac{1}{3} = \frac{8}{3}$

$$\mu = \frac{8}{3}$$

Variance of the Binomial distribution $= npq = 8x\, \frac{1}{3}\, x\, \frac{2}{3}$

$$\sigma^2 = \frac{16}{9}$$

$$\text{Standard deviation } \sigma = \sqrt{\text{variance}} = \sqrt{npq} = \sqrt{\frac{16}{9}} = \frac{4}{3}$$

Problem 10: In a Binomial distribution consisting of 5 independent trials, probabilities of 1 and 2 success are 0.4096 and 0.2048 respectively. Find the parameter p of the distribution. **(April 2004)**

Solution:

Given that $n = 5$

$$P(X = 1) = 0.4096$$

$$P(X = 2) = 0.2048$$

Given that ${}^5C_1\, p^1 (1-p)^4 = 0.4096$

$${}^5C_2\, p^2 (1-p)^3 = 0.2048$$

$$\frac{{}^5C_1\, p^1 (1-p)^4}{{}^5C_2\, p^2 (1-p)^3} = \frac{0.4096}{0.2048}$$

$$\frac{5(1-p)}{10p} = \frac{0.4096}{0.2048}$$

$$1.024(1-p) = 4.096p$$

$$p = 0.2 \quad \Rightarrow q = 1 - 0.2 = 0.8$$

Problem 11: Out of 800 families with 5 children each, how many would you expect to have (*i*) 3 boys (*ii*) either 2 or 3 boys. **(Aug 2007, Nov 2006, Nov 2005, April 2005)**

Solution:

$$\text{Probability of boy} = P(B) = P = \frac{1}{2}$$

$$q = 1 - p = \frac{1}{2}$$

We know that $P(X = x) = {}^{n}C_x\, p^x\, q^{n-x}$

where, n = number of trials = 5

x = number of boys in a family

(*i*) Probability of a family having 3 boys

$$P(X = 3) = {}^{5}C_3 \left(\frac{1}{2}\right)^3 \left(\frac{1}{2}\right)^{5-3}$$

$$= \frac{5}{16} = 0.3125$$

Expected number of families having 3 boys $= 800 \left(\frac{5}{16}\right) = 250$

250 families have 3 boys out of 5 children

(*ii*) Probability of a family having 2 boys

$$P(X = 2) = {}^{5}C_2 \left(\frac{1}{2}\right)^2 \left(\frac{1}{2}\right)^{5-2} = {}^{5}C_2 \left(\frac{1}{2}\right)^5 = 0.3125$$

$$P(X = 3) \cup P(X = 2) = 0.3125 + 0.3125$$

$$= 0.625$$

Expected number of families having either 2 or 3 boys $= 800 \times 0.625 = 500$

Problem 12: 20% of items produced from a factory are defective. Find the probability that in a sample of 5 chosen at random. (*i*) None is defective, (*ii*) One is defective, (*iii*) $P(1 < x < 4)$.

(Nov 2007, Nov 2006, Nov 2005, April 2005)

Solution:

Probability of defective items $p = \frac{20}{100}$, $n = 5$

(*i*) $P(X = 0)$ (none is defective)

$$P(X = 0) = {}^{5}C_0 \left(\frac{1}{5}\right)^0 \left(1 - \frac{1}{5}\right)^{5-0}$$

$$P(X = 0) = \left(\frac{4}{5}\right)^5$$

(*ii*) $P(X = 1)$ (one is defective)

$$P(X = 1) = {}^{5}C_1 \left(\frac{1}{5}\right)^1 \left(1 - \frac{1}{5}\right)^{5-1}$$

$$P(X = 1) = \left(\frac{4}{5}\right)^4$$

(*iii*) $P(1 < X < 4) = P(X = 2) + P(X = 3)$

$$= {}^5C_2 \left(\frac{1}{5}\right)^2 \left(\frac{4}{5}\right)^3 + {}^5C_3 \left(\frac{1}{5}\right)^3 \left(\frac{4}{5}\right)^2$$

$$= 10 \left(\frac{1}{5}\right)^2 \left(\frac{4}{5}\right)^2 \left[\frac{4}{5} + \frac{1}{5}\right]$$

$$= 10 \left(\frac{4^2}{5^4}\right)$$

Problem 13: Determine the Binomial distribution for which the mean is 4 and variance 3 and find its mode.

Solution:

Given that mean = 4 = np

Variance is 3 = npq

$\therefore q = \dfrac{3}{4}$ $\qquad \therefore p = \dfrac{1}{4}$ $\qquad \therefore n = \dfrac{4}{p} = 16$

Binomial distribution is $\left(\dfrac{1}{4} + \dfrac{3}{4}\right)^{16}$

Mode = $(n + 1)p = \dfrac{17}{4}p$

$(n + 1)p$ is not an integer

$\therefore$ Integral part of $(n + 1)p$

i.e. 4 is the mode.

Problem 14: Fit a Binomial distribution to the following data. **(Nov 2004, May 2004)**

x	0	1	2	3	4	5
f	2	14	20	34	22	8

Solution:

In fitting Binomial distribution. First of all the mean and variance of the data are equated to np and npq. Then the expected frequencies are calculated from these values of n and p.

Here n = number of trials = 5

$$N = \text{total frequency} = \sum_{i=0}^{5} f_i = 2 + 14 + 20 + 34 + 22 + 8 = 100$$

Mean of the Binomial distribution

$$\mu = np = \frac{\sum f_i x_i}{\sum f_i}$$

$$= \frac{0(2) + 1(14) + 2(20) + 3(34) + 4(22) + 5(8)}{100}$$

$$= \frac{284}{100} = 2.84$$

$$np = 2.84$$

$$p = \frac{2.84}{5} = 0.568 \Rightarrow q = 1 - 0.568 = 0.432$$

$f(x) = N.P(X = x) = N.\ {}^nC_x\ p^x\ (q)^{n-x}$

$f(0) = N.\ P(X = 0) = N.\ {}^nC_0\ p^0\ (q)^n = 100.\ {}^5C_0\ (0.568)^0\ (0.432)^5 = 1,$

$f(1) = N.\ P(X = 1) = N.\ {}^nC_1\ p^1\ (q)^{n-1} = 100.\ {}^5C_1\ (0.568)^1\ (0.432)^4 = 10,$

$f(2) = N.\ P(X = 2) = N.\ {}^nC_2\ p^2\ (q)^{n-2} = 100.\ {}^5C_2\ (0.568)^2\ (0.432)^3 = 26,$

$f(3) = N.\ P(X = 3) = N.\ {}^nC_3\ p^3\ (q)^{n-3} = 100.\ {}^5C_3\ (0.568)^3\ (0.432)^2 = 34,$

$f(4) = N.\ P(X = 4) = N.\ {}^nC_4\ p^4\ (q)^{n-4} = 100.\ {}^5C_4\ (0.568)^4\ (0.432)^1 = 22,$

$f(5) = N.\ P(X = 5) = N.\ {}^nC_5\ p^5\ (q)^{n-5} = 100.\ {}^5C_5\ (0.568)^5\ (0.432)^0 = 6$

x	0	1	2	3	4	5
f	2	14	20	34	22	8
Expected frequency	1	10	26	34	22	6

Problem 15: Fit a Binomial distribution to the following data: **(Nov 2004)**

x	0	1	2	3	4	5
f	38	144	342	287	164	25

Solution:

Here n = Number of trials = 5

$$N = \text{Total frequency} = \sum_{i=0}^{5} f_i$$

$$= 38 + 144 + 342 + 287 + 164 + 25$$

$$= 1000$$

$$N = 1000$$

Mean of Binomial distribution $\mu = \dfrac{\sum f_i x_i}{\sum f_i}$

$$\mu = \frac{0(38)+1(144)+2(342)+(3)(287)+4(164)+5(25)}{1000}$$

$$\mu = \frac{2470}{1000} = 2.47$$

$$\mu = 2.47$$

$$\mu = np$$

$$p = \frac{\mu}{n} = \frac{2.47}{5} = 0.494$$

$$p = 0.494$$
$$q = 1 - p = 1 - 0.494 = 0.506$$

$f(x) = \text{N. } P(X = x) = \text{N. } {}^nC_x\, p^x\, (q)^{n-x}$

$f(0) = \text{N. } P(X = 0) = \text{N. } {}^nC_0\, p^0\, (q)^n = 1000.\ {}^5C_0\, (0.494)^0\, (0.506)^5 = 0.033$

$f(1) = \text{N. } P(X = 1) = \text{N. } {}^nC_1\, p^1\, (q)^{n-1} = 1000.\ {}^5C_1\, (0.494)^1\, (0.506)^4 = 0.162$

$f(2) = \text{N. } P(X = 2) = \text{N. } {}^nC_2\, p^2\, (q)^{n-2} = 1000.\ {}^5C_2\, (0.494)^2\, (0.506)^3 = 0.316$

$f(3) = \text{N. } P(X = 3) = \text{N. } {}^nC_3\, p^3\, (q)^{n-3} = 1000.\ {}^5C_3\, (0.494)^3\, (0.506)^2 = 0.309$

$f(4) = \text{N. } P(X = 4) = \text{N. } {}^nC_4\, p^4\, (q)^{n-4} = 1000.\ {}^5C_4\, (0.494)^4\, (0.506)^1 = 0.151$

$f(5) = \text{N. } P(X = 5) = \text{N. } {}^nC_5\, p^5\, (q)^{n-5} = 1000.\ {}^5C_5\, (0.494)^5\, (0.506)^0 = 0.0294$

Problem 16: Find the maximum n such that the probability of getting no head in tossing a coin n times is greater than 0.1.

Solution:

Q = Probability of getting a head $= \frac{1}{2}$

Given that

$$P(x = 0) > \frac{1}{10}$$

$$ {}^nC_0\, p^0\, (1 - p)^{n-0} > \frac{1}{10}$$

$$\left(\frac{1}{2}\right)^n > \frac{1}{10}$$

$$\left(\frac{1}{2}\right)^1 > 0.5$$

$$\left(\frac{1}{2}\right)^2 = 0.5$$

$$\left(\frac{1}{2}\right)^3 = 0.125$$

$$\left(\frac{1}{2}\right)^4 = 0.0625$$

For $n \geq 4$ probability is $< \frac{1}{10}$

Hence , $n = 3$

Since $\left(\frac{1}{2}\right)^3 = 0.125 > \frac{1}{10}$

Problem 17: The mean of Binomial distribution is 3 and variance is $\frac{9}{4}$. Find

(*i*) The value of n (*ii*) $P(X \geq 1)$ (*iii*) $P(X \leq 7)$ (*iv*) $P(1 \leq x < 6)$ **(Nov 2007)**

Solution:

(*i*) Mean $= \mu = 3 = np$

$$\text{Variance } \sigma^2 = npq = \frac{9}{4}$$

$$\therefore \qquad q = \frac{9}{4 \times 3} = \frac{3}{4}$$

$$\therefore \qquad p = \frac{1}{4},\ n = 3 \times 4 = 12$$

(*ii*) $P(X \geq 1) = 1 - P(X = 0) = 1 - {}^nC_0\, p^0\, q^{n-0} = 1 - \left(\frac{3}{4}\right)^{12}$

(*iii*) $P(X \leq 7) = P(X = 0) + P(X = 1) + P(X = 2) + P(X = 3) + P(X = 4) + P(X = 5) + P(X = 6) + P(X = 7)$

$$= {}^{12}C_0 \cdot \left(\frac{3}{4}\right)^{12} + {}^{12}C_1 \cdot \frac{1}{4} \cdot \left(\frac{3}{4}\right)^{11} + {}^{12}C_2 \cdot \left(\frac{1}{4}\right)^2 \times \left(\frac{3}{4}\right)^{10} + {}^{12}C_3 \cdot \left(\frac{1}{4}\right)^3 \times \left(\frac{3}{4}\right)^9$$

$$+ {}^{12}C_4 \cdot \left(\frac{1}{4}\right)^4 \times \left(\frac{3}{8}\right)^8 + {}^{12}C_5 \cdot \left(\frac{1}{4}\right)^5 \times \left(\frac{3}{4}\right)^7 + {}^{12}C_6 \cdot \left(\frac{1}{4}\right)^6 \times \left(\frac{3}{4}\right)^6 + {}^{12}C_7 \cdot \left(\frac{1}{4}\right)^7 \times \left(\frac{3}{4}\right)^5$$

$$= \left(\frac{3}{4}\right)^{12} + 12 \times \frac{1}{4} \times \left(\frac{3}{4}\right)^{11} + 66 \times \left(\frac{1}{4}\right)^2 \times \left(\frac{3}{4}\right)^{10} + 220 \times \left(\frac{1}{4}\right)^3 \times \left(\frac{3}{4}\right)^9 + 495$$

$$\times \left(\frac{1}{4}\right)^4 \times \left(\frac{3}{4}\right)^8 + 72 \times \left(\frac{1}{4}\right)^5 \times \left(\frac{3}{4}\right)^7 + 84 \times \left(\frac{1}{4}\right)^6 \times \left(\frac{3}{4}\right)^6 + 72 \times \left(\frac{1}{4}\right)^7 \times \left(\frac{3}{4}\right)^5$$

$$= \frac{3^5}{4^{12}} \left[3^7 + 12 \times 3^6 + 66 \times 3^5 + 220 \times 3^4 + 495 \times 3^3 + 84 \times 3 + 72\right]$$

$$= \frac{3^5}{4^{12}} [6570]$$

(*iv*) $P(1 \leq X < 6) = P(X = 1) + P(X = 2) + P(X = 3) + P(X = 4) + P(X = 5)$

$$= {}^{12}C_1 \cdot \frac{1}{4} \cdot \left(\frac{3}{4}\right)^{11} + {}^{12}C_2 \cdot \left(\frac{1}{4}\right)^2 \times \left(\frac{3}{4}\right)^{10} + {}^{12}C_3 \cdot \left(\frac{1}{4}\right)^3 \times \left(\frac{3}{4}\right)^9 + {}^{12}C_4 \cdot \left(\frac{1}{4}\right)^4 \times \left(\frac{3}{4}\right)^8$$

$$+ {}^{12}C_5 \cdot \left(\frac{1}{4}\right)^5 \times \left(\frac{3}{4}\right)^7$$

$$= 12 \times \frac{1}{4} \times \left(\frac{3}{4}\right)^{11} + 66 \times \left(\frac{1}{4}\right)^2 \times \left(\frac{3}{4}\right)^{10} + 220 \times \left(\frac{1}{2}\right)^3 \times \left(\frac{3}{4}\right)^9 + 495 \times \left(\frac{1}{4}\right)^4 \times \left(\frac{3}{4}\right)^8$$

$$+ 72 \times \left(\frac{1}{4}\right)^5 \times \left(\frac{3}{4}\right)^7$$

$$= \frac{3^7}{4^{12}} [972 + 1782 + 1980 + 1485 + 72] = \frac{3^7}{4^{12}} (6291).$$

3.3 POISSON DISTRIBUTION

Poisson distribution was discovered by the French mathematician and physicist Simeon Denis Poisson (1781–1840) who published it in 1837.

Poisson Distribution as a limiting case of Binomial distribution

(Feb 2007, May 2007, May 2006)

If the parameters n and p of Binomial distribution are known we can find the distribution, but in situation, where n is very large and p is very small, however if we assume that as $n \to \infty$ and $p \to 0$ such that np is always finite say λ, we get the Poisson approximation to the Binomial distribution

$$P(X = x) = P(x) = {}^nC_x\, p^x q^{n-x} = \frac{n(n-1)(n-2)\ldots\ldots(n-x+1)}{x!} p^x (1-p)^{n-x}$$

$$= \frac{n(n-1)(n-2)\ldots\ldots(n-x+1)}{x!}\left(\frac{\lambda}{n}\right)^x\left(1-\frac{\lambda}{n}\right)^{n-x} \quad \text{(Since } np = \lambda, p = \lambda/n)$$

$$= \frac{\lambda^x}{x!}\frac{n}{n}\frac{(n-1)}{n}\frac{(n-2)}{n}\ldots\ldots\ldots\frac{n-x+1}{n}\left(1-\frac{\lambda}{n}\right)^{n-x}$$

$$= \frac{\lambda^x}{x!}\left(1-\frac{1}{n}\right)\left(1-\frac{2}{n}\right)\ldots\ldots\ldots\left(1-\frac{x-1}{n}\right)\frac{\left(1-\frac{\lambda}{n}\right)^n}{\left(1-\frac{\lambda}{n}\right)^x}$$

As $n \to \infty$ then $\left(1-\frac{1}{n}\right)\left(1-\frac{2}{n}\right)\ldots\ldots\ldots\left(1-\frac{x-1}{n}\right) \to 1$ and also $\left(1-\frac{\lambda}{n}\right)^x \to 1$

$$\therefore \quad \frac{\lambda^x}{x!}\underset{n\to\infty}{\text{Lt}}.\left(1-\frac{\lambda}{n}\right)^n = \frac{\lambda^x}{x!}\left[\underset{n\to\infty}{\text{Lt}}\left(1-\frac{\lambda}{n}\right)^{\frac{-n}{\lambda}}\right]^{-\lambda} = e^{-\lambda}$$

$$= \frac{\lambda^x}{x!}e^{-\lambda}$$

$$\therefore \quad P(X = x) = p(x)$$

$$= \frac{\lambda^x}{x!}e^{-\lambda}, \text{ where } \lambda \text{ is a finite number } np.$$

Definition

A random variable X is said to follow Poisson distribution if its probability mass function is given by $P(x) = \frac{\lambda^x}{x!}e^{-\lambda} \quad x = 0, 1, 2,\ldots = 0$ otherwise

Here λ is known as the parameter of the distribution. We shall use the notation $X \sim P(\lambda)$ is denote that X is a Poisson variate with parameter.

3.4 MEAN AND VARIANCE OF POISSON DISTRIBUTION

(April 2007, Nov 2006, Nov 2005)

For the Poisson distribution $P(x) = \frac{\lambda^x}{x!}e^{-\lambda}$.

$$\text{Mean } \mu = \sum_{x=0}^{\infty} xP(x)$$

$$= \sum_{x=0}^{\infty} x \frac{\lambda^x e^{-\lambda}}{x!}$$

$$= \sum_{x=0}^{\infty} \frac{\lambda^x e^{-\lambda}}{(x-1)!}$$

$$= e^{-\lambda} \lambda \sum_{x=0}^{\infty} \frac{\lambda^{x-1}}{(x-1)!}$$

$$= e^{-1} \lambda \left(1 + \frac{\lambda}{1!} + \frac{\lambda^2}{2!} + \frac{\lambda^3}{3!} + \ldots\ldots\ldots\ldots\right)$$

$$= e^{-\lambda} \lambda e^{\lambda} = \lambda$$

$\therefore$ Variance $\sigma^2 = \sum_{x=0}^{\infty} x^2 P(x) - \mu^2$

$$= \sum_{x=0}^{\infty} \left(x^2 + x - x\right) P(x) - \mu^2$$

$$= \sum_{x=0}^{\infty} xp(x) + \sum_{x=0}^{\infty} x(x-1)P(x) - \mu^2$$

$$= \lambda + \sum_{x=0}^{\infty} x(x-1)\frac{\lambda^x e^{-\lambda}}{x!} - \lambda^2$$

$$= \lambda + \sum_{x=0}^{\infty} \frac{\lambda^x e^{-\lambda}}{(x-2)!} - \lambda^2$$

$$= \lambda + \lambda^2 - \lambda^2 = \lambda$$

Hence variance of Poisson distribution is also λ.

Thus mean and variance of Poisson distribution is equal to the parameter λ.

3.5 ADDITIVE PROPERTY OF POISSON VARIATES

If X and Y are two independent Poisson variates with parameters λ_1 and λ_2, then X + Y is also a Poisson variate with parameter $\lambda_1 + \lambda_2$.

3.6 SOLVED PROBLEMS

Problem 1: Average number of accidents on any day on a national high way is 1.8. Determine the probability that the number of accidents are (*i*) at least one (*ii*) at most one.

(Nov 2007, May 2007, Nov 2006, Nov 2004)

Solution: Given $\lambda = 1.8$

$$P(X \geq 1) = 1 - P(X = 0) = 1 - \frac{\lambda^0 e^{-\lambda}}{0!} = 1 - e^{-1.8}$$

$$P(X \le 1) = P(X = 0) + P(X = 1)$$

$$= \frac{\lambda^0 e^{-\lambda}}{0!} + \frac{\lambda^1 e^{-\lambda}}{1!}$$

$$= e^{-\lambda} + e^{-\lambda} \times 1.8$$

$$= e^{-1.8} (1 + 1.8)$$

$$= 2.8 \times e^{-1.8}.$$

Problem 2: If X is a Poisson variate such that P(X = 0) = P(X = 2) + 3 P(X = 4), find (*i*) the mean of x (*ii*) $P(X \le 2)$.

Solution: Given $P(X = 0) = P(X = 2) + 3P(X = 4)$

$$\frac{\lambda^0 e^{-\lambda}}{0!} = \frac{\lambda^2 e^{-\lambda}}{2!} + 3\frac{\lambda^4 e^{-\lambda}}{4!}$$

$$e^{-\lambda} = e^{-\lambda}\left[\frac{\lambda^2}{2} + \frac{\lambda^4}{8}\right]$$

$$1 = \left[\frac{\lambda^2}{2} + \frac{\lambda^4}{8}\right]$$

$$1 = \left[\frac{4\lambda^2 + \lambda^4}{8}\right]$$

$$\lambda^4 + 4\lambda^2 - 8 = 0$$

By solving equation (1), we get $\lambda = 1.21$

(*i*) Mean of $x = \lambda = 1.21$

(*ii*) $P(X \le 2) = P(X = 0) + P(X = 1) + P(X = 2)$

$$= \frac{\lambda^0 e^{-\lambda}}{0!} + \frac{\lambda^1 e^{-\lambda}}{1!} + \frac{\lambda^2 e^{-\lambda}}{2!}$$

$$= e^{-\lambda} [1 + \lambda + \frac{\lambda^2}{2}]$$

$$= e^{-1.21} [1 + 1.21 + \frac{(1.21)^2}{2}]$$

$$= e^{-1.21} \times 2.94205.$$

Problem 3: A distributor of bean seeds determines from extensive tests that 5% of large batch of seeds will not germinate. He sells the seeds in packets of 200 and guarantees 90% germination. Determine the probability that a particular packet will violate the guarantee.

Solution:

We are given

P = Probability that the seeds will not germinate = 5% = 0.05

$n = 200$

λ = Mean number of defective seeds in a packet of 200 = np = 200 × 0.05 = 10

Since p is small we may use Poisson's distribution. Probability of x defective seeds in a packet of 200 is

$$P(X = x) = \frac{\lambda^x}{x!} e^{-\lambda}$$

Probability that a packet will violate the guarantee is [*i.e.*, germination less than 90%]

$$P(X < 180) = \sum_{x=0}^{179} \frac{e^{-10} 10^x}{x!} = e^{-10} \sum_{x=0}^{179} \frac{x^{-10} 10^x}{x!}$$

Problem 4: Suppose 2% of the people on the average are left handed. Find (*i*) the probability of finding 3 or more left handed, (*ii*) the probability of finding ≤ 1 left handed. **(Nov 2006)**

Solution:

Let x denote the number of left handed people

So, $$P(X = x) = \frac{\lambda^x}{x!} e^{-\lambda} \quad x = 0, 1, 2 \ldots \text{ given } \lambda = 2$$

$$\therefore \quad P(X = x) = \frac{(2)^x e^{-2}}{x!}$$

(*i*)
$$\begin{aligned} P(X \geq 3) &= 1 - P(X < 3) \\ &= 1 - [P(X = 0) + P(X = 1) + P(X = 2) \\ &= 1 - e^{-2} [1 + 2 + \frac{2^2}{2!}] \\ &= 1 - e^{-2} [5] \\ &= 0.3233 \end{aligned}$$

(*ii*)
$$\begin{aligned} P(X \leq 1) &= P(X = 0) + P(X = 1) \\ &= \frac{2^0 e^{-2}}{0!} + \frac{2^1 e^{-2}}{1!} \\ &= e^{-2} [1 + 2] \\ &= 3 \times e^{-2}. \end{aligned}$$

Problem 5: If a Poisson distribution is such that $P(X = 1). \frac{3}{2} = P(X = 3)$. Find (*i*) $P(X \geq 1)$ (*ii*) $P(X \leq 3)$. **(May 2007, Nov 2006, May 2005)**

Solution:

Given $$P(X = 1) \frac{3}{2} = P(X = 3)$$

$$\frac{\lambda^1 e^{-\lambda}}{1!} \cdot \frac{3}{2} = \frac{\lambda^3 e^{-\lambda}}{3!}$$

$$\Rightarrow \quad \lambda^2 = 9 \Rightarrow \lambda = \pm 3$$

λ cannot be negative. Hence $\lambda = 3$

(*i*) $P(X \geq 1) = 1 - P(X < 1)$

$= 1 - P(X = 0)$

$$= 1 - \frac{3^0 e^{-3}}{0!} = 1 - e^{-3} = 0.9502$$

(*ii*) $P(X \leq 3) = P(X = 0) + P(X = 1) + P(X = 2) + P(X = 3)$

$$= \frac{3^0 e^{-3}}{0!} + \frac{3^1 e^{-3}}{1!} + \frac{3^2 e^{-3}}{2!} + \frac{3^3 e^{-3}}{3!}$$

$$= e^{-3} [1 + 3 + \frac{3^3}{2!} + \frac{3^3}{3!}]$$

$$= e^{-3}(13) = 0.6472$$

Problem 6: If the probability of a bad reaction from a certain injection is determine the chance that out of 3,000 individuals more than two will get a bad reaction. **(April 2004)**

Solution: Probability of bad reaction $p = 0.001$

$n = 3,000$

Mean $\lambda = np = 3$

Probability of more than two will get bad reaction

$P(X > 2) = 1 - P(X \leq 2)$

$= 1 - [P(X = 0) + P(X = 1) + P(X = 2)]$

$$= 1 - e^{-3} (1 + 3 + \frac{9}{2})$$

$$= 1 - \frac{17}{2} e^{-3} = 0.5619.$$

Problem 7: A hospital switch board receives an average of 4 emergency calls in a 10 minutes interval. What is the probability that (*i*) There are at most 2 emergency calls, (*ii*) Exactly 3 emergency calls in a 10 minutes interval.

Solution:

Given $\lambda = 4$

(*i*) $P(X \leq 2) = P(X = 0) + P(X = 1) + P(X = 2)$

$$= e^{-4} (1 + 4 + \frac{16}{2}) = 13 e^{-4} = 0.238$$

(*ii*) $$P(X = 3) = \frac{e^{-4} 4^3}{3!} = 0.195.$$

Problem 8: Average number of accidents on any day on a national highway is 0.8. Determine the probability that the number of accident is (*i*) at least, (*ii*) atmost one.

Solution:

Given average $\lambda = 0.8$

(*i*) $$P(X \geq 1) = 1 - P(X = 0) = 1 - e^{-0.8} = 1 - 0.449 = 0.551$$

(*ii*) $$P(X \leq 1) = P(X = 0) + P(X = 1) = e^{-0.8} + e^{-0.8}(0.8) = (2.8)\, e^{-0.8} = 0.809.$$

Problem 9: Find the probability that at most 5 defective components will be found in a lot of 200 it experience. Shows that 2% of such components are defective. Also find the probability of more than five defective the probability of more than five defective components.

Solution: Probability of defective components $p = 0.02$

$$n = 200$$

$$\text{Mean } \lambda = np = 200 \times 0.02 = 4$$

Probability that at most 5 defective components

$$P(X \leq 5) = P(X = 0) + P(X = 1) + P(X = 2) + P(X = 3) + P(X = 4) + P(X = 5)$$

$$= e^{-4} + \frac{4}{1!}e^{-4} + \frac{4^2}{2!}e^{-4} + \frac{4^3}{3!}e^{-4} + \frac{4^4}{4!}e^{-4} + \frac{4^5}{5!}e^{-4}$$

$$= \left(1 + 4 + \frac{16}{2} + \frac{64}{6} + \frac{256}{24} + \frac{256}{30}\right)e^{-4}$$

$$= \left(13 + \frac{64}{3} + \frac{128}{15}\right)e^{-4} = 0.7851$$

Probability that more than 5 components are defective is $P(X > 5) = 1 - P(X \leq 5) = 1 - 0.7851 = 0.2149$

Problem 10: In a certain factory turning out razor blades, there is a small chance of 0.002 for any blade to be defective. The blades are supplied in packets of 10, use Poisson distribution to calculate the approximate number of packets containing no defective, one defective and two defective blades respectively in a consignment of 10,000 packets.

Solution:

Given $\lambda = np = 10 \times 0.002 = 0.002$

$$e^{-0.02} = 1 - 0.02 + \frac{(0.02)^2}{2!} \ldots\ldots\ldots\ldots = 0.9802 \text{ approximately}$$

Probability of no defective blade is $e^{-\lambda} = e^{-0.02} = 0.9802$

$\therefore$ No of packets containing no defective blade is $10000 \times 0.9802 = 9802$

Similarly the number of packets containing one defective blade $= 10{,}000 \times \lambda\, e^{-\lambda}$

$$= 10000 \times (0.02) \times 0.9802 = 196$$

Finally the number if packets containing two defective blades

$$= 10000 \times \frac{\lambda^2 e^{-\lambda}}{2!} = 10000 \times \frac{(0.02)^2}{2!} \times 09802 = 2 \text{ approximately.}$$

Problem 11: 20% of the bolts produced in a factory found to be defective. Find the probability that in a sample of 10 bolts chosen at random, exactly two will be defective by using (*i*) Binomial distribution, (*ii*) Poisson distribution.

Solution:

(*i*) Probability of a defective bolt is $p = 0.2$

$\therefore$ $q = 0.8$

$n = 10$

Probability that exactly two bolts are defective is $P(X = 2) = {}^{10}C_2 (p)^2 (q)^8$

$$P(X = 2) = {}^{10}C_2 (0.2)^2 (0.8)^8 = 0.3019$$

(*ii*) In Poisson distribution, mean $\lambda = np$

$$= 10 \times 0.2 = 2$$

From the Poisson distribution

Probability that exactly two bolts are defective is $P(X = 2) = \frac{\lambda^2 e^{-\lambda}}{2!}$

$$= \frac{2^2 e^{-2}}{2!} = 2e^{-2} = 2 \times 0.1353$$

$$= 0.2707.$$

Problem 12: If the probability of a defective fuse from a manufacturing unit is 2%, in a box of 200 fuses, find the probability that (*i*) exactly 4 fuses are defective, (*ii*) more than 3 fuses are defective.

Solution: Probability of defective fuse is $p = 0.02$

$n = 200$

$\therefore$ Mean of the distribution $\lambda = np = 200 \times 0.02 = 4$

(*i*) Probability that exactly 4 fuses are defective is $P(X = 4) = \frac{\lambda^4 e^{-\lambda}}{4!}$

$$= \frac{4^4 e^{-4}}{4!}$$

$$= 10.667 \times 0.0183$$

$$= 0.1952$$

(*ii*) Probability that more than 3 fuses are defective is $P(X > 3) = 1 - P(X \leq 3)$

$$= 1 - \left(\frac{4^0}{0!} e^{-4} + \frac{4^1}{1!} e^{-4} + \frac{4^2}{2!} e^{-4} + \frac{4^3}{3!} e^{-4}\right)$$

$$= 1 - \left(\frac{4^0}{0!} + \frac{4}{1!} + \frac{4^2}{2!} + \frac{4^3}{3!}\right) e^{-4}$$

$$= 1 - \left(1 + 4 + 8 + \frac{32}{3}\right) \times 0.0183$$

$$= 0.5669.$$

Problem 13: A manufacturer of cotter pins known that 5% of his product is defective. If he sells cotter pins in boxes of 100 and guarantees that not more than 10 pins will be defective, what is the approximate probability that a box will fail to meet the guaranteed quality?

Solution:

Given $n = 100$

Probability of a defective pin is $P = 0.05$

Mean $\lambda = np = 100 \times 0.05 = 5$

Probability of x defective pins in a box is $P(X = x) = \dfrac{\lambda^x e^{-\lambda}}{x!}$

Probability that a box will fail to meet the guaranteed quality is $P(X > 10) = 1 - P(X \le 10)$

$$= 1 - \sum_{x=0}^{10} \frac{e^{-5} 5^x}{x!}$$

$$= e^{-5} \sum_{x=0}^{10} \frac{5^x}{x!}$$

Problem 14: The average number of phone calls/minute coming into a switch board between 2 p.m. and 4 p.m. is 2.5. Determine the probability that during one particular minute there will be

(*i*) 4 or fewer (*ii*) more than 6 calls **(Nov 2007, Feb 2007, May 2006)**

Solution:

Mean $\lambda = 2.5$.

Let X be the number of phone calls/minute during 2 : 00 P.M. and 4 : 00 P.M.

For the Poisson distribution $P(X = x) = \dfrac{\lambda^x e^{-\lambda}}{x!}$

(*i*) Probability that the 4 or fewer phone calls/minute is

$P(X \le 4) = P(X = 0) + P(X = 1) + P(X = 2) + P(X = 3) + P(X = 4)$

$$= \frac{\lambda^0 e^{-\lambda}}{0!} + \frac{\lambda^1 e^{-\lambda}}{1!} + \frac{\lambda^2 e^{-\lambda}}{2!} + \frac{\lambda^3 e^{-\lambda}}{3!} + \frac{\lambda^4 e^{-\lambda}}{4!}$$

$$= \frac{(2.5)^0 e^{-2.5}}{0!} + \frac{(2.5)^1 e^{-2.5}}{1!} + \frac{(2.5)^2 e^{-2.5}}{2!} + \frac{(2.5)^3 e^{-2.5}}{3!} + \frac{(2.5)^4 e^{-2.5}}{4!}$$

$$= e^{-2.5}\left[\frac{(2.5)^0}{0!} + \frac{(2.5)^1}{1!} + \frac{(2.5)^2}{2!} + \frac{(2.5)^3}{3!} + \frac{(2.5)^4}{4!}\right]$$

$$= e^{-2.5}\,[10.856]$$

$$= 0.8911.$$

(*ii*) Probability that more than 6 phone calls/minute is $P(X > 6) = 1 - P(X \le 6)$

$$= 1 - [P(X = 0) + P(X = 1) + P(X = 2) + P(X = 3) + P(X = 4) + P(X = 5) + P(X = 6)]$$

$$= 1 - \left(\frac{\lambda^0 e^{-\lambda}}{0!} + \frac{\lambda^1 e^{-\lambda}}{1!} + \frac{\lambda^2 e^{-\lambda}}{2!} + \frac{\lambda^3 e^{-\lambda}}{3!} + \frac{\lambda^4 e^{-\lambda}}{4!} + \frac{\lambda^5 e^{-\lambda}}{5!} + \frac{\lambda^6 e^{-\lambda}}{6!}\right)$$

$$= 1 - \left[\frac{(2.5)^0 e^{-2.5}}{0!} + \frac{(2.5)^1 e^{-2.5}}{1!} + \frac{(2.5)^2 e^{-2.5}}{2!} + \frac{(2.5)^3 e^{-2.5}}{3!} + \frac{(2.5)^4 e^{-2.5}}{4!} + \frac{(2.5)^5 e^{-2.5}}{5!} + \frac{(2.5)^6 e^{-2.5}}{6!}\right]$$

$$= 1 - e^{-2.5}\left[\frac{(2.5)^0}{0!} + \frac{(2.5)^1}{1!} + \frac{(2.5)^2}{2!} + \frac{(2.5)^3}{3!} + \frac{(2.5)^4}{4!} + \frac{(2.5)^5}{5!} + \frac{(2.5)^6}{6!}\right]$$

$$= 1 - e^{-2.5}\,[12.009] = 1 - 0.9858 = 0.0141.$$

Problem 15: Suppose the average number of telephone calls coming into a telephone exchange between 10 A.M. to 11 A.M. is 2, while between 11 A.M. to 12 noon is 6, determine the probability that more than five calls come in between 10 A.M. to 12 noon, assuming that calls are independent.

Solution: Poisson distribution is additive $X = X_1 + X_2 = 2 + 6 = 8$

$$P(X > 5) = 1 - P(X - 5)$$

$$= 1 - [P(X = 0) + P(X = 1) + P(X = 2) + P(X = 3) + P(X = 4) + P(X = 5)]$$

$$= 1 - \left[e^{-8} + \frac{4}{1!}e^{-8} + \frac{4^2}{2!}e^{-8} + \frac{4^3}{3!}e^{-8} + \frac{4^4}{4!}e^{-8} + \frac{4^5}{5!}e^{-8}\right]$$

$$= 1 - \left(1 + 4 + \frac{16}{2} + \frac{64}{6} + \frac{256}{24} + \frac{256}{30}\right)e^{-8}$$

$$= 1 - \left(13 + \frac{64}{3} + \frac{128}{15}\right)e^{-8}$$

$$= 1 - 0.191 = 0.8088.$$

Problem 16: The average rate of phone calls received is 0.6 calls per minute at an office. Determine the probability that (*i*) There will be one or more calls in a minute, (*ii*) There will be at least three calls during 4 minutes.

Solution:

Average rate of phone calls received is =0.6

(*i*)

$$P(X \ge 1) = 1 - P(X < 1)$$

$$= 1 - P(X = 0)$$

$$= 1 - \frac{\lambda^0 e^{-\lambda}}{0!} = 1 - e^{-0.6}$$

$$= 1 - 0.549$$

$$= 0.451.$$

(*ii*) Average of phone calls received during 4 minutes interval is $\lambda = 4 \times 0.6 = 2.4$

$$\begin{aligned} P(X \geq 3) &= 1 - P(X < 3) \\ &= 1 - [P(X = 0) + P(X = 1) + P(X = 2)] \\ &= 1 - \left[e^{-2.4} + \frac{2.4}{1!} e^{-2.4} + \frac{(2.4)^2}{2!} e^{-2.4}\right] \\ &= 1 - e^{-24} [1 + 2.4 + 2.88] \\ &= 1 - 0.570 \\ &= 0.4302. \end{aligned}$$

Problem 17: If a bank receives on an average 6 bad cheques per day, what are the probability that it will receive (*i*) 4 bad cheques on any given day, (*ii*) 10 bad cheques on any two consecutive days. **(JNTU 1999)**

Solution:

(*i*) The average of bad cheques $\lambda = 6$

$\therefore$ Probability that 4 bad cheques are received in any day is

$$\begin{aligned} P(X = 4) &= \frac{\lambda^4 e^{-\lambda}}{4!} \\ &= \frac{6^4 e^{-6}}{4!} = 0.1339 \end{aligned}$$

(*ii*) The average of bad cheques in two days is $\lambda = 2 \times 6 = 12$

$\therefore$ Probability that 10 bad cheques on two consecutive days is

$$\begin{aligned} P(X = 10) &= \frac{\lambda^{10} e^{-\lambda}}{10!} \\ &= \frac{12^{10} e^{-12}}{10!} = 0.105. \end{aligned}$$

3.7 RECURRENCE RELATION

For the Poisson distribution $P(x) = \dfrac{\lambda^x e^{-\lambda}}{x!}$

$$P(x + 1) = \frac{\lambda^{x+1} e^{-\lambda}}{(x + 1)!}$$

$$\frac{P(x + 1)}{P(x)} = \frac{\lambda}{(x + 1)}$$

$$P(x + 1) = \frac{\lambda}{(x + 1)} .P(x).$$

3.8 SOLVED PROBLEMS

Problem 1: If X is Poisson variate such that P(X = 0) = P(X = 1), find P(0) and using recurrence relation, find the probability at $x = 1, 2, 3, 4$ and 5. **(Nov 2004)**

Solution:

Given that

$$P(X = 0) = P(X = 1)$$

$$\frac{\lambda^0 e^{-\lambda}}{0!} = \frac{\lambda^1 e^{-\lambda}}{1!} \Rightarrow \lambda = 1$$

∴ Mean $\lambda = 1$

$$P(X = 0) = P(0) = e^{-1} = 0.368$$

$$P(X = 1) = P(1) = e^{-\lambda} = e^{-1} = 0.368$$

$$P(X = 2) = P(2) = \frac{1}{2}.e^{-1} = 0.184$$

$$P(X = 3) = P(3) = \frac{1}{6} e^{-1} = 0.0613$$

$$P(X = 4) = P(4) = \frac{1}{24} e^{-1} = 0.0153$$

$$P(X = 5) = P(5) = \frac{1}{120} e^{-1} = 0.00306$$

∴ Probabilities at X = 0, 1, 2, 3, 4, 5 are 0.368, 0.368, 0.184, 0.0613, 0.0153, 0.00306.

Problem 2: Fit a Poisson distribution calculates theoretical frequencies for the following data:

x :	0	1	2	3	4
f :	109	65	22	3	1

(Nov 2004)

Solution:

To fit a Poisson distribution, find the parameter λ of the distribution from the given data.

$$\lambda = \text{Arithmetic mean} = \frac{\sum f_i x_i}{\sum f_i} = \frac{122}{200} = 0.61$$

Total frequency $N = \Sigma f_i = 200$

Thus the Poisson distribution that fit to the given data $P(x) = \dfrac{\lambda^x e^{-\lambda}}{x!}$

$$\therefore \quad f(x) = N.P(x)$$

$$f(0) = N.P(0) = 200 \times \frac{(0.61)^0 e^{-0.61}}{0!} = 200 \times 0.543 = 108.6$$

$$f(1) = N.P(1) = 200 \times \frac{(0.61)^1 e^{-0.61}}{1!} = 200 \times 0.331 = 66.3$$

$$f(2) = N.P(2) = 200 \times \frac{(0.61)^2 e^{-0.61}}{2!} = 200 \times 0.1 = 20.2$$

$$f(3) = N.P(3) = 200 \times \frac{(0.61)^3 e^{-0.61}}{3!} = 200 \times 0.02 = 4.1$$

$$f(4) = N.\ P(4) = 200 \times \frac{(0.61)^4 e^{-0.61}}{4!} = 200 \times 0.0022 = 0.64$$

Hence theoretical frequencies are 108.6, 66.3, 20.2, 4.1, 0.64

Problem 3: Fit a Poisson distribution to the following data:

x	0	1	2	3	4	5	6	7	8	9	10	11	12	13	14	15	16
f	0	20	43	53	86	70	54	37	18	10	5	2	2	0	0	0	0

Solution :

To fit a Poisson distribution, find the parameter λ of the distribution from the given data.

$$\lambda = \text{arithmetic mean} = \frac{\sum f_i x_i}{\sum f_i} = \frac{1872}{400} = 4.68$$

Total frequency $N = \Sigma f_i = 400$

Thus the Poisson distribution that fit to the given data $P(x) = \dfrac{\lambda^x e^{-\lambda}}{x!}$

$\therefore$

$$f(x) = N.P(x)$$

$$f(0) = N.P(0) = 400 \times \frac{(4.68)^0 e^{-4.68}}{0!} = 400 \times 0.0183 = 0$$

$$f(1) = N.P(1) = 400 \times \frac{(4.68)^1 e^{-4.68}}{1!} = 29$$

$$f(2) = N.P(2) = 400 \times \frac{(4.68)^2 e^{-4.68}}{2!} = 58$$

$$f(3) = N.P(3) = 400 \times \frac{(4.68)^3 e^{-4.68}}{3!} = 78$$

$$f(4) = N.P(4) = 400 \times \frac{(4.68)^4 e^{-4.68}}{4!} = 78$$

$$f(5) = N.P(5) = 400 \times \frac{(4.68)^5 e^{-4.68}}{5!} = 62$$

$$f(6) = N.P(6) = 400 \times \frac{(4.68)^6 e^{-4.68}}{6!} = 42$$

$$f(7) = N.P(7) = 400 \times \frac{(4.68)^7 e^{-4.68}}{7!} = 24$$

$$f(8) = N.P(8) = 400 \times \frac{(4.68)^8 e^{-4.68}}{8!} = 12$$

$$f(9) = N.P(9) = 400 \times \frac{(4.68)^9 e^{-4.68}}{9!} = 5$$

$$f(10) = N.P(10) = 400 \times \frac{(4.68)^{10} e^{-4.68}}{10!} = 2$$

$$f(11) = N.P(11) = 400 \times \frac{(4.68)^{11}\, e^{-4.68}}{11!} = 1$$

$$f(12) = N.P(12) = 400 \times \frac{(4.68)^{12}\, e^{-4.68}}{12!} = 0.$$

3.9 NORMAL DISTRIBUTION

The normal distribution was first discovered in 1733 by English mathematician De-Moivre, who obtained this continuous distribution as a limiting case of the binomial distribution and applied it to problems arising the game of chance. It was also known to Laplace, no later than 1774 but through a historical error it was credited to Gauss, who first made reference to it in the beginning of 19th century, as the distribution of errors in Astronomy. Gauss used the normal curve to describe the theory of accidental errors of measurements involved in the calculation of orbits of heavenly bodies. Throughout the eighteenth and nineteenth centuries, various efforts were made to establish the normal model as the underlying law ruling all continuous random variables. Thus, the name "normal". These efforts, however, failed because of false premises. The normal model has, nevertheless, become the most important probability model in statistical analysis.

Definition

A random variable X is said to have a normal distribution with parameters μ and σ^2 if its density function is given by the Probability law

$$f(x; \mu, \sigma) = \frac{1}{\sigma\sqrt{2\pi}}\, e^{-\left(\frac{(x-\mu)^2}{2\sigma^2}\right)}, \quad -\infty < x < \infty, \quad -\infty < \mu < \infty \text{ and } \sigma > 0$$

Note:

1. A random variable X with mean μ and variance σ^2 following the normal law is expressed as $X \sim N(\mu, \sigma^2)$.
2. If $X \sim N(\mu, \sigma^2)$, then $Z = \frac{(x-\mu)}{\sigma}$ is a standard normal variate with $E(Z) = 0$ and $V(Z) = 1$ then we write $Z \sim N(0, 1)$.
3. The probability density function of standard normal variate Z is given by

$$\phi(Z) = \frac{1}{\sqrt{2\pi}}.e^{-\left(\frac{z^2}{2}\right)}$$

3.10 NORMAL DISTRIBUTION AS A LIMITING FORM OF BINOMIAL DISTRIBUTION

The probability function of the binomial distribution with parameters n, p is given by

$P(X = x) = {}^nC_x,\ p^x q^{n-x} = 0, 1, 2, 3, \ldots\ldots\ldots, n$ ...(1)

Define $$Z = \frac{(x-\mu)}{\sigma} = \frac{(x-np)}{\sqrt{npq}}$$ ...(2)

When $x = 0 \Rightarrow Z = \dfrac{-np}{\sqrt{npq}} = -\sqrt{\dfrac{np}{q}}$

$x = n \Rightarrow Z = \dfrac{(n - np)}{\sqrt{npq}} = \dfrac{(np)}{\sqrt{npq}} = \sqrt{\dfrac{nq}{p}}$

thus, in this limit as $n \to \infty$, Z takes the values $-\infty$ to ∞.

$$\underset{x\to 0}{\text{Lt}}\ P(x) = \underset{x\to 0}{\text{Lt}}\ \frac{n!}{x!(n-x)!} p^x q^{n-x}$$

$$\underset{x\to 0}{\text{Lt}}\ P(x) = \left[\frac{\sqrt{2\pi}e^{-n} n^{n+\frac{1}{2}} p^x q^{n-x}}{\sqrt{2\pi}e^{-x} x^{x+\frac{1}{2}} \sqrt{2\pi}e^{-(n-x)} (n-x)^{n-x+\frac{1}{2}}}\right]$$

$$= \underset{x\to 0}{\text{Lt}}\ \frac{1}{\sqrt{2\pi}} \frac{1}{\sqrt{npq}} \frac{(np)^{x+\frac{1}{2}} (nq)^{n-x+\frac{1}{2}}}{x^{x+\frac{1}{2}} (n-x)^{n-x+\frac{1}{2}}}$$

$$= \underset{x\to 0}{\text{Lt}} \left[\frac{1}{\sqrt{2\pi}\sqrt{npq}} \left(\frac{np}{x}\right)^{x+\frac{1}{2}} \left(\frac{nq}{n-x}\right)^{n-x+\frac{1}{2}}\right] \qquad ...(3)$$

From (2), we get $X = np + Z\sqrt{npq} \Rightarrow \dfrac{x}{np} = 1 + Z\dfrac{n-x}{nq}$

Further $n - X = n - np - Z\sqrt{npq} = nq - Z\sqrt{npq} \Rightarrow \dfrac{n-x}{nq} = 1 - Z\sqrt{\dfrac{p}{nq}}$

Also $dz = \dfrac{1}{\sqrt{npq}} dx$

Hence the probability differential of the distribution of Z, in the limit is

$$dG(z) = g(z)\, dz = \left(\frac{1}{\sqrt{2\pi}} \times \frac{1}{N}\right) dz,$$

where $N = \left(\dfrac{x}{np}\right)^{x+\frac{1}{2}} \left(\dfrac{n-x}{nq}\right)^{n-x+\frac{1}{2}}$

$$\Rightarrow \log N = \left(x + \frac{1}{2}\right) \log (x / np) + \left(n - x + \frac{1}{2}\right) \log \{(n-x)/nq\} \qquad ...(4)$$

$$= (np + z\sqrt{npq} + \frac{1}{2}) \log \{1 + z\sqrt{\frac{q}{np}}\} + (nq - z\sqrt{npq} + \frac{1}{2}) \log \left\{1 - z\sqrt{\frac{p}{nq}}\right\}$$

$$= \left(np + z\sqrt{npq} + \frac{1}{2}\right) \left\{z.\sqrt{\frac{q}{np}} - \frac{1}{2} z^2 \left(\frac{q}{np}\right) + \frac{1}{3} z^3 \left(\frac{q}{np}\right)^{3/2} - \cdots\cdots\right\} +$$

$$\left(np - z\sqrt{npq} + \frac{1}{2}\right)\left\{-z\sqrt{\frac{p}{np}} - \frac{1}{2}z^2\left(\frac{p}{nq}\right) - \frac{1}{3}z^3\left(\frac{p}{nq}\right)^{3/2} - \ldots\ldots\right\}$$

$$= \left[\left[\left\{z\sqrt{npq} - \frac{1}{2}qz^2 + \frac{1}{3}z^3\frac{q^{3/2}}{\sqrt{np}} + z^2, -\frac{1}{2}z^3\frac{q^{3/2}}{\sqrt{np}} + \frac{1}{2}z\sqrt{\frac{q}{np}} - \frac{1}{4}z^2\frac{q}{np} + \ldots\right\}\right.\right. +$$

$$\left.\left\{-z\sqrt{npq} - \frac{1}{2}z^2 p - \frac{1}{3}z^3\frac{p^{3/2}}{\sqrt{nq}} + z^2 p + \frac{1}{2}z^3\frac{p^{3/2}}{\sqrt{np}} - \frac{1}{2}z\sqrt{\frac{p}{\sqrt{np}}} - \frac{1}{4}z^2\frac{p}{np} + \ldots\right\}\right]$$

$$= \left[-\frac{1}{2}z^2(p+q) + z^2(p+q) + \frac{z}{2\sqrt{n}}\left(\frac{q}{p} + \frac{p}{q}\right) + 0(n)^{-\frac{1}{2}}\right]$$

$$= \frac{z^2}{2} + 0\left(n^{-\frac{1}{2}}\right) \to \frac{z^2}{2} \text{ as } n \to \infty$$

$$\therefore \operatorname*{Lt}_{n\to\infty} \log N = \frac{z^2}{2} \Rightarrow \operatorname*{Lt}_{n\to\infty} N = e^{\frac{z^2}{2}}$$

Substituting in equation number (4), we get

$$dG(z) = g(z)dz = \frac{1}{\sqrt{2\pi}}e^{\frac{-z^2}{2}}dz, -\infty < z < \infty \qquad \ldots[4(a)]$$

Hence the probability function of Z is : $g(z) = \frac{1}{\sqrt{2\pi}}e^{\frac{-z^2}{2}}, -\infty < z < \infty$ ...[4(*b*)]

This is the probability density function of the normal distribution with mean 0 and unit variance.

If X is normal variate with mean μ and s.d. σ, then $Z = (X - \mu)/\sigma$ is standard normal variate Jacobian of transformation is $1/\sigma$. Hence substituting in [4(*b*)] the p.d.f. of a normal variance X with

E(X) = μ, Var(X) = σ^2 is given by $f_x(x) \begin{cases} \frac{1}{\sigma\sqrt{2\pi}}e^{-(x-\mu)^2/2\sigma^2}, -\infty < x < \infty \\ 0, \text{otherwise} \end{cases}$

Graph of the Normal Distribution

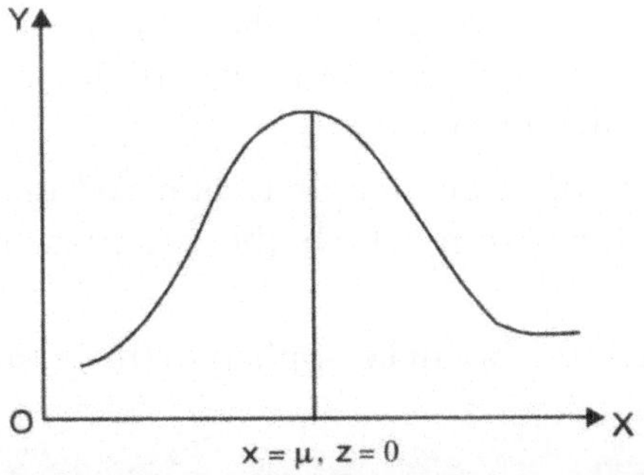

Graph of the normal distribution is called the normal curve. It is bell shaped and symmetrical about the mean μ the two tails of the curve extended to $-\infty$ to ∞ respectively and gradually approach to X-axis . The curve is unimodel and mode of the normal distribution is coincide with its mean. The like X = μ divides the area under the normal curve above the X-axis into two equal parts. Thus the median of the distribution is also coincide with its mean, mode. They are under the normal curve between the two given coordinates $X = x_1$ and $X = x_2$ represents the probability of values failing into the given internal. The total area under the normal curve above the X-axis is unity.

Chief Characteristics of Normal Distribution

The normal probability curved with mean μ and standard deviation σ is given by the equation $f(x) = \frac{1}{\sigma\sqrt{2\pi}} e^{-\left(\frac{(x-\mu)^2}{2\sigma^2}\right)}$ has the following properties:

1. The curve is bell shaped and symmetrical about the line X = μ
2. Mean, Median and Mode of the distribution coincide.
3. Normal probability curve with same mean different variance differ in maximum ordinate.
4. X-axis is an asymptote to the curve.
5. The curve has two points of inflection at $x = \mu \pm \sigma$. Both these points are equidistance from mean.
6. The mean deviation about the mean of Normal Distribution is 4/5 of its standard deviation.
7. Total area under the curve above X-axis is unity.
8. The area of the normal curve between $\mu \pm \sigma$ is 68.2%
9. The area of the normal curve between $\mu \pm 2\sigma$ is 95.4%
10. The area of the normal curve between $\mu \pm 2\sigma$ is 99.73%

Importance of Normal Distribution

Normal distribution plays a very important role in statistical theory

1. Most of the distributions occurring in practice example Binomial, Poisson, Hypergeometric distributions etc., can be approximated by normal distribution moreover, many of the sampling distributions students χ, χ^2 distributions etc., tend to normally for large samples.
2. Many of the distribution curved of sample statistic, the distributions of sample mean, sample variance etc., tend to normally for large samples and as such they can best be studied with the help of the normal curve.
3. The entire theory of small sample tests based on the fundamental assumption that parent populations from which the samples have been drawn follow normal distribution.
4. Theory of normal curves can be applied to the graduation of the curves which are not normal.

5. Normal distribution finds large application in statistical quality control in industry for setting control limits.

Mean Deviation from the Mean of Normal Distribution (Nov 2006)

$$\text{Mean deviation} = \int_{-\infty}^{\infty} |x-\mu|\, f(x)dx$$

$$= \int_{-\infty}^{\infty} |x-\mu| \frac{1}{\sigma\sqrt{2\pi}} e^{-(x-\mu)^2/2\sigma^2} dx$$

$$= \frac{1}{\sigma\sqrt{2\pi}} \int_{-\infty}^{\infty} \sigma |z| e^{-z^2/2} dz \quad \text{put } \frac{x-\mu}{\sigma} = z \Rightarrow x-\mu = z\sigma$$

$$\Rightarrow \quad dx = \sigma dz$$

$$= \frac{\sigma}{\sqrt{2\pi}} \int_{-\infty}^{\infty} |z| e^{-z^2/2} dz \quad \frac{z^2}{2} = t \Rightarrow zdz = dt$$

$$= \frac{2\sigma}{\sqrt{2\pi}} \int_{0}^{\infty} e^{-t} dz$$

$$= \sqrt{\frac{2}{\pi}}.\sigma \left[-e^{-t}\right]_0^{\infty}$$

$$= \sqrt{\frac{2}{\pi}}.\sigma = \frac{4}{5}\sigma$$

$\therefore$ Mean deviation from the mean of Normal distribution of $\frac{4}{5}$ its standard deviation.

Mode of the Normal Distribution

Mode is the value of x for which $f(x)$ is maximum. *i.e.*, mode is the solution of $f^1(x) = 0$ and $f^{11}(x) < 0$. For the normal distribution with mean μ and variance σ^2

$$f(x) = \frac{1}{\sigma\sqrt{2\pi}} e^{-(x-\mu)^2/2\sigma^2}$$

$$f^1(x) = \frac{1}{\sigma\sqrt{2\pi}} e^{-(x-\mu)^2/2\sigma^2} \left(\frac{1}{\sigma^2}(x-\mu)\right)$$

$$= -\frac{1}{\sigma^2}(x-\mu)\, f(x)$$

$$f^{11}(x) = \frac{1}{\sigma^2}\left\{f(x)+(x-\mu)f^1(x)\right\} \quad ...(1)$$

Now $\quad f^1(x) = 0 \Rightarrow x-\mu = 0 \Rightarrow x = \mu \quad ...(2)$

At $x = \mu$, from (1) $f^1(x) < 0$. Hence $x = \mu$ is mode of Normal distribution.

Median of Normal Distribution

$f(x)$ is probability density function $\int_{-\infty}^{\infty} f(x)dx = 1$

Let M is the median of the normal distribution, we have $\int_{-\infty}^{M} f(x)dx = \frac{1}{2}$

$$\Rightarrow \qquad \frac{1}{\sigma\sqrt{2\pi}} \int_{-\infty}^{M} e^{-(x-\mu)^2/2\sigma^2} dx = \frac{1}{2}$$

$$\frac{1}{\sigma\sqrt{2\pi}} \left[\int_{-\infty}^{\mu} e^{-(x-\mu)^2/2\sigma^2} dx + \int_{\mu}^{M} e^{-(x-\mu)^2/2\sigma^2} dx \right] = \frac{1}{2}$$

But
$$\frac{1}{\sigma\sqrt{2\pi}} \int_{-\infty}^{\mu} e^{-(x-\mu)^2/2\sigma^2} dx + \frac{1}{\sqrt{2\pi}} \int_{\mu}^{0} e^{-z^2/2} dx = \frac{1}{2}$$

$$\Rightarrow \qquad \int_{\mu}^{M} e^{-(x-\mu)^2/2\sigma^2} dx = 0$$

$$\Rightarrow \qquad \mu = M$$

$\therefore$ M = μ is the median of Normal distribution.

Hence for Normal distribution Mean = Median= Mode. **(Feb 2007, May 2007)**

Standard form of Normal Distribution

If X is normal random variable with mean μ and variance σ^2 than the random variable $z = \frac{x-\mu}{\sigma}$ has the normal distribution with mean $\mu = 0$ and Standard deviation is 1 the random variable z is called standard normal random variable.

$\therefore$ The probability density function for the normal distribution in standard form is

$\phi(z) = \frac{1}{\sqrt{2\pi}} e^{-\frac{z^2}{2}}$. If $f(z)$ is probability density function of normal distribution, then

$$P(Z_1 \le Z \le Z_2) = F(Z_2) - F(Z_1)$$

3.11 SOLVED PROBLEMS

Problem 1: If x is normally distributed, and the mean of x is 12 and standard deviation is 4 then

(*i*) Find the probabilities of (*a*) $x \ge 20$ (*b*) $x < 20$ (c) $0 \le x \le 12$

(*ii*) Find x_1 when $P(x > x_1) = 0.24$

(*iii*) Find x_1 and x_2 when $P(x_1 < x < x_2) = 0.5$ and $P(x > x_2) = 0.25$

Solution:

(*i*) Given mean $\mu = 12$ and standard deviation $\sigma = 4$

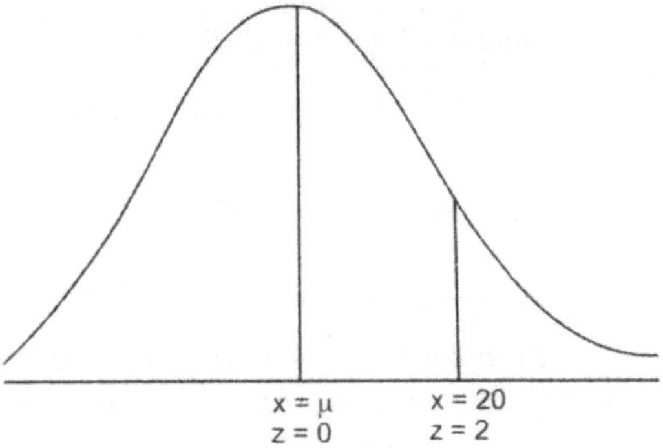

(*a*) When $x = \mu \Rightarrow Z = \dfrac{x - \mu}{\sigma}$

When $x = 20 \Rightarrow Z = \dfrac{20-12}{4} = 2$

$\therefore\ P(x \geq 20) = P(Z > 2)$

$= 0.5 - P(0 < Z < 2)$

$= 0.5 - 0.4772 = 0.0228$

(*b*) $P(x < 20) = 1 - P(x \geq 20)$

$= 1 - 0.0228 = 0.9772$

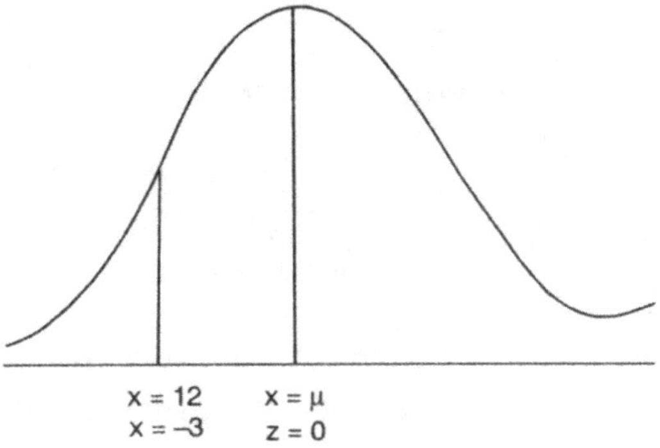

(*c*) When $x = 0 \Rightarrow Z = \dfrac{0-12}{4} = -3$

When $x = 12 \Rightarrow Z = \dfrac{12-12}{4} = 0$

$\therefore\ P(0 \leq x \leq 12) = P(-3 \leq Z \leq 0)$

$= P(0 \leq Z \leq 3) = 0.4986$

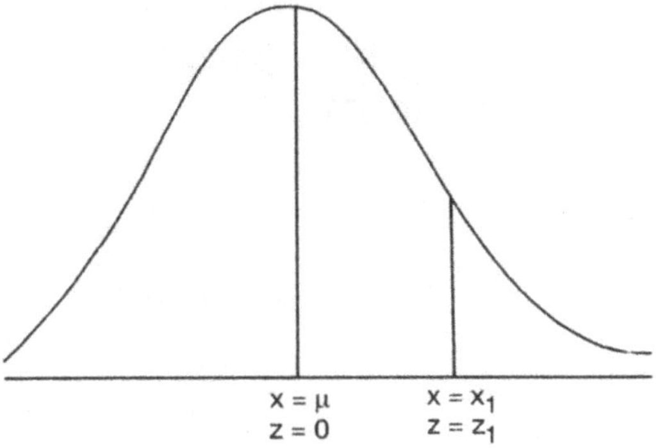

(*ii*) When $x = x_1$, $Z = \dfrac{x_1 - 12}{4} = z_1$ (say)

Then we are given $P(x > x_1) = 0.24$

$\Rightarrow P(Z > z_1) = 0.24 \Rightarrow P(0 < Z < z_1)$

$= 0.26$

$\therefore$ From Normal Tables, $z_1 = 0.71$ (approx)

Hence $\dfrac{x_1 - 12}{4} = 0.71 \Rightarrow x_1 = 12 + 4$

$\times\ 0.71 = 14.84$

(*iii*) We are given: $P(x_1 < x < x_2) = 0.5$ and $P(x > x_2) = 0.25$ ...(1)

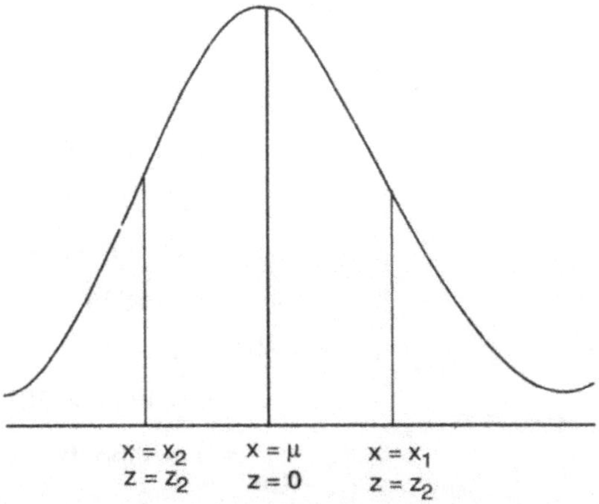

From (1) obviously the points x_1 and x_2 are located as shown in following adjoining.

When X= x_1, $Z = \dfrac{x_2 - 12}{4} = z_1$

We have $P(Z > z_1) = 0.25$

$\Rightarrow P(0 < Z < z_1) = 0.25$

$\therefore\ z_1 = 0.67$ (from tables)

Hence $\dfrac{x_2 - 12}{4} = 0.67$

$\Rightarrow x_2 = 12 + 4 \times 0.67 = 14.68$

And $\dfrac{x_1 - 12}{4} = -0.67$

$\Rightarrow x_1 = 12 - 4 \times 0.67 = 9.32.$

Problem 2: x is a normal variate with mean 30 and standard deviation 5. Find the probabilities that (*i*) $26 \le x \le 40$, (*ii*) $x \ge 45$ and (*iii*) $|x - 30| > 5$

Solution: Here $\mu = 30$ and $\sigma = 5$

(*i*) When $x = 26$, $Z = \dfrac{x-\mu}{\sigma} = \dfrac{26-30}{5}$

$= -0.8$

and when $x = 40$, $z = \dfrac{40-30}{5} = 2$

$\therefore P(26 \le x \le 40) = P(-0.8 \le Z \le 2)$

$= P(-0.8 \le Z \le 0) + P(0 \le Z \le 2)$

$= P(0 \le Z \le 0.8) + (0 \le Z \le 2)$

$= 0.2881 + 0.4772$

$= 0.7653$

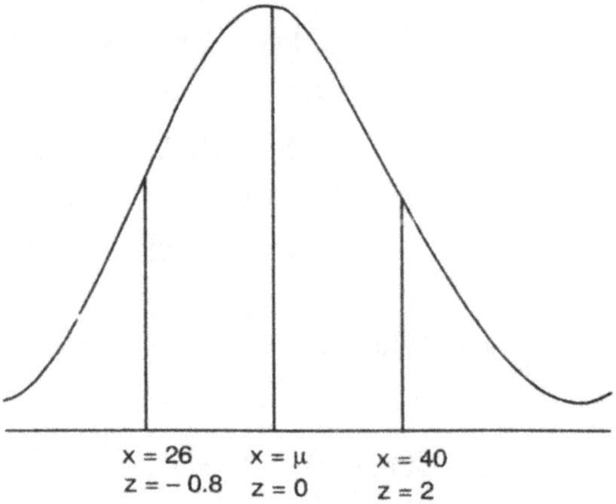

(*ii*) When $x = 45$ $z = \dfrac{45-30}{5} = 3$

$\therefore P(x \ge 45) = P(Z \ge 3)$

$= 0.5 - P(0 \le Z \le 3)$

$= 0.5 - 0.49865$

$= 0.00135$

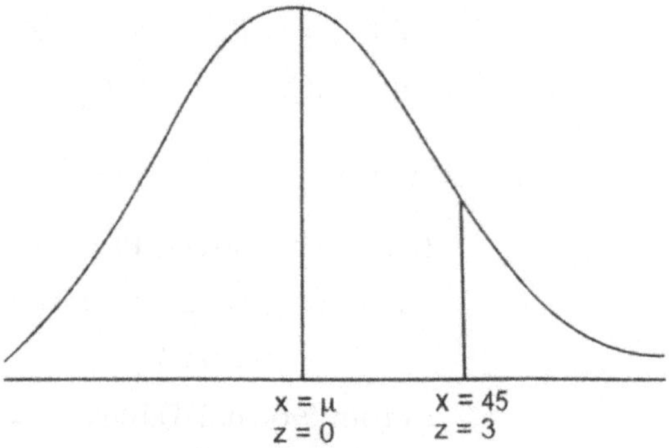

(*iii*) $P(|x - 30| \le 5) = P(25 \le x \le 35)$

$= P(-1 \le Z \le 1)$

$= 2P(0 \le Z \le 1)$

$= 2 \times 0.3413$

$= 0.6826$

$\therefore P(|x - 30| > 5)$

$= 1 - P(|x - 30|) \le 5)$

$= 1 - 0.6826 = 0.3174$

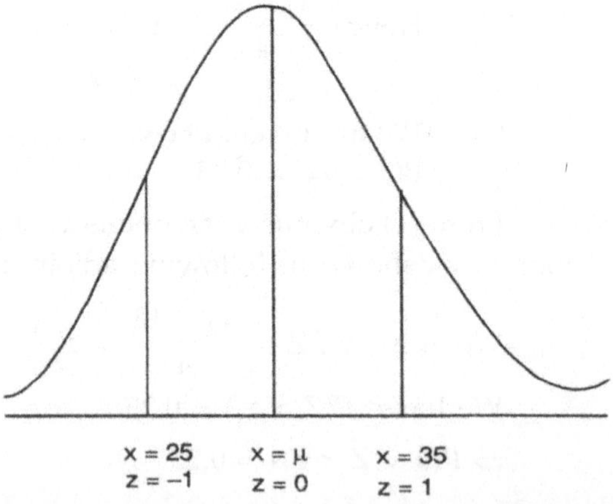

Problem 3 : The mean and standard deviation of a normal variate are 8 and 4 respectively find, (*i*) $P(50 \le x \le 10)$, (*ii*) $P(x \ge 5)$. **(Nov 2006, April 2008)**

Solution: Given $\mu = 8$, $\sigma = 4$

We know that $z = \dfrac{x - \mu}{\sigma} = \dfrac{x - 8}{4}$

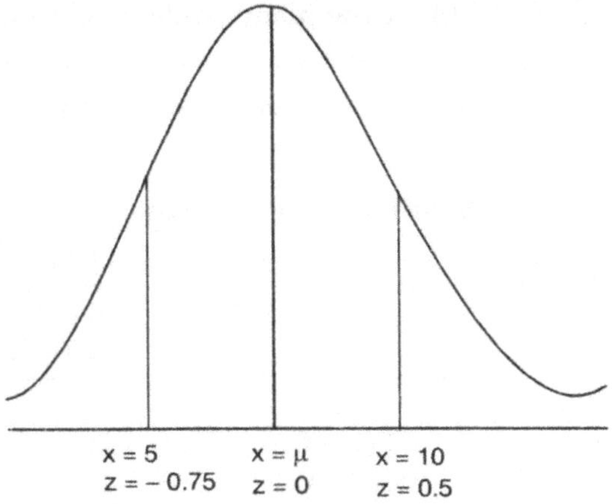

When $x = 5$; $z = \dfrac{5 - 8}{4} = -\ 0.75$

When $x = 10$; $z = \dfrac{10 - 8}{4} = 0.5$

(*i*) $P(5 \le x \le 10) = P(-\ 0.75 \le z \le 0.5)$

$= P(-\ 0.75 \le z \le 0) + P(0 \le z \le 0.5)$

$= P(0 \le z \le 0.75) + P(0 \le z \le 0.5)$

= Area from (0 to 0.75) + Area from (0 to 0.5)

$= 0.2734 + 0.1916 = 0.465$

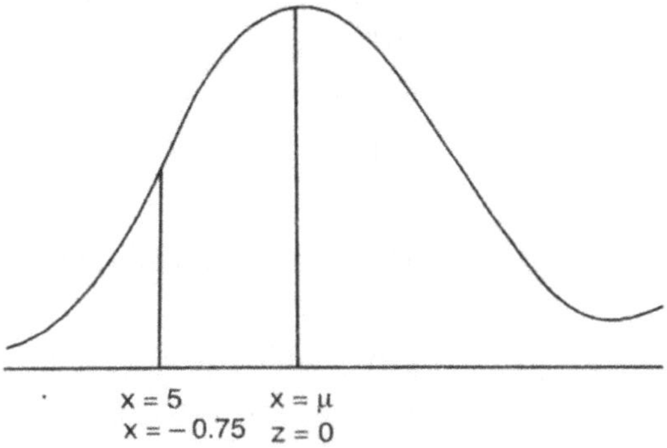

(*ii*) $P(x \ge 5)$

$\Rightarrow P(z \ge -\ 0.75) \Rightarrow P(z \le 0.75)$

$= 0.5 + P(0 \le z \le 0.75)$

$= 0.5 + 0.2734$

$= 0.7734$

Problem 4: If the mean and Standard deviation of normal distribution are 70 and 16, find $P(38 < x < 46)$. **(Nov 2006)**

Solution: Given $\mu = 70$, $\sigma = 16$

We know $z = \dfrac{x - \mu}{\sigma} = \dfrac{x - 70}{16}$

When $x = 38 \Rightarrow z = \dfrac{38 - 70}{16} = -\ 3$

When $x = 46 \Rightarrow z = \dfrac{46 - 70}{16} = -\ 1.5$

$P(38 < x < 46) = P(3 < z < -1.5)$

$= P(0 < z < -3) - P(0 < z < 1.5)$

= Area from (0 to –3) –Area from (0 to 15)

$= 0.4772 - 0.4332$

$= 0.044.$

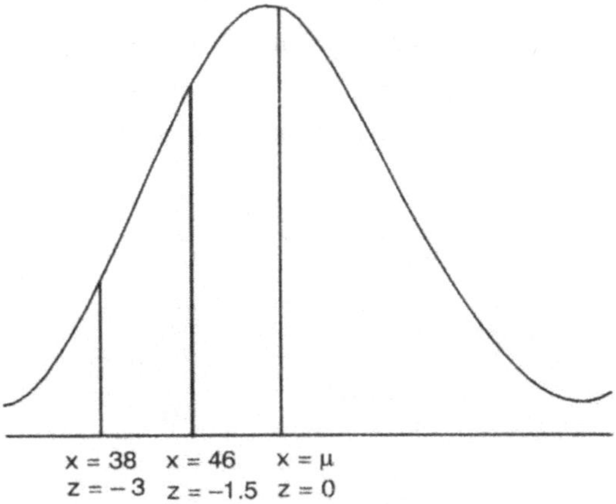

Problem 5: The marks obtained in mathematics by 1000 students are normally distributed with mean 78% and standard deviation 11%. Determine

(*i*) How many students got marks above 90%?

(*ii*) What was the highest mark obtained by the lowest 10% of the student?

(*iii*) Within what limits did the middle of 90% of the student lie?

(April 2006, Feb 2007, Nov 2007)

Solution: We know that

$$z = \frac{x - \mu}{\sigma} = \frac{x - 0.78}{0.11}$$

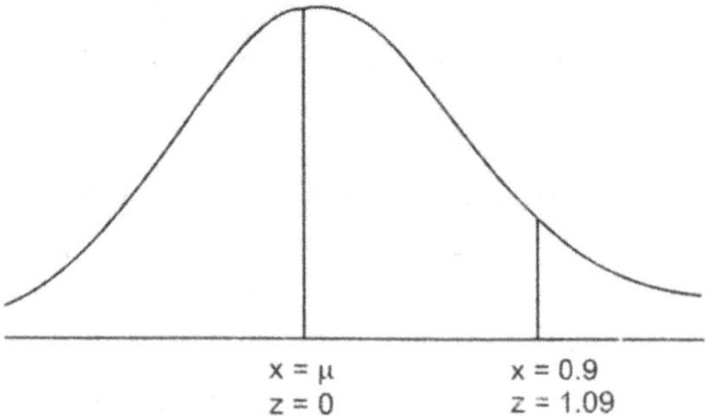

(*i*) For $x = 0.9 \Rightarrow z = \frac{0.9 - 0.78}{0.11} = 1.09$

$P(x > 0.9) = P(z > 1.09) = 0.5 - 0.3621$
$= 0.1379$

Number of students with marks above 90%

$= 1000 \times P(x > 0.9)$

$= 1000 \times 0.1379$

$= 138$

(*ii*) The lowest 10% students constitute 0.1 area (< 1/2) of extreme left tail. So z_1 must be negative.

From table 0.4 = 0.5 – 0.1

= 0.5 – area 0.1 from 0 to z_1

So $z_1 = -1.28$

Thus, $-1.28 = \frac{x - 0.78}{0.11}$

$\therefore x = 0.6392$

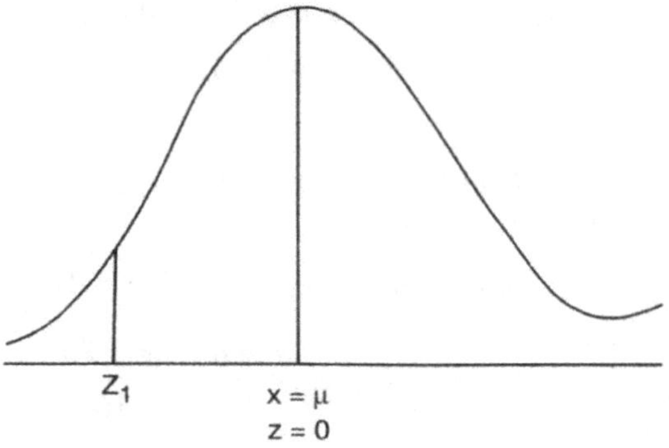

Thus

The highest mark obtained by the lowest 10% of students is 63.92=64%

(*iii*) Middle 90% correspond to 0.9 area, leaving 0.05 area on both sides. Then the corresponding z's are ± 1.64.

$1.64 = z_2 = \frac{x_2 - 0.78}{0.11}$

$\Rightarrow x_2 = 96.04\%$

$-1.64 = z_1 = \frac{x_1 - 0.78}{0.11}$

$\Rightarrow x_1 = 59.96\%$

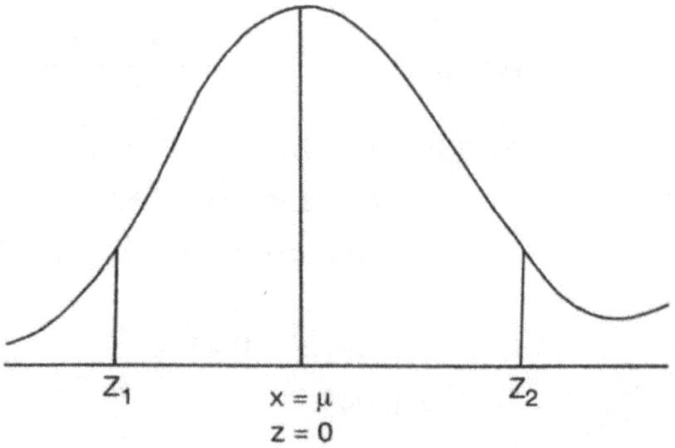

Thus the middle 90% have marks in between 60 to 90.

Problem 6: Suppose the weights of 80 male students are normally distributed with mean $\mu = 140$ pounds and standard deviation 10 pounds. Find the number of students whose weights are

(*i*) between 138 and 148 pounds

(*ii*) more than 152 pounds **(May 2007, May 2006)**

Solution : Mean $\mu = 140$ pounds

Standard deviation $\sigma = 10$ pounds

$$n = 800$$

Standard normal variate

$$z = \frac{x-\mu}{\sigma} = \frac{x-140}{10}$$

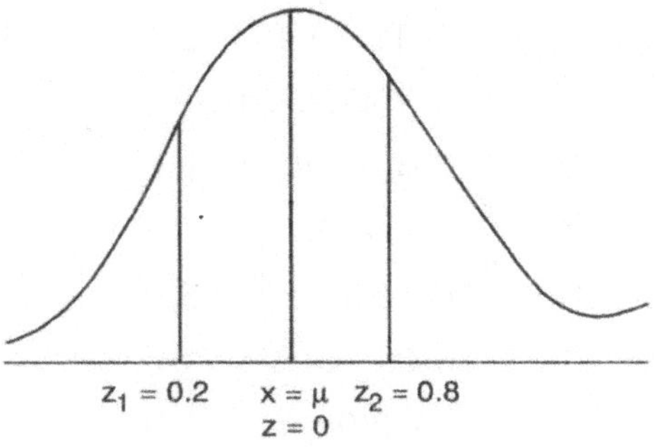

(*i*) Between 138 and 148 pounds

$P(138 \le x \le 148)$

$$x_1 = 138 \Rightarrow z_1 = \frac{138-140}{10} = 0.2$$

$$x_2 = 148 \Rightarrow z_2 = \frac{148-140}{10} = 0.8$$

$P(138 \le x \le 148)$

$= P(-0.2 \le z \le 0.8)$

$= P(-0.2 \le z \le 0) + P(0 \le z \le 0.8)$

$= P(0 \le z \le 0.2) + P(0 \le z \le 0.8)$

$= 0.0793 + 0.2881$

$= 0.3674$

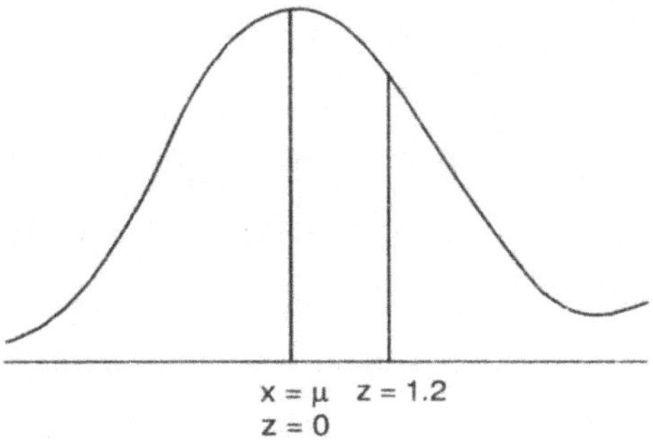

(*ii*) More than 152 pounds

$P(x > 152)$

$$x = 152 \Rightarrow z = \frac{152-140}{10} = 1.2$$

$P(x > 152)$

$= P(z > 1.2)$

$= 0.5 - P(0 \le z \le 1.2)$

$= 0.5 - 0.3849 = 0.1151$

Problem 7: If z is normal variate, find the probability that

(*i*) to the left of $z = -1.78$

(*ii*) to the right of $z = -1.45$

(*iii*) corresponding to $-0.80 \le z \le 1.53$

(*iv*) to the left of $z = -2.52$ and to the right of $z = 1.83$

(Nov 2007, Feb 2007, May 2007, Nov 2006, Nov 2004)

Solution:

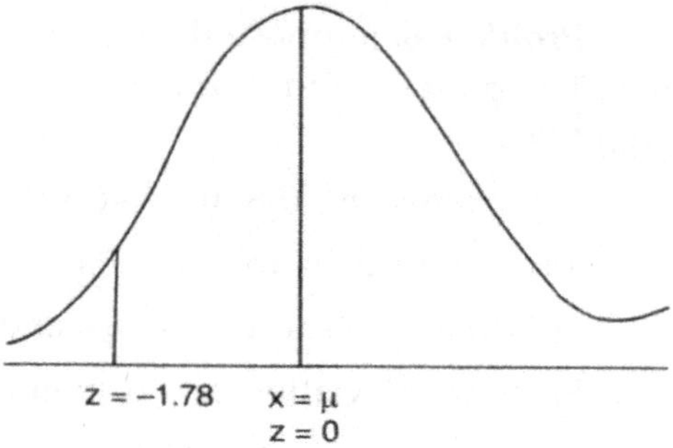

(*i*) P(z < − 1.78)

= P(z > 1.78)

= 0.5 − P(0 < z < 1.78)

= 0.5 − 0.4625

(from the table)

= 0.0375

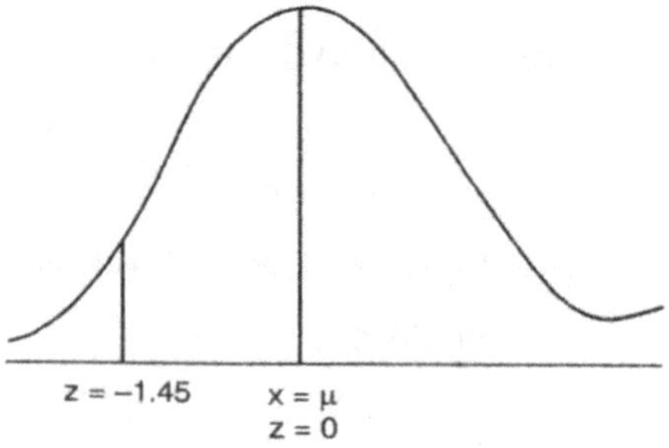

(*ii*) P(z > − 1.45)

= P(− 1.45 < z < 0) + 0.5

= 0.4256 + 0.5

(from the table)

= 0.9256

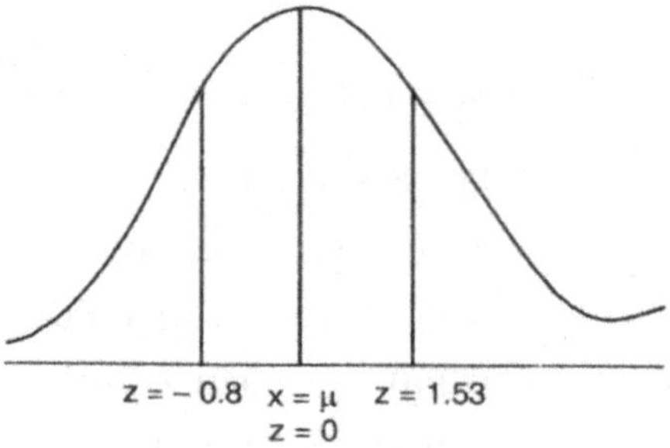

(*iii*) P(− 0.8 ≤ z ≤ 1.53)

= P(− 0.8 ≤ z ≤ 0)

+ P(0 ≤ z ≤ 1.53)

= P(0 ≤ z ≤ 0.8)

+ P(0 ≤ z ≤ 1.53)

= 0.2881 + 0.4370

(from the table)

= 0.7251

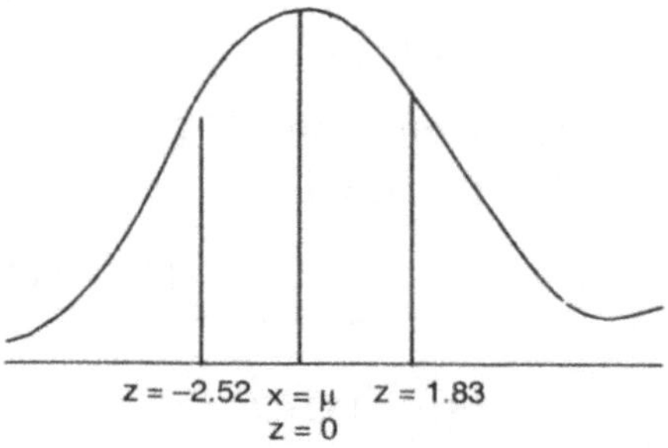

(*iv*) P(z ≤ 2.52) + P(z ≥ 1.83)

= P(z ≥ 2.52) + P(z ≥ 1.83)

= [0.5 − P(0 ≤ z ≤ 2.52)]

+ [0.5 − P(0 ≤ z ≤ 1.83)]

= [0.5 − 0.4941] + [0.5 − 0.4664]

= 0.0059 + 0.0336

= 0.0395.

Problem 8 : In a normal distribution 31% of the items are under 45 and 8% are over 64. Find the mean and variance of the distribution. **(May 2003)**

Solution:

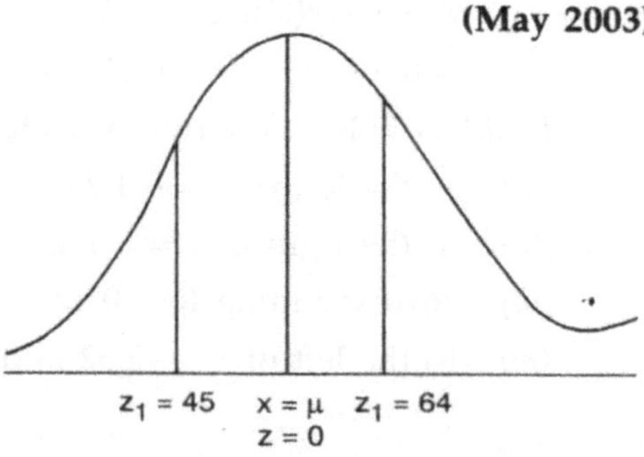

Let μ be the mean and σ be the standard deviation

$$\therefore\ z = \frac{x-\mu}{\sigma}$$

∴ P (*x* < 45) = P(z < z_1) = 0.31, where

$$z_1 = \frac{45-\mu}{\sigma}$$

From tables, we get $z_1 = -0.5$

$$\therefore \quad \frac{45-\mu}{\sigma} = -0.5$$

$$\Rightarrow \quad 45 = \mu - 0.5\sigma \qquad ...(1)$$

$P(x > 64) = P(z > z_2) = 0.8$ where

$$z_2 = \frac{64-\mu}{\sigma}$$

From tables $z_2 = 1.4$

$$\therefore \quad \frac{64-\mu}{\sigma} = 1.4$$

$$64 = \mu + 1.4\sigma \qquad ...(2)$$

Solving equations (1) and (2), we get $\mu = 50$, $\sigma = 10$

Problem 9: In a normal distribution 7% of the items are under 35 and 89% are under 63. Find the mean and standard deviation of the distribution.

Solution:

$P(x < 63) = 0.89$

$\Rightarrow P(x < 63) = 0.11$

Also $P(x < 35) = 0.07$

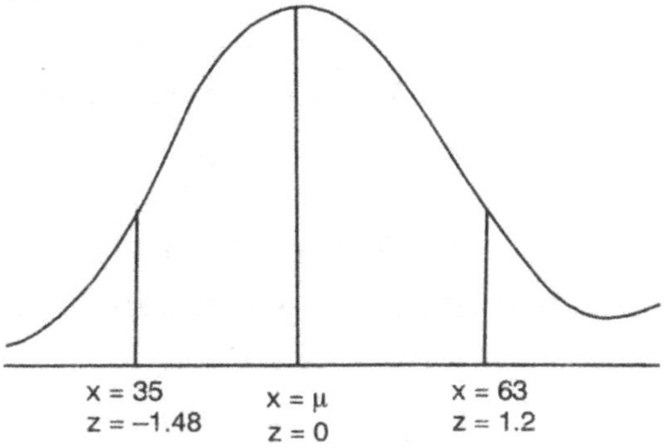

When $x = 35$, $z = \frac{35-\mu}{\sigma} = -z_1$ ...(1)

(Since 35 is to the left of the mean)

When $x = 63$, $z = \frac{63-\mu}{\sigma} = z_2$. ...(2)

$\therefore$ From tables $z_1 = 1.48$, $z_2 = 1.23$

Substituting in (1) and (2)

$$\frac{35-\mu}{\sigma} = -1.48$$

$$\mu - 1.48\sigma = 35 \qquad ...(3)$$

$$\frac{63-\mu}{\sigma} = 1.23$$

$$\therefore \quad \mu + 1.23\sigma = 63 \qquad ...(4)$$

Solving (3) and (4), we get $\mu = 50.3$ and $\sigma = 10.33$.

Problem 10: A sales tax officer has reported that the average sales of the 500 business that he has to deal with during a year is Rs. 36,000 with a standard deviation of 10,000. Assuming that the sales in these business are normally distributed, find

(*i*) the number of business as the sales of while are Rs. 40,000

(*ii*) the percentage of business the sales of while are likely to range between Rs. 30,000 and Rs. 40,000. **(May 2007, Nov 2006, May 2005)**

Solution:

Mean $\mu = 36{,}000$, $N = 500$, $\sigma = 10{,}000$

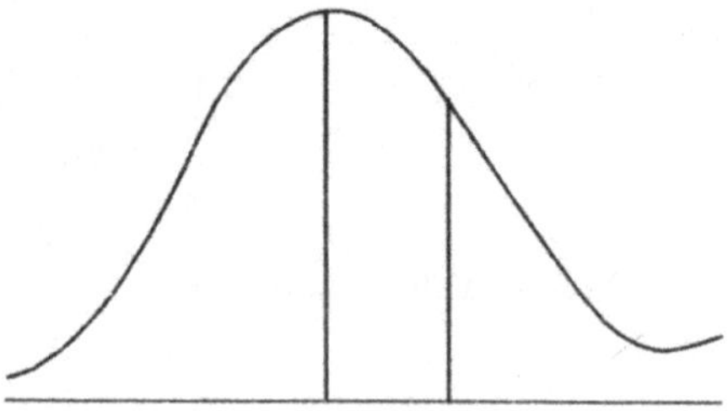

We know that

$$z = \frac{x-\mu}{\sigma} = \frac{x - 36{,}000}{10{,}000}$$

When, $x = 40{,}000$

$$\Rightarrow z = \frac{40{,}000 - 36{,}000}{10{,}000} = \frac{4{,}000}{10{,}000} = 0.4$$

When, $x = 30{,}000 \Rightarrow z = \dfrac{30{,}000 - 36{,}000}{10{,}000} = \dfrac{-6000}{10{,}000} = -0.6$

(*i*) $P(x > 40{,}000) = P(z > 0.4)$

$= 0.5 - P(0 \le z \le 0.4)$

$= 0.5 - 0.1554$ (from table)

$= 0.3446$

$\therefore$ The number of business as the sales is $0.3446 \times 500 = 172$

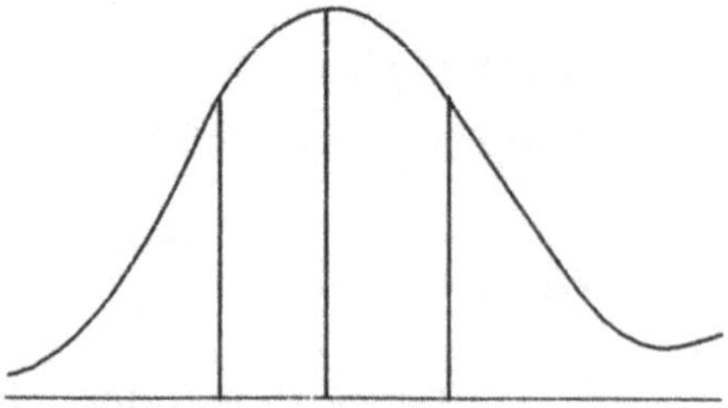

(*ii*) $P(30{,}000 < x < 40{,}000)$

$= P(-0.6 < z < 0.4)$

$= P(-0.6 < z < 0) + P(0 < z < 0.4)$

$= P(0 < z < 0.6) + P(0 < z < 0.4)$

$= 0.2258 + 0.1554$ (from the table)

$= 0.3812$

$\therefore$ The percentage of business sales is 38.12%.

Problem 11: Given that the mean heights of students in a class is 158 cms with standard deviation of 20 cms. Find how many students heights lie between 150 cms and 170 cms if there are 100 students in the class. **(Nov 2007, April 2007)**

Solution:

Mean $\mu = 158$ cms

Standard deviation $\sigma = 20$ cms

Probability of students whose heights lie between 150 cms and 170 cms $= P(150 \le x \le 170)$

$$x_1 = 150, \quad x_2 = 170$$

$$z = \frac{x-\mu}{\sigma}$$

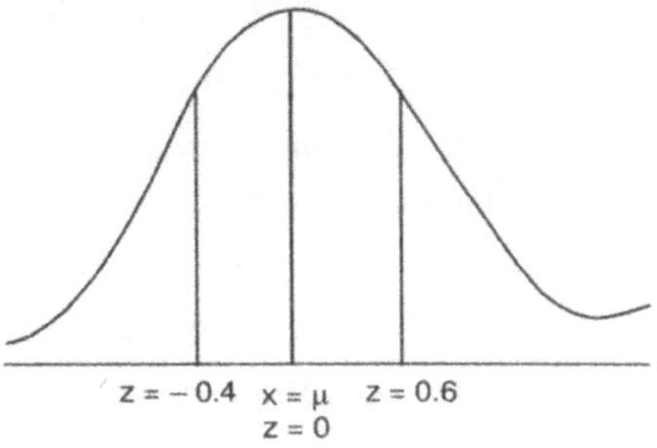

$$z_1 = \frac{150-158}{20} = -0.4$$

$$z_2 = \frac{170-158}{20} = 0.6$$

$$P(150 \le x \le 170) = P(-0.4 \le z \le 0.6)$$
$$= P(-0.4 \le z \le 0) + P(0 \le z \le 0.6)$$
$$= P(0 \le z \le 0.4) + P(0 \le z \le 0.6)$$
$$= 0.1554 + 0.2257 = 0.3811$$

Number of students who height lie between 150 cms and 170 cms

= probability × total number of students

= 0.38 × 100 = 38 (number of people should be integers)

Problem 12: The marks obtained in statistics in a certain examination are found to be normally distributed. If 15% of the candidates ≥ 60 marks, 40% < 30 marks, find the mean and standard deviation of marks

Solution:

When $x = 60$

$$\Rightarrow z = \frac{60-\mu}{\sigma} = z_1 \quad ...(1)$$

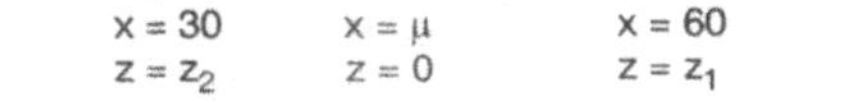

$P(x \ge 60) = P(z \ge z_1) = 0.15$

$= P(0 \le z \le z_1)$

$= 0.35$

$\Rightarrow z_1 = 1.04$ (from table)

From (1), $\dfrac{60-\mu}{\sigma} = 1.04$

$\Rightarrow 1.04\,\sigma + \mu = 60 \quad ...(2)$

When $x = 30$

$$\Rightarrow z = \frac{30-\mu}{\sigma} = z_2 \quad ...(3)$$

$\Rightarrow P(x < 30) = P(z \le z_2) = 0.4 = P(0 < z < z_2) = 0.1$

$\Rightarrow z_2 = -0.25$

From (3), $\dfrac{30-\mu}{\sigma} = -0.25$

$\Rightarrow -0.25\,\sigma + \mu = 30$ ...(4)

Solve (2) and (4), we get

$\mu = 34.55$ and $\sigma = 18.2$

Problem 13: If the masses of 300 students are normally distributed with mean 68 kgs and standard deviations 3 kgs how many students have masses?

(*i*) Greater than 72 kg

(*ii*) Less than or equal to 64 kg

(*iii*) Between 65 and 71 kg inclusive. **(JNTU 2002, April 2005)**

Solution: Mean $\mu = 68$ kgs

Standard deviation $\sigma = 3$ kgs

Standard normal variate $z = \dfrac{x-\mu}{\sigma}$

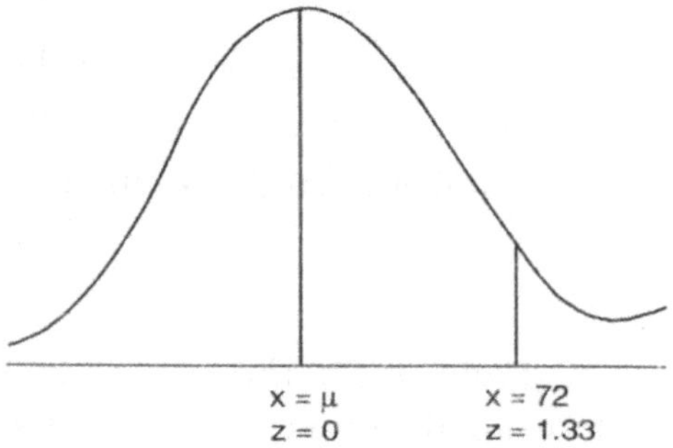

(*i*) When $x = 72 \Rightarrow z = \dfrac{72-68}{3} = 1.33$

$\therefore P(x > 72) = P(z > 1.33)$

$= 0.5 - P(0 < z < 1.33)$

$= 0.5 - 0.4082$

$= 0.0918$

$\therefore$ Number of students whose mass is greater than 72 kgs is

$0.0918 \times 300 = 28$

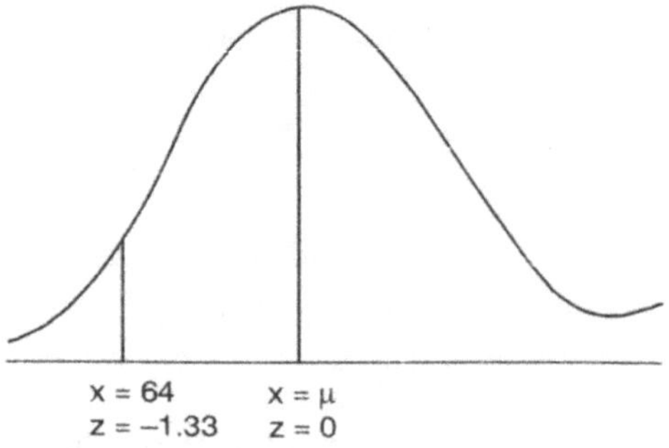

(*ii*) When $x = 64 \Rightarrow z = \dfrac{64-68}{3} = -1.33$

$\therefore P(x \le 64) = P(z \le -1.33)$

$= 0.5 - P(0 < z < 1.33)$

$= 0.5 - 0.4082$

$= 0.0918$

$\therefore$ Number of students whose mass is less than or equal to 64 kgs is $0.0918 \times 300 = 28$

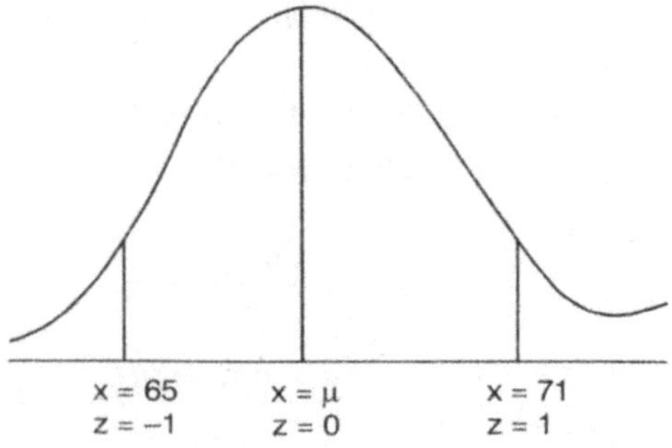

(*iii*) When $x = 65 \Rightarrow z = \dfrac{65-68}{3} = -1$

When $x = 71 \Rightarrow z = \dfrac{71-68}{3} = 1$

$\therefore P(65 \le x \le 71) = P(-1 \le z \le 1)$

$= 2 \times P(0 \le z \le 1)$

$= 2 \times 0.3413$

$= 0.6826$

$\therefore$ Number of students whose mass lies between 65 and 75 kgs are $0.6826 \times 300 = 205$.

Problem 14: A random variable has normal distribution with $\mu = 62.4$. Find the standard deviation if the probability is 0.20 that it will take on a value greater than 79.2. **(Jan 2003)**

Solution:

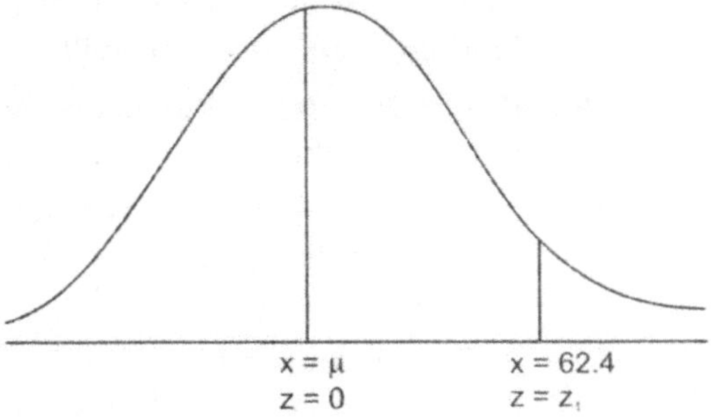

Mean $\mu = 62.4$

Standard normal variate $z = \dfrac{x - \mu}{\sigma}$

When $x = 79.2 \Rightarrow z = \dfrac{79.2 - 62.4}{\sigma} = z_1$ ...(1)

$$P(x > 79.2) = P(z > z_1)$$
$$= 0.5 - P(0 < z < z_1) = 0.2$$
$$= P(0 < z < z_1) = 0.3$$

$\Rightarrow$ $z_1 = 0.84$

From (1), $\dfrac{79.2 - 62.4}{!} = 0.84 \Rightarrow \dfrac{16.84}{!} = 0.84$

$\therefore$ $\sigma = 20.$

3.12 EXERCISE

1. If A and B play games of chess of which 6 are won by A, 4 are won by B and end 2 end in draw. Find the probability that (*i*) A and B win alternately, (*ii*) B wins at least one game, (*iii*) Two games end in draw. **(Jan 2003)**
2. If the chance that one of the ten telephone lines is busy as an instant is 0.3. (*i*) What is the chance that 5 of the lines are busy? (*ii*) What is the most probable number of busy lines and what is the probability of this number? (*iii*) What is the probability that all the lines are busy? **(May 2004)**
3. Assume that 60% of all Engineering students are good in Mathematics. Determine the probability that among 18 Engineering students. (*i*) Exactly 10, (*ii*) At least 10, (*iii*) At most 8, (*iv*) At least 2 and at most 9 are good in English. **(May 2004, May 2005, April 2007)**
4. The probability that an entering student will graduate is 0.4. Determine the probability that out of 5 students (*i*) none, (*ii*) one, (*iii*) at least one will graduate. **(May 2004)**
5. A burger alarm system has six fail safe components. The probability of each failing is 0.05. Find these probabilities (*i*) Exactly three will fairs, (*ii*) Fewer than two will fail, (*iii*) None will fail. **(May 2004)**
6. The probability that a man aged 50 year will die with in a year is 0.01125. What is the probability that of 12 such men at least eleven with reach their 51st birthday? **(Nov 2003)**
7. In eight through of a die 5 or 6 is considered a success find the mean number of success and standard deviation. ($\mu = 2.7$, $\sigma = 1.3$) **(April 2004, Nov 2003)**

8. The incident of occupational disease in an industry is such that the workers have a 20% chance of suffering from it. What is the probability that out of 6 works chosen at random, four or more will suffer the disease.

9. If 3 of 20 tyres are defective and 4 of then are randomly chosen for inspection. What is the probability that only one of the defective tyre will be included?

10. A shipment of 20 tape recorders contains 5 defective, find the standard deviation of the probability distribution of the number of defectives in a sample of 10 randomly chosen for inspection.

11. A die is thrown 8 times it is required to find the probability that 3 will show (*i*) Exactly 2 times, (*ii*) at least, (*iii*) at most once.

12. Seven unbiased coins are tossed and number of heads noted. The experiment is repeated 128 times and the following distribution is obtained.

No. of heads :	0	1	2	3	4	5	6	7
Frequency :	7	6	19	35	30	23	7	1

Fit a Binomial distribution assuming (*i*) The coin is unbiased, (*ii*) The nature of the coin is not known.

13. Using recurrence formula find the probabilities when X = 0, 1, 2, 3, 4 and 5. If the mean Poisson distribution is 3. **(April 2005)**

14. Out of 2000 families with 4 children each, how many families would *u* expect to have (*i*) at least one boy, (*ii*) exactly two boys, (*iii*) no girl. **(Nov 2006)**

15. The probabilities of a Poisson variate taking the values of 1 and 2 are equal. Find (*i*) m, (*ii*) $P(x \geq 1)$, (*iii*) $P(1 < x < 4)$. **(April 2007)**

16. In a sample of 1000 cases, the mean of certain test is 14 and standard deviation is 2.5. Assuming the distribution to be normal find (*i*) how many students score between 12 and 15? (*ii*) how many score above 18? (*iii*) how many score below 8?

(Aug 2007, April 2007)

17. If x is a Poisson variate such $3P(x = 4) = \frac{1}{2} P(x = 2) + P(x = 0)$. Find (*i*) The mean of x (*ii*) $P(x \leq 2)$. **(Feb 2007, May 2007)**

18. If the variance of a Poisson variate is 3. Find the probability that

(*i*) $x = 0$, (*ii*) $1 < x < 4$, (*iii*) $0 < x < 3$. **(Nov 2007, Aug 2007, May 2007)**

19. 2% of the items of a factory are defective. The items are packed in boxes. What is the probability that there will be (*i*) 2 defective items, (*ii*) at least 3 defective items.

(April 2007)

20. Let x denote the number of heads in a single toss of 4 fair coins. Determine (*i*) $P(x < 2)$, (*ii*) $P(1 < x \leq 3)$.

21. Fit a Poisson distribution to the following data and calculate the theoretical frequencies.

X:	0	1	2	3	4	5
F:	142	156	69	27	5	1

22. Find the probability that at most 5 defective components will be found in a lot of 200 if experience shows that 2% of such components are defective. Also find the probability of more than five defective components. **(Nov 2007)**

23. In 100 set of ten tosses of an unbiased coin, in how many cases should we expect (*i*) Seven heads and three tails, (*ii*) At least seven heads. **(April 2007)**

24. The mark of 1000 students in a university are found to be normally distributed with mean 70 and standard deviation 5. Estimate the number of students whose marks will be (*i*) between 60 and 75, (*ii*) more than 75, (*iii*) less than 68. **(April 2007)**

25. In 256 sets of 12 tosses of a coin, in how may case one can expect eight heads and 4 tails? **(April 2007)**

26. The mean voltage of a battery is 15 volts and standard deviation is 2 volts. What is the probability that 4 such batteries connected in series will have a combined voltage of 60.8 or more volts. **(Nov 2003)**

27. The mean weight of 500 male students at certain college is 75 kg and the standard deviation is 7 kg. Assuming that the weights are normally distributed. Find how many students weigh (*i*) Between 60 and 78 kg, (*ii*) More than 92 kg? **(Jan 2003, Dec 2002)**

28. A shipment of 20 tape recorders contain 5 defectives, find the standard deviation of the probability distribution of number of defectives in sample of 10 randomly chosen for inspection.

29. Find the probability of getting 3 and 6 heads inclusive in 10 tosses of fair coin by using (*i*) Binomial distribution, (*ii*) The normal approximation to the binomial distribution.

30. The probability that John hits a target is $\frac{1}{2}$. He firs 6 times. Find the probability that he hits target (*i*) exactly 2 times, (*ii*) more than 4 times, (*iii*) at least once. **(Aug 2007)**

31. A Poisson distribution has a double mode at $x = 2$ and $x = 3$. Find the maximum probability and also find $P(x \geq 2)$. **(Aug 2007)**

32. The weekly wages of 1000 workers are normally distributed around a mean of Rs. 70 and standard deviation of Rs. 5. Estimate the number of workers whose weekly wages will be (*i*) between 70 and 72, (*ii*) between 69 and 72, (*iii*) more than 80. **(Aug 2007)**

33. Given that the mean heights of students in a class is 158 cms with standard deviation of 20 cms. Find how many students heights lie between 150 cms and 170 cms if there are 100 students in the class. **(Aug 2007)**

34. The probability that the life of a bulb is 100 days is 0.05. Find the probability that one of 6 bulbs (*i*) at least one, (*ii*) greater than 4, (*iii*) none, will be having a life of 100 days. **(Aug 2007)**

35. 1000 students had written an examination the mean of test is 35 and standard deviation is 5. Assuming the distribution to be normal find (*i*) how many students marks lie between 25 and 40? (*ii*) how many students get more than 40? (*iii*) how many students get below 20?

ANSWERS

2. 0.2688, 0.2688, $(0.3)^{10}$

3. 0.1734, 0.7368, 0.1347, 0.2632

4. 0.078, 0.259, 0.922

5. 0.0214, 0.02320, 0.7351

6. 0.1192

14. 625, 750, 630

16. 444, 55, 116

18. 0.0498, 0.5976, 0.5976

19. 0.272, 0.320

22. 0.784, 0.216

23. 12, 17

24. 818, 159, 345

25. 31

27. 325, 4

29. 0.773, 0.6337

31. 0.2241, 0.8008

32. 155, 235, 23

33. 38

35. 819, 159, 1

3.13 QUIZ

1. The mean of the Binomial distribution with parameters 'n' and 'p' is....

(a) np (b) npq

(c) $\sqrt{npq}$ (d) $\sqrt{np}$

2. The variance of the Binomial distribution with parameters n and p is....

(a) np (b) npq

(c) $\sqrt{npq}$ (d) $\sqrt{np}$

3. The standard deviation of the Binomial distribution with parameters n and p is....

(a) np (b) npq

(c) $\sqrt{npq}$ (d) $\sqrt{np}$

4. A shipment of 20 record players contains 5 defective. In a sample of 10 record players the standard deviation of the number of defectives is....

(a) $\frac{8}{5}$ (b) $\frac{3}{5}$

(c) $\frac{3}{8}$ (d) $\frac{5}{8}$

5. The mean of the distribution of getting heads in tossing a coin 5 times is............

(a) $\frac{1}{2}$ (b) $\frac{3}{2}$

(c) $\frac{7}{2}$ (d) $\frac{5}{2}$

6. The probability of getting a spade in drawing three cards with replacement is.........

(a) $\frac{1}{64}$ (b) $\frac{1}{32}$

(c) $\frac{3}{64}$ (d) $\frac{3}{32}$

7. Mean of Binomial distribution is 4 and variance is 2 then, p = ...

(a) $\frac{1}{3}$ (b) $\frac{1}{2}$

(c) $\frac{1}{4}$ (d) $\frac{3}{4}$

8. Mean of Binomial distribution is 6 and variance is 2 mode = ...

(a) 5 (b) 6

(c) 8 (d) 9

9. A coin is tossed 7 times. The mode of distribution of getting head is ....

(a) 2 (b) 3

(c) 4 (d) 5

10. The probability of getting one boy in a family of 4 children is ....

(a) $\frac{1}{2}$ (b) $\frac{2}{3}$

(c) $\frac{1}{3}$ (d) $\frac{1}{4}$

11. In a village three are 96 families of 4 children each, the number of families of getting at least one boy are ....

(a) 80 (b) 90

(c) 70 (d) 60

12. A Poisson distribution is such that probability at $x = 1$ = probability at $x = 2$ then mean is ....

(a) 1 (b) 2

(c) 3 (d) 4

13. X is a Poisson variate such that $P(X = 3) = P(X = 5)$ then μ = ...

(a) $\sqrt{12}$ (b) $\sqrt{15}$

(c) $\sqrt{24}$ (d) $\sqrt{29}$

14. If X is a Poisson variate which is a bimodal at $x = 3$ and at $x = 4$. Then μ = ...

(a) 1 (b) 3

(c) 6 (d) 8

15. If X is a Poisson variate such that $P(X = 1) = 2$ and $P(X = 2) = 1$, then μ = ...

(a) 1 (b) 2

(c) 3 (d) 4

16. Among the items manufactured in a factory 5% are defective. The probability of getting one defective blade in a pack of five blades is ....

(*a*) 0.01 (*b*) 0.0224

(*c*) 0.03 (*d*) 0.048

17. If X is Poisson variate such that $\mu = 2$, $P(X = 1) =$ ....

(*a*) 0.9 (*b*) 0.27

(*c*) 0.36 (*d*) 0.4

18. If X is a Poisson variate such that $\mu = 3$, then $P(X \geq 1) =$ ....

(*a*) 0.233 (*b*) 0.1233

(*c*) 0.5123 (*d*) 0.6723

19. If probability density function $f(x) = \frac{x^2}{9}$ in $0 < x < 3$, mean = ...

(*a*) $\frac{1}{4}$ (*b*) $\frac{3}{4}$

(*c*) $\frac{5}{4}$ (*d*) $\frac{9}{4}$

20. The probability density function $f(x) = \frac{1}{2}(2 - x)$ in $0 \leq x \leq 2$, $p(0 \leq x \leq a) = \frac{3}{4}$, then $a =$ ....

(*a*) 1 (*b*) 2

(*c*) 3 (*d*) $\frac{1}{2}$

21. If X is a random variable assigned to the number of heads obtained in tossing two coins and $P(|X - 1|) \geq 2 \leq 2$...............

(*a*) $\frac{1}{8}$ (*b*) $\frac{1}{4}$

(*c*) $\frac{1}{3}$ (*d*) $\frac{1}{2}$

22. A coin is tossed n times the probability at $x = 2 = \frac{3}{8}$ and probability at $x = 3 = \frac{1}{4}$, then $n =$ ....

(*a*) 1 (*b*) 2

(*c*) 4 (*d*) 5

23. If the X is Poisson variate $P(X = 1) = 0.3679$, $P(X = 2) = 0.184$, then $\mu =$ ....

(*a*) 1 (*b*) 2

(*c*) 3 (*d*) 4

24. If x is a standard normal variate $P(0 \leq z \leq 1.2) =$ ....

(*a*) 0.3849 (*b*) 0.21

(*c*) 0.48 (*d*) 0.57

25. If $P(0 \leq z \leq z_1) = 0.3770$, then z_1 = ...

(*a*) ± 1.16 (*b*) ≤ 2.26

(*c*) ≤ 3.12 (*d*) ± 4.16

26. If mean = 5 variance = $\frac{10}{3}$ of a Binomial distribution, then n = ...

(*a*) 0 (*b*) 3

(*c*) 5 (*d*) 7

27. One of the modes of Poisson distribution is 2 and $P(x = 2) = 0.612$, then μ = ...

(*a*) 1 (*b*) 3

(*c*) 2 (*d*) 4

28. If a Binomial distribution is bimodal at $x = 4$ and 5 and $p = q$, then n = ...

(*a*) 1 (*b*) 3

(*c*) 6 (*d*) 9

29. If on an average one ship in every ten is wrecked. The probability that only 2 arrive safely one of 5 ships is..................

(*a*) 0.81 (*b*) 0.081

(*c*) 0.9 (*d*) 0.0081

30. A die is thrown 8 times. The probability that 3 turns up exactly 2 times is ....

(*a*) 0.25 (*b*) 0.41

(*c*) 0.3 (*d*) 0.59

31. Assuming that 20% of the population of a city is literate. Assuming that 100 investigators each take 10 individuals to see whether they are literate, the number of people to report at least one literate is ....

(*a*) 60 (*b*) 50

(*c*) 40 (*d*) 90

32. If the variance of a Poisson variate is 2 and $P(X = 1) = 0.2706$, then $P(X = 2)$ =

(*a*) 0.1353 (*b*) 0.2706

(*c*) 0.1804 (*d*) 0.0902

33. If an ace of spades is drawn from a pack of well shuffled cards once in 104 trials, then the mean is

(*a*) 1 (*b*) 2

(*c*) 3 (*d*) 4

34. If the mean pf a Poisson variate is 1.32 the probability when $x = 3 = 0.1025$, then $P(X = 4)$ =

(*a*) 0.012 (*b*) 0.0338

(*c*) 0.1028 (*d*) 0.205

35. If X is a normal variate, then the mean deviation about the mean μ is

(*a*) $\frac{1}{2}\sigma$ (*b*) σ

(*c*) $\frac{3}{4}\sigma$ (*d*) $\frac{4}{5}\sigma$

36. The mode of normal distribution is..................

(*a*) μ (*b*) $\frac{\mu}{2}$

(c) $x - \mu$ (*d*) $\mu\,\sigma$

If the mean of a Binomial distribution is 4 and the variance is 3, then

37. $p =$

(*a*) 1 (*b*) $\frac{1}{2}$

(c) $\frac{1}{4}$ (*d*) $\frac{3}{4}$

38. $q =$

(*a*) 1 (*b*) $\frac{1}{2}$

(c) $\frac{1}{4}$ (*d*) $\frac{3}{4}$

39. $n =$

(*a*) 4 (*b*) 16

(c) 8 (*d*) 23

40. mode =

(*a*) 4 (*b*) 16

(c) 8 (*d*) 32

41. The probability of getting at least 4 heads in 6 tosses of a fair coin

(*a*) $\frac{11}{32}$ (*b*) $\frac{19}{32}$

(c) $\frac{21}{32}$ (*d*) $\frac{13}{32}$

42. The probability of guessing at least 6 out of 10 answers in a true-false examination is

(*a*) $\frac{193}{256}$ (*b*) $\frac{193}{1024}$

(c) $\frac{193}{512}$ (*d*) none of these

43. The lower bound on $P(-3 < x < 3)$, where $\mu = E(x) = 0$ and variance $\sigma^2 = 1$ is

(*a*) $\frac{2}{9}$ (*b*) $\frac{4}{9}$

(c) $\frac{7}{9}$ (*d*) $\frac{8}{9}$

44. The number of parameters in a Poisson distribution is (*a*) 1 (*b*) 2

(*c*) 3 (*d*) 4

45. The distribution in which mean is equal to the variance is

(*a*) Binomial distribution (*b*) Poisson distribution

(*c*) Normal distribution (*d*) Uniform distribution

46. The distribution in which mean, median and mode coincide is

(*a*) Binomial (*b*) Poisson

(*c*) Normal (*d*) Uniform

47. In a normal distribution with mean 12 and standard deviation 4, $P(0 \le x \le 12)$ is

(*a*) 0.52 (*b*) 0.4986

(*c*) 0 (*d*) none of these

48. If the mean of a normal distribution is 12, then its mode is

(*a*) 6 (*b*) 4

(*c*) 8 (*d*) none

49. If the mean of a normal distribution is 25, then its median is

(*a*) 30 (*b*) 28

(*c*) 25 (*d*) none

50. In a Binomial distribution mean is 48 and the SD is 4, then p =

(*a*) $\frac{1}{3}$ (*b*) $\frac{2}{3}$

(*c*) $\frac{1}{12}$ (*d*) $\frac{11}{12}$

51. In a Poisson distribution $P(X = 2) = 9P(X = 4) + 90\ P(X = 6)$

(*a*) 3 (*b*) 2

(*c*) 1 (*d*) none of these

52. The daily wages of 1000 workers are normally distributed with a mean of Rs. 70 and S.D. of Rs. 5. Then the number of workers with wages between Rs. 70 and Rs. 72 is

(*a*) 235 (*b*) 470

(*c*) 236 (*d*) 472

53. The mean of a Binomial distribution is 2.4 and variance is 1.44 the value of n is ....

(*a*) 6 (*b*) 5

(*c*) 4 (*d*) 3

54. In a Poisson distribution, the probability $P(X = 0)$ is twice the probability $P(X = 1)$, then the mean of the distribution is ....

(*a*) 1/2 (*b*) 1/3

(*c*) 1/4 (*d*) 1/5

55. The area under the normal curve between $z = 0$ and $z = 1.5$ is 0.4332, then the area which lies to the left of $= -1.5$ is ....

(*a*) 0.001 (*b*) 0.0668

(*c*) 0.0778 (*d*) 0.0787

56. If the value of the standard normal variable at $x = 34$ is -2 and mean $\mu = 50$, then σ is ....

(*a*) 5 (*b*) 6

(*c*) 7 (*d*) 8

57. If the mean and standard deviation of the sampling distribution of 80 samples each of size 25 with replacement from a normal population of 3000 with mean 68 and standard deviation 3 respectively, then the expected number of samples whose mean lies between 66.8 and 68.3 is ....

(*a*) 51 (*b*) 52

(*c*) 53 (*d*) 54

58. The value of k for a random sample of size 24 for a normal distribution such that $P(k < t < 2.807) = 0.095$ is ....

(*a*) 1.215 (*b*) 1.319

(*c*) 2.319 (*d*) none

59. The mean voltage of a battery is 1.5 V and standard deviation is 0.2 V. The probability that 5 such batteries are connected in series will have a combined voltage of 75.8 or more voltage is ....

(*a*) 0.0228 (*b*) 0.0118

(*c*) 0.228 (*d*) 0.118

60. The probability that in family of children, there will be at least one boy is...............

(*a*) 1/16 (*b*) 2/16

(*c*) 15/16 (*d*) 17/16

61. The mean and variance of a Binomial variate x are respectively 7/3 and 14/9, then $P(X = 6 \text{ or } 7)$ is ....

(*a*) 12/729 (*b*) 13/729

(*c*) 14/729 (*d*) 15/729

62. The marks scored by 1600 candidates in a test follow a normal distribution with mean 60 and standard deviation scored marks less than 40, the number of candidates who have scored between 60 and 80 is

(*a*) 500 (*b*) 400

(*c*) 300 (*d*) 200

63. Probability of student is not swimmers is 1/5. Out of 5 students chosen at random the probability that 4 are swimmers is ....

(*a*) 0.108 (*b*) 0.308

(*c*) 0.408 (*d*) 0.208

64. If x is Poisson variate such that $E(x^2) = 6$, then $E(x) =$ ....

(*a*) 1 (*b*) 2

(*c*) 3 (*d*) 4

65. The asymptote of normal distribution curve is...............

(*a*) Y = 0 (*b*) both axis

(*c*) $x = 0$ (*d*) none

66. The value of k for a random sample of size 18 for a normal distribution such that $P(-2.110 < t < k) = 0.965$ is ....

(*a*) 3.5 (*b*) 0.567

(*c*) 1.567 (*d*) 2.567

67. Out of bolts produced by a factory 2% are defective. In a shipment of 3600 bolts from the factory, the expected number of defective bolts is ....

(*a*) 70 (*b*) 71

(*c*) 72 (*d*) 73

68. If x is normal variate, then mean deviation about the mean ....

(*a*) 4/5 (*b*) 4/5 σ

(*c*) σ (*d*) μ

69. If $P(0 < z < 1.88) = 0.19699$, the area under the standard curve which is the left of $z = 1.88$ is ....

(*a*) 1.9 (*b*) 1

(*c*) 0.9699 (*d*) 9.699

70. The mean and standard deviation of a normal distribution of x are 120 and 40. The probability that x is over 120 is ....

(*a*) 1 (*b*) 0.5

(*c*) 0.1 (*d*) 0.2

71. If x is a Poisson variate such that the probability of itself taking a positive value is $1 - e^2$, then $P(-1.5 < x \leq 0) =$ ....

(*a*) 1 (*b*) e

(*c*) e^3 (*d*) e^2

72. In a Binomial distribution $n = 400$ and $p = 1/5$, its standard deviation.................

(*a*) 8 (*b*) 6

(*c*) 4 (*d*) 2

73. In a normal distribution the maximum ordinate at $x = \mu$ is ....

(*a*) $1/\sigma$ (*b*) $1/\mu$

(*c*) $\dfrac{1}{\sqrt{2\pi\sigma}}$ (*d*) $\dfrac{1}{\sqrt{2\pi}}$

74. In a Poisson distribution $P(X = 0) = 0.2$, then $V(x)$ ....

(*a*) log 2 (*b*) log 3

(*c*) log 4 (*d*) log 5

75. The mean of Poisson distribution is 2.25, then standard deviation is ....

(*a*) 1 (*b*) 1.5

(*c*) 2 (*d*) 2.5

76. If x is a Poisson variate such that probability of itself takes a positive value is $e^{1.5}$, then its mean is................

(*a*) 1.5 (*b*) 2

(*c*) 2.5 (*d*) 1

77. The variance of B (6, 1/2) is................

(*a*) 0 (*b*) 0.5

(*c*) 1 (*d*) 1.5

78. Out of 1000 families with 4 children each, the probable number of families all of those children are daughters is.................

(*a*) 500 (*b*) 550

(*c*) 625 (*d*) 650

79. For a Poisson distribution x, if P(1) = P(2), then P(0) = ...

(*a*) e (*b*) e^{-2}

(*c*) e^3 (*d*) e^2

80. A book contains 100 misprints distributed randomly through out its 100 pages. The probability that a page observed at random contain at least 2 misprints (Assume Poisson Distribution) is ....

(*a*) $1/e$ (*b*) $1 - (2/e)$

(*c*) $2/e$ (*d*) none

81. A box containing 100 bulbs, 10 are defective, the probability is that out of sample of 5 bulbs, none is defective is ....

(*a*) $(9/10)^5$ (*b*) $(10/9)^5$

(*c*) $(1/9)^5$ (*d*) $(1/10)^5$

82. The probability that a normal variate x lies outside the range $\mu \pm 3\sigma$ is.................

(*a*) 0.89 (*b*) 0.9973

(*c*) 0.785 (*d*) 1

83. In an experiment the success is twice that of failure. If the experiment is repeated 6 times the probability of at least 4 success is ....

(*a*) 0.1888 (*b*) 0.2888

(*c*) 0.6888 (*d*) 0.4888

84. If the mean and variance of a Binomial variate x are 2 and 1 respectively, then the probability that x takes a value greater than one is ....

(*a*) 8/16 (*b*) 9/16

(*c*) 10/16 (*d*) 11/16

85. The height of 800 men form a normal distribution with mean 165 cm and standard deviation 25 cm. If 150 men have their height between 160 cm and 170 cm. The number of men whose height is less than 160 cm is ....

(*a*) 325 (*b*) 350

(*c*) 300 (*d*) 400

86. The area under normal distribution curve between $z = 0$ and $z = \infty$ is ....

(*a*) 1 (*b*) 0.5

(*c*) 1.5 (*d*) 2

87. For a Binomial distribution $\mu = 10/3$ and the sum of mean and variance is 40/9, then the parameter P is

(*a*) 2/3 (*b*) 1/3

(*c*) 1 (*d*) none

ANSWERS

1. (*a*)	**2.** (*b*)	**3.** (*c*)	**4.** (*b*)	**5.** (*d*)	**6.** (*c*)	**7.** (*b*)	**8.** (*b*)	**9.** (*c*)	**10.** (*d*)
11. (*b*)	**12.** (*b*)	**13.** (*c*)	**14.** (*c*)	**15.** (*a*)	**16.** (*d*)	**17.** (*b*)	**18.** (*b*)	**19.** (*d*)	**20.** (*a*)
21. (*a*)	**22.** (*c*)	**23.** (*a*)	**24.** (*a*)	**25.** (*a*)	**26.** (*c*)	**27.** (*b*)	**28.** (*d*)	**29.** (*d*)	**30.** (*a*)
31. (*d*)	**32.** (*b*)	**33.** (*b*)	**34.** (*b*)	**35.** (*d*)	**36.** (*a*)	**37.** (*c*)	**38.** (*d*)	**39.** (*b*)	**40.** (*a*)
41. (*a*)	**42.** (*c*)	**43.** (*d*)	**44.** (*a*)	**45.** (*b*)	**46.** (*c*)	**47.** (*b*)	**48.** (*d*)	**49.** (*c*)	**50.** (*b*)
51. (*c*)	**52.** (*a*)	**53.** (*a*)	**54.** (*a*)	**55.** (*b*)	**56.** (*d*)	**57.** (*b*)	**58.** (*b*)	**59.** (*a*)	**60.** (*c*)
61. (*d*)	**62.** (*a*)	**63.** (*c*)	**64.** (*b*)	**65.** (*a*)	**66.** (*d*)	**67.** (*c*)	**68.** (*d*)	**69.** (*c*)	**70.** (*b*)
71. (*d*)	**72.** (*a*)	**73.** (*c*)	**74.** (*d*)	**75.** (*b*)	**76.** (*a*)	**77.** (*d*)	**78.** (*c*)	**79.** (*b*)	**80.** (*b*)
81. (*a*)	**82.** (*b*)	**83.** (*c*)	**84.** (*d*)	**85.** (*a*)	**86.** (*b*)	**87.** (*a*)			

Chapter 4

SAMPLING DISTRIBUTION

4.0 RANDOM SAMPLES

Population: A collection of objects or individuals is said to be population.

Ex: Number of students in a college, population is two types.

(*i*) Finite population

(*ii*) Infinite population.

Finite Population: If the elements in the population are finite then that population is said to be a finite population.

Ex: Students in a College.

Infinite Population: If the elements in the population are infinite then that population is said to be an infinite population.

Ex: Stars in the sky.

Sample: A small part from the population is said to be a sample. The number of students in a class room of the number of students of particular college.

Sample size: The number of elements in a sample is called sample size. The sample size is above 30, then that sample is said to be large sample. If the sample size is ≤ 30, then that sample is said to be small sample.

Parameters: A statistical constants obtain from the population is said to be parameters those parameters are denoted by

Population Size : N

Population Mean : μ

Population Variance : σ^2

Population Proportion : P

Population Standard Deviation: σ

Statistics: A statistical constants obtain from the sample is said to be statistics represented by

Sample Size: n

Sample Mean : $\bar{x}$

Sample Variance : S^2

Sample Proportion : p

Sample Standard Deviation : S

Sample Mean and Variance: Let $x_1, x_2, x_3, \ldots\ldots, x_n$ be the sample taken from a population of size N.

$$\therefore \qquad \text{Sample Mean } \bar{x} = \frac{1}{n}\sum_{i=1}^{n} x_i$$

$$\text{Sample Variance } S^2 = \frac{1}{(n-1)}\sum_{i=1}^{n}(x_i - \bar{x})^2$$

Sampling distribution. If we selected a large number of sample say k of size 'n', from the same finite population of size N and if we calculate some statistics ; in particular mean $\bar{x}$, Variance S^2 etc., of these k samples then we have a series of k means, k variances as given below.

Sample number	*Statistics*	
	Mean	*Variance*
1	$\bar{x}_1$	S_1^2
2	$\bar{x}_2$	S_2^2
3	$\bar{x}_3$	S_3^2
.	.	.
.	.	.
k	$\bar{x}_k$	S_k^2

The set of values of a statistics constitute is called the sampling distribution of that statistics. The set of values $\bar{x}_1, \bar{x}_2, \bar{x}_3, \ldots\ldots\ldots, \bar{x}_k$ of the statistic mean is known as sampling distribution of mean. Similarly, the series of values $S_1^2, S_2^2, S_3^2, \ldots\ldots\ldots, S_k^2$ of the statistic variance is known as sampling distribution of variance.

4.1 THE SAMPLING DISTRIBUTION OF THE MEAN (σ-KNOWN)

Suppose we take a sample of size n from a population to test. If the mean is $\bar{x}$. Suppose we take another sample of size n. The mean of 2nd sample may not be equal to $\bar{x}$, we can calculate different samples. All samples will be having different means.

Suppose all possible samoles of size n taken from a finite population N. Let $(x_1, x_2, \ldots, x_n)$ be the sample taken from the finite population of size N. The mean of the sample can be evaluated using the formula

$$\bar{x} = \frac{1}{n}\sum_{i=1}^{n} x_i \text{ and the variance } s^2 = \frac{1}{n}\sum_{i=1}^{n}(x_i - \bar{x})^2$$

Let $S_i = (x_{1i}, x_{2i}, x_{3i}, \ldots\ldots, x_{ni})$ be the ith sample, the set of all samples is called the sampling distribution. We can calculate the mean and variance of the sampling distribution.

Mean of the Sampling distribution of Mean: Let $\bar{x}_1, \bar{x}_2, \bar{x}_3, \ldots\ldots\ldots, \bar{x}_k$ be the means of samples. Let μ and σ^2 be the mean and variance of finite population of size N.

Let $\bar{x}$ be the mean of the sample of size 'n'

$$E(\bar{x}) = E\left(\frac{x_1 + x_2 + \ldots\ldots x_n}{n}\right)$$

$$= \frac{1}{n} E\,(x_1 + x_2 + \ldots\ldots x_n)$$

$$= \frac{1}{n} [E(x_1) + E(x_2) + \ldots\ldots E(x_n)]$$

Since sample is random. We can assume that

$$E(x_1) = E(x_2) = \ldots\ldots\ldots\ldots\ldots\ldots E(x_n) = \mu$$

$$\therefore \quad E(\bar{x}) = \mu$$

$\therefore$ Mean of the sampling distribution of means = Mean of the finite population of size N.

Let $V(\bar{x})$ be the variance of sampling distribution of means.

$$\sigma_{\bar{x}}^2 = V(\bar{x}) = \frac{V(x_1 + x_2 + \ldots\ldots x_n)}{n}$$

$$= \frac{1}{n^2} V(x_1 + x_2 + \ldots\ldots\ldots x_n)$$

$$= \frac{1}{n^2} \{V(x_1) + V(x_2) + \ldots\ldots\ldots + V(x_n)\}$$

We assume that

$$V(x_1) = V(x_2) = \ldots\ldots\ldots\ldots = V(x_n) = \sigma^2$$

$$\therefore \quad V(\bar{x}) = \frac{n!^2}{n^2} = \frac{!^2}{n}$$

$\therefore$ Variance of sampling distribution of Means $\neq$ Variance of population.

Standard Error: The standard deviation of sampling distribution of a statistics is known as standard error.

$$\therefore \quad \text{Standard error of means} = \frac{!}{\sqrt{n}}$$

Note:

1. The formula $\sigma_{\bar{x}}^2 = V(\bar{x})$ is applicable each draw was independent of other or the sampling is simple and with replacement.
2. In case of the sampling is without replacement.

$$\sigma_{\bar{x}}^2 = V(\bar{x}) = \left(\frac{N-n}{N-1}\right)\frac{\sigma^2}{n}$$

where N is the size of population and n is sample size $n \leq N$

$\left(\frac{N-n}{N-1}\right)$ is called the finite population correction factor.

4.2 SOLVED PROBLEMS

Problem 1: Find the value of the finite population correction factor for $n = 10$ and $N = 1000$.

Solution:

$$\text{Correction factor for finite population} = \left(\frac{N-n}{N-1}\right)$$

where N = number of items in a population

n = number of items in the sample

Given that $N = 1000$, $n = 10$

$$\therefore \text{ Correction factor} = \left(\frac{1000-10}{1000-1}\right) = \frac{990}{999} = 0.99.$$

Problem 2: A sample is collected from the items produced by a factory. The sample size is 81. The standard deviation of the population is 0.3. Find the standard error of the mean of sampling distribution.

Solution:

n = sample size = 81

σ = standard deviation of the population = 0.3

Standard error = standard error of the mean of sampling distribution

$$\frac{!}{\sqrt{n}} = \frac{0.3}{\sqrt{81}} = 0.333$$

Problem 3: A normal population has a standard deviation of 1. A sample of size 400 is collected for testing. Find the standard error of the mean of sampling distribution.

Solution:

n = sample size = 400

σ = standard deviation of the population = 1

Standard error = standard error of the mean of sampling distribution

$$= \frac{!}{\sqrt{n}} = \frac{1}{\sqrt{400}} = 0.05$$

Problem 4: Take 30 slips of paper and label 5 each – 4 and 4, four each – 3 and 3, three each – 2 and 2 and each – 1, 0 and 1. If each slip of paper has the same probability of being drawn find the probabilities of getting – 4, – 3, – 2, – 1, 0, 1, 2, 3, 4 and find the mean and variance of this distribution. **(Feb 2007, May 2007, Nov 2006)**

Solution:

The thirty slips are

– 4, – 4, – 4, – 4, – 4, 4, 4, 4, 4, 4, – 3, – 3, –3, – 3, 3, 3, 3, 3, – 2, – 2, – 2, 2, 2, 2, – 1, – 1, 1, 1, 0, 0

Samples are one each collected

(– 4, 4, – 3, 3, – 2, 2, 1, 0, 1)

Probability of getting

$$p(-4) = \frac{5}{30} = \frac{1}{6}$$

$$p(4) = \frac{5}{30} = \frac{1}{6}$$

$$p(-3) = \frac{4}{30} = \frac{2}{15}$$

$$p(3) = \frac{4}{30} = \frac{2}{15}$$

$$p(-2) = \frac{3}{30} = \frac{1}{10}$$

$$p(2) = \frac{3}{30} = \frac{1}{10}$$

$$p(-1) = \frac{2}{30} = \frac{1}{15}$$

$$p(1) = \frac{2}{30} = \frac{1}{15}$$

$$p(0) = \frac{2}{30} = \frac{1}{15}$$

∴ Probability distribution of the sample

x	–4	4	–3	3	–2	2	–1	1	0
$p(x)$	$\frac{1}{6}$	$\frac{1}{6}$	$\frac{2}{15}$	$\frac{2}{15}$	$\frac{1}{10}$	$\frac{1}{10}$	$\frac{1}{15}$	$\frac{1}{15}$	$\frac{1}{15}$

$$\text{Mean } \mu_{\bar{x}_i} = \sum xp(x) = \frac{-4}{6} + \frac{4}{6}\frac{-6}{15} + \frac{6}{15}\frac{-4}{10} + \frac{4}{10}\frac{-1}{15} + \frac{1}{15} + 0 = 0$$

$$\text{Variance } \sigma^2_{\bar{x}}\,\bar{x} = \sum (x - \mu)^2 P(x)$$

$$= 1\left[\frac{16}{2} \times 2 + \frac{9 \times 2 \times 2}{15} + \frac{4 \times 2}{10} + \frac{2 \times 1}{15}\right] = \left[\frac{32}{6} + \frac{36}{15} + \frac{8}{10} + \frac{2}{15}\right] = \frac{260}{30} = \frac{26}{3}$$

Mean of population

$$\mu = \frac{-4+(-4)-4-4-4+4+4+4+4+4-3-3-3-3+3+3+3+3-2-2-2+2+2+2-1-1+1+1+0+0}{30}$$

$= 0$

$$\sigma^2 = \frac{1}{30}\sum x^2 = \frac{1}{30}.260 = \frac{26}{3}$$

In this case $\sigma^2 = \sigma^2_{\bar{x}}$

Problem 5: The weights of 1500 ball bearings are normally distributed with a mean of 635 grams and standard deviation of 3.136 gms. If 300 random samples of size 36 are drawn from this population. Determine the expected mean and standard deviation of the sampling distribution of means if sampling is done (*a*) with replacement (*b*) without replacement.

Solution:

$N = 1500,\ n = 36$

$\mu = 635 \qquad \sigma = 3.136$

(*a*) With replacement

Expected mean $= \mu_x = \mu = 635$

Expected Standard Deviation $= \sigma_x = \sigma = \frac{1.36}{\sqrt{36}}$

$= 0.227$

(*b*) Without replacement

Expected mean $\mu_x = \mu = 635$

Expected Standard Deviation $= \sigma_x = \frac{!}{\sqrt{n}}\left(\frac{\sqrt{N-n}}{\sqrt{N-1}}\right)$

$= \frac{1.36}{\sqrt{36}}\left(\frac{\sqrt{1500-36}}{\sqrt{1500-1}}\right)$

$= \frac{(1.36)\times(0.988)}{6} = 0.224.$

Problem 6: A population consists of six numbers 4, 8, 12, 16, 20, 24. Consider all samples of size two which can be drawn without replacement from this population, find

(*i*) population mean
(*ii*) population standard deviation
(*iii*) mean of the sampling distribution of means
(*iv*) standard deviation of the sampling distribution of means

(Feb 2007, Aug 2007, Nov 2007)

Solution: Given population: 4, 8, 12, 16, 20, 24

(*i*) Mean of the population $\mu = \frac{4+8+12+16+20+24}{6}$

$= 14$

(*ii*) Variance of the population $\sigma^2 = \frac{1}{n}\sum_{i=1}^{6}(x_i-\mu)^2$

$= \frac{(4-14)^2+(8+14)^2+...(24-14)^2}{6}$

$= \frac{280}{6} = 46.61$

Standard deviation of population is $\sigma = \sqrt{46.67} = 6.83$

Population size = 6

Sample size = 2

∴ ${}^6C_2$ samples can be drawn from the population without replacement.

∴ Samples with their respective means are given below.

Sample	*Mean*	*Sample*	*Mean*
4, 8	26	8, 24	16
4, 12	8	12, 16	14
4, 16	10	12, 20	16
4, 20	12	12, 24	18
4, 24	14	16, 20	18
8, 12	10	16, 24	20
8, 16	12	20, 24	22
8, 20	14		

which is sampling distribution

(*iii*) Mean of the sampling distribution of means $\mu_{\bar{x}} = \dfrac{210}{15} = 14$

∴ Mean of population = Mean of the sampling distribution of means

Variance of the sampling distribution of means $\sigma_{\bar{x}}^2 = \dfrac{1}{n}\sum_{i=1}^{6}(\bar{x}_i - \mu_{\bar{x}})^2$

$$= \frac{1}{15}[(6-14)^2 + (8-14)^2 + \ldots\ldots + (22-14)^2]$$

$$= 18.667$$

(*iv*) Standard deviation of sampling distribution $= \sqrt{18.667} = 4.32$

$$\text{S.E.} = \left(\frac{N-n}{N-1}\right)\frac{\sigma}{\sqrt{n}}$$

$$= \left(\frac{6-2}{6-1}\right)\frac{6.83}{\sqrt{2}} = 4.32$$

∴ Standard deviation of sampling distribution of mean = Standard Error of Means

Problem 7: If the population is 3, 6, 9, 15, 27

(*a*) List all possible samples of size 3 that can be taken without replacement from the finite population.

(*b*) Calculate the mean of each of the sampling distribution of means.

(*c*) Find the standard deviation of sampling distribution of means.

(Feb 2007, May 2007, Aug 2007, Nov 2004)

Solution:

Mean of the population $\mu = \dfrac{3 + 6 + 9 + 15 + 27}{5} = \dfrac{60}{5} = 12$

Standard deviation of the population

$$\sigma = \sqrt{\frac{(3-12)^2+(6-12)^2+(9-12)^2+(15-12)^2+(27-12)^2}{4}}$$

$$= \sqrt{\frac{360}{4}} = 9.486$$

(*a*) Sampling without replacement (finite population) the total number of samples without replacement in ${}^{N}C_{n} = {}^{5}C_{3} = 10$, the 10 samples are

(3, 6, 9) (3, 6, 15) (3, 6, 27) (3, 9, 15) (3, 9, 27) (3, 15, 27) (6, 9, 15) (6, 9, 27) (6, 15, 27) (6, 15, 27) (9, 15, 27)

Compute the static A.M. for each of these samples.

$\bar{x}_i$	6	8	12	9	13	15	10	14	16	17
f	6	8	9	10	12	13	14	16	15	17

(*b*) $\mu_{\bar{x}_i}$ = mean of sampling distribution of means

$$= \frac{6+8+9+10+12+13+14+15+16+17}{10} = \frac{120}{10} = 12$$

(*c*) $$\sigma^2_{\bar{x}_i} = \frac{(6-12)^2+(8-12)^2+(9-12)^2+...+(15-12)^2+(16-12)^2+(17+12)^2}{9}$$

$$= \frac{36+16+9+4+0+1+4+9+16+25}{9} = \frac{120}{9} = 13.3$$

$$\sigma_{\bar{x}_i} = \sqrt{13.3} = 3.651.$$

Problem 8: A population consists of a four numbers 3, 7, 11, 15. Consider all possible samples of a size two which can be drawn with replacement from their population. Find (*i*) The population mean, (*ii*) The population standard deviation, (*iii*) The mean of the sampling distribution of means, (*iv*) The standard deviation of the sampling distribution of means.

Verify (3) and (4) directly from (1) and (2) using a suitable formulae. Solve this problem if sampling is without replacement.

Solution:

With 4 numbers 16 samples (*i.e.*, 4 × 4) of size 2 can be drawn with replacement. The samples with their respective means are shown below.

Sample	*Mean*	*Sample*	*Mean*	*Sample*	*Mean*	*Sample*	*Mean*
3, 3	3	7, 3	5	11, 3	7	15, 3	9
7, 3	5	7, 7	7	11, 7	9	15, 7	11
11, 3	7	7, 11	9	11, 11	11	15, 11	13
15, 3	9	7, 15	11	15, 15	13	15, 15	15
Total	24		32		40		48

Means of the sampling distribution of means = $\frac{24 + 32 + 40 + 48}{16} = \frac{144}{16} = 9$

The mean of the population = $\frac{3 + 7 + 11 + 15}{4} = 9$

Hence the mean of sampling distribution of means is equal to the population mean.

Standard deviation of sampling distribution:

Mean	*Deviation(d)*	d^2	*Mean*	*Deviation(d)*	d^2
3	3–9 = – 6	36	7	7–9 = –2	4
5	5–9 = – 4	16	9	9–9 = 0	0
6	7–9 = –2	4	11	11–9 = 2	4
9	9–9 = 0	0	13	13–9 = 4	16
5	5 –9 = – 4	16	9	9–9 = 0	0
7	7–9 = –2	4	11	11–9 = 2	4
9	9–9 = 0	0	13	13–9 = 4	16
11	11–9 = 2	4	15	15–9 = 6	36
		$\Sigma d^2 = 80$			$\Sigma d^2 = 80$

Standard deviation of sampling distribution $\sqrt{\frac{\sum(X-\mu)^2}{n-1}} = \sqrt{\frac{160}{15}} = 3.265$

Standard deviation of population:

X	*x* = (X–9)	x^2
3	3–9 = – 6	36
7	7–9 = – 2	4
11	11–9 = 2	4
15	15–9 = 6	36
	$\Sigma x = 0$	$\Sigma x^2 = 80$

$$\sigma = \sqrt{\frac{1}{n}\sum x^2}$$

$$= \sqrt{\frac{1}{40} \times 80} = 4.472$$

The standard deviation of sampling distribution of means is equal to standard error of mean,

i.e., standard error = $\frac{\sigma}{\sqrt{n}} = \frac{4.472}{2}$ = 2.236, where n = sample size

If the sampling is without replacement the following six samples can be drawn:

(3, 7), (3, 11), (3, 15), (7, 11), (7, 15), (11, 15)

Their respective means are 5, 7, 9, 9, 11, 13

Mean of sampling distribution of means = $\frac{5+7+9+9+11+13}{6} = \frac{54}{6} = 9$

It is the same as population mean.

Variance of sampling distribution of means

$$\sigma_{\bar{x}}^2 = \frac{(5-9)^2+(7-9)^2+(9-9)^2+(9-9)^2+(11-9)^2+(13-9)^2}{6}$$

$$\sigma_{\bar{x}}^2 = \frac{16+4+4+16}{6} = \frac{40}{6} = 6.67$$

$$\sigma_{\bar{x}}^2 = \sqrt{6.67} = 2.58$$

It is equal to $\sqrt{\frac{N-n}{N-1} \cdot \frac{\sigma^2}{n}} = \sqrt{\frac{4-2}{4-1} \cdot \frac{20}{2}}$

$= 2.58.$

4.3 CENTRAL LIMIT THEOREM

Statement: If $\bar{x}$ is the mean of random sample of size n taken from a normal population having the mean μ and variance σ^2, then

$Z = \frac{\bar{x}-\mu}{\sigma/\sqrt{n}}$ is a random variable whose distribution function approaches that of the standard normal distribution.

4.4 SOLVED PROBLEMS

Problem 1: A random sample of size 100 is taken from an infinite population having mean 76 and variance 256, what is the probability that $\bar{x}$ will be between 75 and 78.

(April 2007, Nov 2006)

Solution:

Given n = 100, μ = 76, σ^2 = 256

$\therefore \quad Z = \frac{\bar{x}-\mu}{\sigma/\sqrt{n}}$

When $\bar{x} = 75 \Rightarrow \quad Z = \frac{75-76}{16/10} = -0.625$

Then $\bar{x} = 78 \Rightarrow \quad Z = \frac{78-76}{16/10} = 1.25$

$$P(75 \le \bar{x} \le 78) = P(-0.625 \le Z \le 1.25)$$
$$= 0.2324 + 0.3944$$
$$= 0.6268.$$

Problem 2: If the distribution of the weights of all men travelling between Hyderabad and Chennai has a mean of 163 pounds and Standard deviation of 18 pounds. What is the probability that the consigned gross weight of 36 men travelling between these two litres is more than 6000 pounds?

Solution:

Given $\mu = 163$

$\sigma = 18$

Average weight of man $\bar{x} = \dfrac{6000}{36} = 166.7$

$n = 36$

$$Z = \frac{166.7 - 163}{18/6} = 1.23$$

$$P(\bar{x} > 166.7) = P(Z > 1.23)$$
$$= 0.5 - P(0 < 2 < 1.23)$$
$$= 0.5 - 0.3907 = 0.1093.$$

Problem 3: A random sample is taken from N(30, 12). How large sample should be taken if the sample mean is to lie between 25 and 35 with probability 0.95?

Solution:

Given $\mu = 30$

$\sigma^2 = 12$

$\sigma = 3.464$

$$P(25 < \bar{x} < 35) = 0.95$$

When $\bar{x} = 25$

$$\Rightarrow \quad Z = \frac{25 - 30}{\frac{3.464}{\sqrt{n}}} = Z_1 = \frac{5\sqrt{n}}{3.464}$$

$$\therefore \quad P(25 < \bar{x} < 35) = P(-Z_1 < Z < Z_1) = 0.95$$

$$\Rightarrow \quad 2P(0 < Z < Z_1) = 0.95$$

$$P(0 < Z < Z_1) = 0.475$$

$$Z_1 = 1.96$$

$$\frac{5\sqrt{n}}{3.464} = 1.96$$

$$\therefore \quad n = 2$$

Problem 4: A random sample of size 64 is taken from normal population with $\mu = 51.4$ and $\sigma = 6.8$ what is the probability that the mean of the sample will

(i) Exceed 52.9

(ii) Between 50.5 and 52.3

(iii) < 50.6

Solution:

Given

$$n = 64$$
$$\mu = 51.4$$
$$\sigma = 6.8$$

$$\frac{s}{\sqrt{n}} = \frac{6.8}{\sqrt{64}} = \frac{6.8}{8} = 0.85$$

(i) When $\bar{x} = 52.9 \Rightarrow Z = \dfrac{52.9 - 51.4}{0.85} = 1.76$

$$P(\bar{x} > 52.9) = P(0 < Z < 1.76)$$
$$= 0.5 - P(0 < Z < 1.76)$$
$$= 0.5 - 0.4608$$
$$= 0.0392$$

(ii) When $\bar{x} = 50.5 \Rightarrow Z = -1.06$

$$\bar{x} = 52.3 \Rightarrow Z = 1.06$$

$$P(50.5 < \bar{x} < 52.3) = P(-1.06 < Z < 1.06)$$
$$= 2P(0 < Z < 1.06)$$
$$= 2 \times 0.3554$$
$$= 0.7108$$

(iii) When $\bar{x} = 50.6$

$\Rightarrow$ $Z = -0.94 = 0.1736.$

Problem 5: The mean height of students in a college is 155 cms and standard deviation is 15. What is the probability that the mean weight of 36 students is less than 157 cms?

Solution:

μ = mean of the population = mean height of students of a college is 155 cms.

σ = standard deviation of population = 15 cms

$$Z = \frac{\bar{x} - \mu}{\sigma / \sqrt{n}} = \frac{157 - 155}{\frac{15}{\sqrt{36}}} = \frac{40}{6} = 0.8$$

$$P(\bar{x} \leq 157) = P(Z \leq 0.8)$$
$$= 0.5 + P(0 \leq z \leq 0.8)$$
$$= 0.5 + 0.2881$$
$$= 0.7881$$

$\therefore$ The probability that the mean height of 36 students is less than 157 = 0.7881.

Problem 6: A normal population has a mean of 0.1 and a standard deviation of 2.1. Find the probability that the mean of simple sample of 900 members will be negative.

(Feb 2007, Aug 2007, May 2007, May 2006)

Solution:

$$Z = \frac{\bar{x}-\mu}{\sigma/\sqrt{n}}$$

$$\mu = \text{mean of population} = 0.1$$

$$\sigma = \text{standard deviation of the population is } 2.1$$

$$Z = \frac{\bar{x}-0.1}{2.1/\sqrt{900}} = \frac{x-0.1}{0.07}$$

$$= \frac{x}{0.07} - 1.43$$

$\bar{x}$ is negative if $Z < -1.43$

$$P(\bar{x} < 0) = P(Z < -1.43)$$

$$P(Z < -1.43) = P(1.43 \le Z < \infty)$$

$$= 0.5 - 0.4236 = 0.0764$$

$\therefore$ The probability that $\bar{x}$ is $-$ve is 0.764.

Problem 7: A research worker wishes to estimate mean of a population by using sufficiently large sample. The probability is 95% that sample mean will not differ from the true mean by more than 25 percentage of the standard deviation. How large a sample should be taken?

Solution:

From the table the area under the normal curve between the values -1.96 and 1.96 is 0.95.

$$\therefore \quad \left|\frac{\bar{x}-\mu}{\sigma/\sqrt{n}}\right| = 1.96$$

$$\Rightarrow \quad |\bar{X}-\mu| = 1.96\,\frac{!}{\sqrt{n}} \qquad \text{...(1)}$$

$$\text{Given that} \quad |\bar{X}-\mu| < \frac{!}{4} \qquad \text{...(2)}$$

From (1) and (2)

$$1.96\,\frac{!}{\sqrt{n}} < \frac{!}{4}, \quad 19.6 < \frac{\sqrt{n}}{4}$$

$$n > 16 \times (19.6) = 62 \text{ (approximately)}$$

4.5 THE SAMPLING DISTRIBUTION OF MEAN (σ - UNKNOWN)

In the case of sampling distribution of the mean with known standard deviation the information about population standard deviation s was required. If the sample size n is large then even population standard deviation is not known it does not make any problem, since in such situation sample standard deviation s will be used as a substitute. However, when it comes to the random variable whose values are given by $\frac{\bar{x}-\mu}{s/\sqrt{n}}$.

Statement: If $\bar{x}$ is the mean of a random sample of size n taken from a normal population having the mean μ and standard deviation s. Then $t = \frac{\bar{x}-\mu}{s/\sqrt{n}}$, where $s^2 = \frac{1}{n-1}\sum_{i=1}^{n}(x_i - \bar{x})^2$ is a random variable having the t – distribution with $(n-1)$ degree freedom.

4.6 SOLVED PROBLEMS

Problem 1: In 16 one hour test runs, the gasoline consumption of an engine averaged 16.4 gallons with a standard deviation of 2.1 gallons. Test the claim that the average gasoline consumption of this engine is 12.0 gallons per hour.

Solution:

Given $n = 16$, $\mu = 12.0$, $\bar{x} = 16.4$ and $s = 2.1$

From t-distribution,

We have
$$t = \frac{\bar{x}-\mu}{s/\sqrt{n}}$$
$$= \frac{16.4 - 12.0}{2.1/\sqrt{16}}$$
$$= \frac{4.4\sqrt{16}}{2.1}$$
$$= \frac{17.6}{2.1} = 8.38.$$

The table of t-distribution shows that the probability of getting a value of t greater than 2.947 is 0.005 for 15 degrees of freedom, then the probability of getting a value greater than 8 must be negligible. Hence it would seem reasonable to conclude that the true average hourly gasoline consumption of the engine exceeds 12.0 gallons.

Problem 2: A manufacturer of fuses claims that with a 20% overload the fuses will blow in 12.40 minutes on the average. To test this claim, a sample of 20 of the fuses was subjected to a 20% of over load, and the times it took them to blow had a mean of 10.63 minutes and a standard deviation of 2.48 minutes. If it can be assumed that the data constitute a random sample from a normal population, do they claim to support or refuse the manufacturer's claim.

Solution:

Given $n = 20$, $\mu = 12.40$, $\bar{x} = 10.63$, $s = 2.48$

From t-distribution

We have
$$t = \frac{\bar{x}-\mu}{s/\sqrt{n}}$$

$$= \frac{10.63 - 12.40}{2.48/\sqrt{20}}$$

$$= \frac{-1.77\sqrt{20}}{2.48}$$

$$= -3.19$$

This value of t-distribution has 19 degrees of freedom. From the table of t-distribution for $v = 19$ we find that the probability that t will exceed 2.86 is 0.005 and hence the probability that t will be less than –2.861 is also 0.005. since $t = -3.19$ is less than –2.861 and 0.005 is very small probability, we conclude that the data claim to refuse the manufacturers claim that the mean blowing time of his fuses with a 20% overload is 12.40 minutes.

Hence the mean blowing time of his fuses with a 20% overload is said to be less than 12.40 hours.

Problem 3: A random sample of size 25 from a normal population has the mean $\bar{x} = 47.5$ and the standard deviation $s = 8.4$. Does this information tend to support or refute the claim that the mean of the population is $\mu = 42.1$?

Solution:

Given $n = 25$, $\bar{x} = 47.5$, $\mu = 42.1$, $s = 8.4$

We have from t-distribution,

$$t = \frac{\bar{x} - \mu}{s/\sqrt{n}}$$

$$= \frac{47.5 - 42.1}{8.4\sqrt{25}}$$

$$= \frac{5.4\sqrt{25}}{8.4} = 3.21$$

This value of t has 24 degrees of freedom. From the table of t-distribution for $v = 24$, we get the probability that t will exceed 2.797 is 0.005, then the probability of getting a value greater than 3.21 is negligible. Hence we conclude that the information given in the data of this example tend to refute the claim that the mean of the population is $\mu = 4.21$.

4.7 EXERCISE

1. A random sample of size 144 is taken from an infinite population having the mean 75 and variance 225. What is the probability that $\bar{x}$ will lie between 72 and 77? **(Aug 2007, May 2006)**
2. Determine the probability that the sample mean area covered by a sample of 40 of 1 litre paint boxes will be between 510 to 520 square feet given that 1 litre of such paint box covers on the average 513.3 square feet with S.D. of 31.5 square feet.
3. There are 25 cards written on four of them written 5, on six of them written 10, on five of them written 15, on three of them written 20, on four of them written 25 and on three of them written 30. One are each of 5, 10, 15, 20, 25, 30 are selected without replacement.

Find (*i*) mean of the population, (*ii*) standard deviation of the population, (*iii*) mean of the sampling distribution of means, (*iv*) standard deviation of the sampling distribution of means verify that (*i*) = (*iii*) and (*ii*) ≠ (*iv*). **(Aug 2007)**

4. A population consists of 5, 10, 14, 18, 13 and 24. Consider all possible samples of size two which can be drawn without replacement from the population. Find (*i*) the mean of population, (*ii*) standard deviation of population, (*iii*) mean of the sampling distribution of means, (*iv*) standard deviation of the sampling distribution of means. **(Feb 2007, Nov 2006)**

5. The mean of certain normal population is equal to the standard error of the mean of the samples of 64 from that distribution. Find the probability that the mean of the sample of size 36 will be negative. **(May 2006)**

6. A population consists of the five numbers 2, 3, 6, 8, 11. consider all possible samples of size two which can be drawn with replacement from this population. Find (*a*) The mean of the population, (*b*) Standard deviation of the population, (*c*) The mean of the sampling distribution of means, (*d*) The standard error of the means. Verify (*c*) and (*d*) directly from (*a*) and (*b*) . **(Jan 2003, April 2005)**

7. In a normal distribution with mean 375 and standard deviation 48, how large a sample must be taken so that the probability will be at least 0.95 that the sample mean falls between 370 and 380.

8. The average cost of a studio condominium in the cedar lakes development is Rs. 62,000 and the standard deviation is Rs. 4200. What is the probability that a condominium in this development will cost at least Rs. 65,000.

9. A random sample of size 25 from a normal population has the mean $\bar{x} = 47$ and the standard deviation $s = 7$. Basing our decision on the statistic of *t*-distribution, can we say that the given information supports the conjecture that the mean of the population is $\mu = 42$?

10. If the size of the sample is 20. The size of population is 1000. Then find the population correction factor of the population.

11. If the sample size is 25 and population size 10000. Then find the population correction factor.

12. A random sample of size 81 is taken from an infinite population having the mean 65 and standard deviation 10. What is the probability that $\bar{x}$ will lie between 66 and 68? **(Nov 2007)**

13. The guaranteed average life of a certain type of electric light bulbs is 1000 hours with a standard deviation of 125 hours. It is decided to sample the output so as to ensure that 90 percent of the bulbs do not full short of the guaranteed average by more than 2.5 percent. What must be the minimum size of the sample?

14. A survey is proposed to be conducted to know the annual earnings of the old statistics graduates of Delhi university. How large should the sample be taken in order to estimate the mean annual earnings within plus and minus Rs. 1000 at 95% confidence level. The standard deviation of the annual earnings of the entire population is known to be Rs. 3000.

15. Find the probability that a random variable having the standard normal distribution will take on a value. (*a*) between 0.87 and 1.28 (*b*) between – 0.34 and 0.62.

16. The mean weekly sales of soap bars in departmental stores was 146.3 bars per store. After an advertising campaign the mean weekly sales in 22 stores for a typical week increased to 153.7 and showed a standard deviation of 17.2 was the advertising compaign successful?

17. A process of making certain bearings is under control if the diameters of the bearings have a mean of 0.5000 cm. What can we say about this process if a sample of 10 of these bearings has a mean diameter of 0.5060 cm and a standard deviation of 0.0040 cm?

18. The claim that the variance of a normal population is $\sigma^2 = 21.3$ is rejected if the variance of a random sample of size 15 exceeds 39.74. What is the probability that the claim will be rejected even through $\sigma^2 = 21.3$?

ANSWERS

1. 0.9368 **2.** 0.6553 **5.** 0.7734

7. at least 355 **8.** $P(Z \geq 0.71) = 0.5000 - 0.2611 = 0.2389$

9. $t = 3.57$, the data do not support the conjecture) **10.** 0.98

11. 0.999 **12.** 0.1806.

4.8 QUIZ

1. A random sample of size 2 is drawn from the population 3, 7, 11, 12. Consider the sampling distribution of sample mean with replacement, then the sample mean is.........
   (*a*) 8 (*b*) 9
   (*c*) 10 (*d*) 11

2. If the mean and standard deviation of the sampling distribution of 80 samples each of size 25 with replacement from a normal population of 3000 with mean 68 and standard deviation 3 respectively, then the expected number of samples whose mean lies between 66.8 and 68.3 is
   (*a*) 53 (*b*) 54
   (*c*) 55 (*d*) 56

3. The value of k for a random sample of size 24 for a normal distribution such that $P(k < t < 2.807) = 0.095$ is................
   (*a*) 2.319 (*b*) 0.319
   (*c*) 1.319 (*d*) none

4. The value of k for a random sample of size 18 for a normal distribution such that $P(2.110 < t < k) = 0.965$ is.................
   (*a*) 3.567 (*b*) 1.567
   (*c*) 0.567 (*d*) 2.567

5. Given $n = 12$ and $t_{0.1} = 1.356$ and $t_{0.025} = 2.179$ then $P(-1.356 < t < 2.179)$ is...........
   (*a*) 0.875 (*b*) 0.785
   (*c*) 0.574 (*d*) 0.857

6. If a sample of size 64 is taken from a population whose standard deviation is 4 then probable error is

(*a*) 0.733 (*b*) 0.337

(*c*) 0.373 (*d*) none

7. Given $n = 30$ and $t_{0.1} = 1.310$ then $P(t > 1.310)$ is..............

(*a*) 0.3 (*b*) 0.2

(*c*) 0.1 (*d*) 0

8. The finite population correction factor for $n = 1000$ and $N = 10000$ is ....

(*a*) 0.5 (*b*) 0.6

(*c*) 0.7 (*d*) 0.9

9. For small samples ($n < 30$) drawn from a normal distribution and with σ unknown a natural static for inference on population mean μ is................

(*a*) $\dfrac{x - \mu}{s/\sqrt{n}}$ (*b*) $\dfrac{\overline{x} - \mu}{s^2/\sqrt{n}}$

(*c*) $\dfrac{\overline{x} + \mu}{s/\sqrt{n}}$ (*d*) none

10. A normal population has a standard deviation of a sample of size 400 is collected for testing then the standard error of a mean of sampling distribution is....................

(*a*) 0.05 (*b*) 0.04

(*c*) 0.03 (*d*) 0.02

11. A random sample of size 2 is drawn form the population 1, 2, 3, 4, 5. consider sample distribution of sample mean with replacement then the population mean is..............

(*a*) 1 (*b*) 2

(*c*) 3 (*d*) 4

12. If the size of sample is 200 and of population is 2000 then population correction factor is...........

(*a*) 0.8 (*b*) 0.7

(*c*) 0.9 (*d*) 0.99

13. If size of the sample 25 and standard deviation of the population is 0.1. Then standard error of the mean of sampling distribution is ....

(*a*) 0 (*b*) 0.01

(*c*) 0.02 (*d*) 0.03

14. If a sample is taken from an infinite population and sample size is increased from 25 to 100. The effect on standard error is ....

(*a*) same value (*b*) multiplied by 2

(*c*) divided by two (*d*) multiplied by 4

15. If a sample is taken from an infinite population and if sample size is increased from 400 to 900. The effect on standard error is................

(a) multiplied by 3/2 (b) divided by 2

(c) multiplied by 2 (d) divided by 4

16. If a sample is taken from an infinite population and the sample size is decreased from 1000 to 160. The effect on standard error is................

(a) multiplied by 2 (b) divided by 2

(c) multiplied by 3/2 (d) multiplied by 5/2

17. The size of the samples is 121. Mean and standard deviation of the population are 70 and 10. The probability that $\bar{x}$ is greater than 72 is.............

(a) 0.0139 (b) 0.02

(c) 0.03 (d) 0.04

18. $\mu = 30.5$, $n = 100$, $\bar{x} = 28.8$, $\sigma = 8.35$, then $|z| =$ ....

(a) 1.68 (b) 2.68

(c) 3.68 (d) 4.68

19. If the population is 2, 4, 6, 8, 10. If samples of size 2 are taken with replacement. Then the mean of the means of sample distribution is......................

(a) 1 (b) 2

(c) 5 (d) 6

20. Samples of size 2 taken from the population 62, 57, 63, 58 variance of the mean of sampling distribution is................

(a) 1.17 (b) 2.17

(c) 3.1 (d) 4.17

21. The marks of 5 students in one subject are 45, 47, 49, 61, 48 and mean of the population is 52 then $t =$ ....

(a) 0.1 (b) 0.3

(c) 0.5 (d) 0.7

22. The finite population correction factor is

(a) $\dfrac{N-n}{N-1}$ (b) $\sqrt{\dfrac{N-n}{N-1}}$

(c) $\dfrac{N-1}{N-n}$ (d) $\sqrt{\dfrac{N-1}{N-n}}$

23. If $n = 10$, $N = 1000$, then the finite population correction factor is

(a) 0.991 (b) $\sqrt{0.991}$

(c) $\dfrac{1}{0.991}$ (d) $\dfrac{1}{\sqrt{0.991}}$

24. A random sample of size 64 is taken from a normal population with $\mu = 51.4$ and $\sigma = 6.8$ then the standard error is

(*a*) 0.8 (*b*) 0.85

(*c*) 0.95 (*d*) none

25. For large samples, sample size *n* is.................

(*a*) < 30 (*b*) $n \le 30$

(*c*) $n > 30$ (*d*) $n \ge 30$

26. If $n < 30$, then $\frac{\bar{x} - \mu}{s / \sqrt{n}}$ is..................

(*a*) *z*-variate (*b*) F-variate

(*c*) *t*-variate (*d*) none

27. A sample of size 25 from a normal population has the mean $\bar{x} = 47.5$ and standard deviation $s = 8.4$ then $t =$ ....

(*a*) 1/3.21 (*b*) 1/2.96

(*c*) 2.96 (*d*) 3.21

28. There are two independent random samples of sizes 13 and 7 from normal population. given that the variance of first sample is 4 times the variance of the second sample then the value of F is................

(*a*) 16 (*b*) 4

(*c*) 2 (*d*) none

29. The number of different samples of size 2 that can be chosen from a finite population of size 25

(*a*) 30 (*b*) 35

(*c*) 100 (*d*) 300

30. The number of different samples of size 2 that can be chosen from a finite population of size 6 is.............

(*a*) 6 (*b*) 12

(*c*) 15 (*d*) none

31. For large samples, $\frac{\bar{x} - \mu}{s / \sqrt{n}}$ is..........

(*a*) *t*-variate (*b*) *z*-variate

(*c*) F-variate (*d*) χ^2-variate

32. In 16 one hour test runs, the gasoline consumption of an engine average 16.4 gallons with a Standard deviation of 2.1 gallons. If the average gasoline consumption of this engine prescribed by the company is 12 than $t =$ ....

(*a*) 8.38 (*b*) 3.88

(*c*) 8.83 (*d*) none

33. The square of a standard normal variate is called

(*a*) χ^2 variate (*b*) F-variate

(*c*) *t*-variate (*d*) none

34. If X is a standard normal variate then X^2 is a chi-square variate with

(*a*) *n* degree of freedom (*b*) 1 degree of freedom

(*c*) 2 degree of freedom (*d*) none

35. For N = 800, *n* = 200 then the finite population correction factor

(*a*) 0.8361 (*b*) 0.9799

(*c*) 0.7509 (*d*) none of these

36. A random sample of size 36 is taken from an infinite population having mean 63 and standard deviation 9. The probability that $\bar{x}$ will be more than 66.75 is....................

(*a*) 0.0062 (*b*) 0.062

(*c*) 0.62 (*d*) 0.4938

ANSWERS

1. (*b*)	**2.** (*a*)	**3.** (*c*)	**4.** (*d*)	**5.** (*a*)	**6.** (*b*)	**7.** (*c*)	**8.** (*d*)	**9.** (*a*)	**10.** (*a*)
11. (*c*)	**12.** (*c*)	**13.** (*c*)	**14.** (*c*)	**15.** (*a*)	**16.** (*d*)	**17.** (*a*)	**18.** (*b*)	**19.** (*d*)	**20.** (*b*)
21. (*d*)	**22.** (*a*)	**23.** (*a*)	**24.** (*b*)	**25.** (*d*)	**26.** (*c*)	**27.** (*d*)	**28.** (*b*)	**29.** (*d*)	**30.** (*c*)
31. (*b*)	**32.** (*a*)	**33.** (*a*)	**34.** (*b*)	**35.** (*c*)	**36.** (*a*)				

Chapter 5

ESTIMATIONS

5.0 INTRODUCTION

The purpose of most statistical investigation is to generalize from information contained in random samples about the populations from which the samples were obtained. In classical approach the methods of statistical inference we must choose a value of a parameter from a possible continuum of alternatives, while in the testing of hypothesis is we must decide whether to accept or reject one specified value or a set of values of a parameter.

Estimate: To find an unknown population parameter, a judgment or a statement is made which is called an estimate.

Estimator: The method or rule to determine an unknown population parameter is called an estimator.

Example: Sample mean is an estimator of population mean because sample mean is a method of determining the population mean.

Let $x_1, x_2, \ldots, x_n$ is a sample taken from a population whose probability density function is $f(x, t)$, where t is the unknown parameter we have to construct some functions, which are estimates of 0 to find the estimate in terms of sample values.

The estimation can be done in two ways

(*i*) Point estimation

(*ii*) Interval estimation

(*iii*) Bayesian estimation

5.1 POINT ESTIMATION

If from the observation in sample, a single value is calculated a sample estimate form an unknown population parameter. The procedure to find the parameter is called a point estimation.

Properties of good estimator: An estimator is said to be good estimator if it is

(*i*) unbiased

(*ii*) consistent

(*iii*) efficient and sufficient

Unbiased estimator: A statistic $\bar{\theta}$ is said to be an unbiased estimator or its value an unbiased estimate if and only if the mean of the sampling distribution of the estimator is equal to θ if $E(\bar{\theta}) = \theta$

Consistency: If an estimator $\bar{\theta}$ approaches the parameter θ, closer and closer as the sample size n increases then $\bar{\theta}$ is said to be consistent estimator of θ. In case of large samples consistency is a desirable property for an estimator to posses. However in small samples, consistency is of little importance.

Efficient and sufficient: If $\bar{\theta}_1$ and $\bar{\theta}_2$ be two estimators of a certain parameter θ, then $\bar{\theta}_1$ is said to be more efficient than $\bar{\theta}_2$ for all sample sizes if

$$\text{Var}(\bar{\theta}_1) < \text{Var}(\bar{\theta}_2)$$

An estimator is said to be sufficient for a parameter, if it contains all the information in the sample regarding the parameter.

Theorem: Let $x_1, x_2, \ldots, x_n$ be random sample drawn from a given population with mean μ and variance σ^2, show that sample mean $\bar{x}$ is an unbiased estimator of population mean μ and variance σ^2, show that sample mean $\bar{x}$ is an unbiased estimator of population mean μ *i.e.*, $E(\bar{x}) = \mu$.

Solution:

$$\bar{x} = \frac{1}{n}\sum_{1}^{\infty} x_i$$

$$E(\bar{x}) = E\left(\frac{1}{n}\sum x_i\right)$$

$$= \frac{1}{n}E\left(\sum_{1}^{\infty} x_i\right)$$

$$= \frac{1}{n} E\,(x_1 + x_2 + \ldots\ldots x_n)$$

$$= \frac{1}{n}(\mu + \mu + \ldots\ldots + \mu)$$

$$= \frac{1}{n}(n\,\mu)$$

$$= \mu$$

$$\therefore \quad E(\bar{x}) = \mu \;\therefore\; \bar{x} \text{ is an unbiased estimator of } \mu.$$

Maximum error: We know that for large 'n'

$Z = \dfrac{\bar{x} - \mu}{\sigma/\sqrt{n}}$ is a random variable having the standard normal distribution *i.e.*, $\left|\dfrac{\bar{x} - \mu}{\sigma/\sqrt{n}}\right| < Z_{\alpha/2}$

$$\Rightarrow \quad -Z_{\alpha/2} < \left|\frac{\bar{x} - \mu}{\sigma/\sqrt{n}}\right| < Z_{\alpha/2}$$

where $Z_{\alpha/2}$ is such that the normal curve area to its right equal to $\alpha/2$.

Let E be the maximum error of $|\bar{x} - \mu|$

$\therefore$ Maximum error $E = Z_{\alpha/2} \frac{\sigma}{\sqrt{n}}$.

Sample size. (Large sample)

$$E = Z_{\alpha/2} \cdot \frac{\sigma}{\sqrt{n}}$$

$$\Rightarrow \qquad \sqrt{n} = \left(\frac{Z_{\frac{\alpha}{2}} . \sigma}{E} \right)$$

$$\therefore \text{ Sample size} \qquad n = \left(\frac{Z_{\frac{\alpha}{2}} . \sigma}{E} \right)^2$$

where σ is standard deviation of population

E is maximum error

$Z_{\alpha/2}$ level

Small sample:

$$t = \frac{\bar{x} - \mu}{\sigma / \sqrt{n}}$$

$$\text{Maximum error} \qquad E = t_{\alpha/2} \cdot \frac{S}{\sqrt{n}}$$

With $(n-1)$ degree freedom

S is standard deviation of sample

n = sample size

5.2 INTERVAL ESTIMATION

In general a point estimator does not coincide with a true value of the parameter. So it is preferred to obtain arrange of values or an interval, in which the parameter value lies. Supported a is the probability that the interval does not include the parameter value.

$P(\theta, a, b) = P(a < \theta < b) = 1 - \alpha$, where θ is the parameter and a and b are random variables depending upon some estimator $\bar{\theta}$ of θ and are the lower and upper limits of the interval. The interval (a, b) is called the confidence interval with $(b-a)$ as the measure of precision of the estimate and $(1-\alpha)$ is the measure of reliability. Confidence interval for mean Let μ and σ be the mean and standard deviation of normal population

Let $Z = \frac{\bar{x} - \mu}{\sigma / \sqrt{n}}$ then the confidence limits for μ are

$$\bar{x} - Z_{\alpha/2} \frac{\sigma}{\sqrt{n}} < \mu < \bar{x} + Z_{\alpha/2} \frac{\sigma}{\sqrt{n}}$$

Small sample. Suppose a random sample of size n taken from a normal population then

$$t = \frac{\bar{x} - \mu}{S/\sqrt{n}}$$

The confidence limits for small sample are $\bar{x} - t_{\alpha/2}\frac{S}{\sqrt{n}} < \mu < \bar{x} + t_{\alpha/2}\frac{S}{\sqrt{n}}$

Confidence α	99%	98%	95%	90%
$Z_{\alpha/2}$	2.58	2.33	1.96	1.645

5.3 SOLVED PROBLEMS

Problem 1: In a study of maximal aerobic capacity 12 women were used as subjects and measurement that made blood plasma volume. The following data give their blood plasma volumes in leters: 3.15, 2.99, 2.77, 3.12, 2.45, 3.85, 2.99, 3.87, 4.06, 2.94, 3.53, 3.20. Assume that these are observations of a normally distributed random variable X that has mean fit and standard deviation σ.

(*i*) Give the value of a point estimate of μ

(*ii*) Determine point estimates of σ^2 and σ

(*iii*) Find a 90% confidence interval for μ.

Solution:

(*i*) $\mu = \dfrac{3.15 + 2.99 + 2.77 + ... + 3.20}{12} = 3.24$

(*ii*) $\sigma^2 = \dfrac{(3.15 - 3.24)^2 + ... + (3.20 - 3.24)^2}{12 - 1} = 0.2372$

$\sigma = 0.4870$

(*iii*) $\bar{x} - Z_{\alpha/2}\dfrac{\sigma}{\sqrt{n}} < \mu < \bar{x} + Z_{\alpha/2}\dfrac{\sigma}{\sqrt{n}}$

$\Rightarrow \bar{x} = \mu = 3.54$

$\sigma = 0.487$

$n = 12$

$3.24 \pm 1.645.\dfrac{0.4870}{3.464}$

3.24 ± 0.2312

$\therefore 3.0088 < \mu < 3.4712$

Problem 2: Assume that the yield per acre for a particular variety of soyabeans is $N(\mu, \sigma^2)$. For a random sample of 5 plots, the yield in bushes per acre were 37.4, 48.8, 46.9, 55 and 44. (*i*) Give a point estimate for μ, (*ii*) Find a 90% confidence interval for μ.

Solution: (*i*) $\bar{x} = \mu = \dfrac{37.4 + 48.8 + 46.9 + 55 + 44}{5} = 46.42$

$$\sigma^2 = \frac{(37.4 + 46.42)^2 + ...(44 - 46.42)^2}{5-1} = 41.68$$

(*ii*) $\bar{x} - Z_{\alpha/2}\frac{\sigma}{\sqrt{n}} < \mu < \bar{x} + Z_{\alpha/2}\cdot\frac{\sigma}{\sqrt{n}}$

$$46.42 \pm 1.645 \times \frac{6.4562}{\sqrt{5}}$$

$$46.42 \pm 4.7509$$

$$41.6681 < \mu < 51.1709$$

Problem 3: Let X and Y equai respectively, the blood volumes in millilitres for a male who is a paraplegic and participate in vigorous physical activities and a male who is ablebodies and participates in normal activities. Assume that X is $N(m_x, S_x^2)$ and Y is $N(\mu_y, \sigma_y^2)$ respectively. Using the following $n = 7$ observations of X: 1612, 1352, 1456, 1222, 1560, 1456, 1924 and $m = 10$ observations of Y: 1082, 1300, 1092, 1040, 910, 1248, 1092, 1040, 1092, 1288.

(*i*) Give a point estimate for $\mu_x - \mu_y$

(*ii*) Find a 95% confidence interval for $\mu_x - \mu_y$. Since the variance σ_x^2 and σ_y^2 might not be equal.

Solution:

(*i*) For x ; $n = 7$

$$\mu_x = \frac{1612 + 1352 + ... + 1924}{7}$$

$$= 1511.71$$

For Y; $n = 10$

$$\mu_y = \frac{1082 + 1300 + ...1288}{10} = 1118.4$$

Point estimate for μ_x = 1511.71

Point estimate for μ_y = 1118.4

Point estimate for $\mu_x - \mu_y$ = 1511.71 – 1118.4

$$\sigma_x^2 = \frac{(1612 - 1151.71)^2 + ... + (1924 - 1151.71)^2}{6} = 158956.12$$

$$\sigma_y^2 = \frac{(1082 - 1118.4)^2 + (1300 - 1118.4)^2 + ... + (1288 - 1118.4)^2}{9} = 13763.6$$

(*ii*) $\bar{x} - Z_{\alpha/2}\frac{\sigma_x}{\sqrt{n}} < \mu < \bar{x} + Z_{\alpha/2}\frac{\sigma_x}{\sqrt{n}} = 1511.71 \pm 1.96\frac{398.7}{\sqrt{7}}$

Confidence interval for X are $1807.06 < \mu < 1216.34$

Confidence interval for Y are $1118.4 \pm 1.96\ \frac{117.31}{\sqrt{10}} = (1045.69,\ 1191.11)$

Problem 4: What is the maximum error one each expect to make with probability 0.90 when using the mean of a random sample of size 64 to estimate the mean of population with $\sigma^2 = 2.56$?

Solution:

Sample size $n = 64$

$\sigma = 1.6$

At 90% confidence level $Z_{\alpha/2} = 1.645$

$\therefore$ Maximum error $E = 1.645 \times \dfrac{1.6}{\sqrt{64}} = 0.329.$

Problem 5: Inspecting ceramic tiles prior to their shipment, a quantity control engineer detects 2, 3, 6, 0, 4 and 9 defectives in six cartons. Each containing 144 tiles. What can we assert with a probability of 0.99 about the possible size of his error, if he uses the mean of this sample to estimate the true average number of defective per carton?

Solution:

Sample mean $\bar{x} = \dfrac{2+3+6+0+4+9}{6} = 4$

Variance $S^2 = 10$

$S = 3.16277$

For 99% $\alpha = 0.01 \Rightarrow t_{\sigma/2} = t_{0.005}$ with 5 degree freedom $= 4.032$

Maximum error $E = t_{\alpha/2}\cdot\dfrac{S}{\sqrt{n}} = 4.032 \times \dfrac{3.16277}{\sqrt{6}} = 5.20529.$

Problem 6: A random sample of size 16 values from a normal population showed a mean of 41.5 inches and the sum of the squares of deviations from means is 135 sq. inches. Find the maximum error with 95% confidence.

Solution:

Given $n = 16$

$\Sigma(x_i - \bar{x})^2 = 135$ $S^2 = \dfrac{135}{15} = 9 \Rightarrow$ $S = 3$

For 95% $\alpha = 0.05 \Rightarrow t_{\alpha/2} = t_{0.025}$ with 15 degree freedom is 2.131

$\therefore$ Maximum error $E = t_{\alpha/2}\cdot\dfrac{S}{\sqrt{n}} = 1.598.$

Problem 7: The mean and standard deviation of a population are 11,795 and 14,054 respectively, (*i*) The maximum error if $\bar{x} = 11{,}795$ and $n = 50$, (*ii*) Constructs 95% confidence interval for the true mean.

Solution:

(*i*) Sample size $n = 50$

Standard deviation of population $\sigma = 14054$

$Z_{\alpha/2}$ at 95% confidence is 1.96

Maximum error $E = \sigma_{\alpha/2}\cdot\dfrac{\sigma}{\sqrt{n}}$

$$= 1.96 \times \frac{14054}{\sqrt{50}}$$

$$= 3896.$$

(*ii*) 95% confidence interval

$$\bar{x} - Z_{\alpha/2}\frac{\sigma_x}{\sqrt{n}} < \mu < \bar{x} + Z_{\alpha/2}\frac{\sigma_x}{\sqrt{n}}$$

$$11795 - 3896 < \mu < 11795 + 3896$$

$$= 7899 < \mu < 15694.$$

Problem 8: It is desired to estimate the mean number of hours of continuous use until a certain computer will first require repairs. If it can be assumed that $\sigma = 48$ hours how large a sample be needed so that one will be able to assert with 90% confidence that the sample mean is off by at most 10 hours.

Solution: Maximum error $E = 10$

Standard deviation $\sigma = 48$

The value of Z at 90% confidence level is $Z_\alpha = 1.645$

$\therefore$ Sample size $n = \left(\frac{Z_\alpha \sigma}{E}\right)^2$

$$= \left(\frac{1.645 \times 48}{10}\right)^2$$

$$= 62.$$

Problem 9: A random sample of 100 teachers in a large metropolitan area revealed a mean weekly salary of Rs. 487 with standard deviation Rs. 48. With what degree of confidence can we assert that the average weekly salary of all teachers in the metropolitan area is between 472 to 502? **(April 2006)**

Solution: Population mean $\mu = 487$

Population standard deviation $\sigma = 48$

Sample size $n = 100$

$$Z = \frac{x - \mu}{\sigma / \sqrt{n}}$$

When $\bar{x} = 472 \Rightarrow$ $Z = \frac{472 - 487}{28/100} = -3.125$

When $\bar{x} = 502 \Rightarrow$ $Z = \frac{502 - 487}{48/100} = 3.125$

$$P(472 < x < 502) = P(-3.125 < Z < 3.125)$$

$$= 2P(0 < Z < 3.125)$$

$$= 0.9982.$$

Thus, we can ascertain with 99.82% confidence.

Problem 10: In a study of an automobile insurance a random sample of 80 body repair costs had a mean of Rs. 472.36 and standard deviation of Rs. 62.35. If $\bar{x}$ is used as a point estimate to the true average repair costs, with what confidence we can assert that the maximum error doesn't exceed Rs. 10?

Solution:

Maximum error $E = 10$

Standard deviation $\sigma = 62.35$

Sample size $n = 80$

Confidence $Z_\alpha = \frac{E\sqrt{n}}{\sigma} = \frac{10 \times \sqrt{80}}{62.35}$

$= 1.43$

From the table $\alpha = 84.72\%$

Problem 11: The mean of random sample is an unbiased estimator of the mean of the population.

Population = 3, 6, 9, 15, 27

(*a*) List of all possible samples of size 3 that can be taken without replacement from the finite population.

(*b*) Calculate the mean of each of the samples listed in (*a*) and assigning each sample a probability of 1/10, verify that the mean of these $\bar{x}$ is equal to 12. Which is equal to the mean of the population θ *i.e.* Prove that $\bar{x}$ is an unbiased estimate of θ.

Solution:

List of possible samples of size 3 taken from 3, 6, 9, 15, 17 without replacement

(${}^5C_3 = 10$ ∴ There are 10 samples)

= [(3, 6, 9), (3, 6, 15), (3, 6, 27), (6, 9, 15), (6, 9, 27), (3, 9, 15), (3, 9, 27),.(9, 15, 27), (6, 15, 27), (3, 15, 27)]

Mean of the population $\theta = \frac{3+6+9+15+27}{5} = 12$

Mean of samples = 6, 8, 12, 10, 14, 9, 13, 17, 16, 15

Probability assigned is 1/10 each.

$\bar{x}$	6	8	12	10	14	9	13	17	16	15
$p(\bar{x})$	$\frac{1}{10}$	$\frac{1}{10}$	$\frac{1}{10}$	$\frac{1}{10}$	$\frac{1}{10}$	$\frac{1}{10}$	$\frac{1}{10}$	$\frac{1}{10}$	$\frac{1}{10}$	$\frac{1}{10}$

$$E(\bar{x}) = 6 \times \frac{1}{10} + 8 \times \frac{1}{10} + 12 \times \frac{1}{10} + 10 \times \frac{1}{10} + 14 \times \frac{1}{10} + 9 \times \frac{1}{10} + 13 \times \frac{1}{10} + 17 \times \frac{1}{10} + 16 \times \frac{1}{10} + 15 \times \frac{1}{10}$$

$$= \frac{1}{10} \times 120 = 12 = \theta$$

$\therefore$ $E(\bar{x}) = \theta$

$\therefore$ $\bar{x}$ is an unbiased estimate of θ.

i.e., the mean of a random sample is an unbiased estimator of the mean of the population.

Problem 12: To estimate the average time it takes to assemble a certain computer component, the industrial engineer at an electronics firm timed 40 technicians in the performance of the task, getting a mean of 12.73 minutes and a standard deviation of 2.06 minutes.

(*a*) What can we say with 99% confidence about the maximum error if $\bar{x} = 12.73$ is used as point estimate of the actual average time required to do the job.

(*b*) Use the given data to construct a 99% confidence interval.

Solution:

(*a*) Maximum error $= Z_{\alpha/2} \dfrac{\sigma}{\sqrt{n}}$

n = sample size = 40

σ = standard deviation of the population = 2.06

At 99% confidence level $Z_{\alpha/2} = 2.575$

$\therefore$ Maximum error $E = 2.575 \times \dfrac{2.06}{\sqrt{40}} = 0.842$

(*b*) Confidence interval $= \bar{x} - Z_{\alpha/2}\dfrac{\sigma}{\sqrt{n}}, \bar{x} + Z_{\alpha/2}\dfrac{\sigma}{\sqrt{n}}$

$\bar{x} = 12.73$

$E = Z_{\alpha/2} \dfrac{\sigma}{\sqrt{n}} = 0.842$

$\therefore$ Confidence interval = (11.888, 13.572).

Problem 13: A random sample of 400 items is found to have mean of 82 and Standard deviation of 18. Find 95% confidence limits for the mean of the population from which the sample is drawn.

Solution:

σ = standard deviation=18

n = sample size=400

at 95% confidence level $Z_{\alpha/2} = 1.96$

$\bar{x}$ = sample mean = 82

Confidence limits are $\bar{x} - Z_{\alpha/2}\dfrac{\sigma}{\sqrt{n}}, < \mu < \bar{x} + Z_{\alpha/2}\dfrac{\sigma}{\sqrt{n}}$

$$Z_{\alpha/2}\frac{\sigma}{\sqrt{n}} = 1.96 \times \frac{18}{20} = 1.764$$

∴ Confidence limits are 80.236 and 83.764.

Problem 14: Find 95% confidence limits for the mean of a normality distributed population from which the following sample was taken 15, 17, 10, 18, 16, 9, 7, 11, 13, 14.

Solution:

$$\bar{x} = \frac{15+17+10+18+16+9+7+11+13+14}{10} = 13$$

$$S^2 = \frac{(x_i - x)^2}{n-1}$$

$$= \frac{(15-13)^2+(17-13)^2+.....+(13-13)^2+(14-13)^2}{2} = 13.3$$

$$t_{\alpha/2} = 2.26$$

$$S^2 = 40/3$$

$$t_{\alpha/2}\sqrt{\frac{s^2}{n}} = 2.26\sqrt{\frac{40}{3}\times\frac{1}{10}} = 2.26 \times \frac{2}{\sqrt{3}} = 2.6$$

Confidential limits are (10.4 and 15.6)

Problem 15: A sample of size 10 was taken from a population Standard deviation of sample is 0.03. Find the maximum error with 99% confidence.

Solution:

S = standard deviation = 0.03

Sample size n = 10

at 99% level with 9 degree freedom is $t_{\alpha/2} = 3.25$

∴ $$E = t_{\alpha/2}\frac{s}{\sqrt{n}} = 3.25 \times \frac{0.03}{3}$$

$$= 0.0325$$

Problem 16: Measurements of the weights of a random sample of 200 ball bearings made by a certain machine during one week showed a mean of 0.824 and a standard deviation of 0.042. Find 95% confidence limits for the mean weight of all the ball bearings.

Solution:

Confidence limits $\bar{x} - Z_{\alpha/2}\frac{\sigma}{\sqrt{n}}$ and $\bar{x} + Z_{\alpha/2}\frac{\sigma}{\sqrt{n}}$

$\bar{x}$ = mean of the sample = 0.824

$Z_{\alpha/2}$ = z value for 95% level (from tables) = 1.96

σ = standard deviation = 0.042

n = size of the sample = 200

$$Z_{\alpha/2}\frac{\sigma}{\sqrt{n}} = 1.96 \times \frac{0.042}{\sqrt{200}} = 0.0059$$

$$\bar{x} - Z_{\alpha/2}\frac{\sigma}{\sqrt{n}} = 0.824 - 0.0059 = 0.8181$$

$$\bar{x} + Z_{\alpha/2}\frac{\sigma}{\sqrt{n}} = 0.824 + 0.0059 = 0.8299$$

∴ Confidence limits are 0.8181 and 0.8299.

5.4 BAYESIAN ESTIMATION

To estimate the mean of a population, μ is treated as a random variable whose distribution is indicative of the strong feelings or assumption of a person about the possible value of μ. Let μ_0 and σ_0 be the mean and standard deviation of prior distribution.

Combining the feelings about the possible values of μ with direct sample evidence, the posterior distribution of μ is Bayesian estimation is approximated by normal distribution with

$$\mu_1 = \frac{n\bar{x}\sigma_0^2 + \mu_0\,\sigma^2}{n\sigma_0^2 + \sigma^2} \quad \text{and}$$

$$\sigma_1 = \sqrt{\frac{\sigma^2\,\sigma_0^2}{n\sigma_0^2 + \sigma^2}}$$

Here μ_1 and σ_1 are known as the mean and standard deviation of the posterior distribution. In the computation of μ_1 and σ_1 , σ^2 is assumed to be known when σ^2 is unknown, which is generally the vase, σ^2 is replaced by sample variance S^2 provided $n \geq 30$.

5.5 EXERCISE

1. What is the size of the sample required to estimate an unknown proportion to with a maximum error of 0.06 with at least 5% confidence given that standard deviation is 2?

(April 2006, Nov 2004)

2. Ten bearings made by a certain process have a mean diameter of 0.5060 cm with a standard deviation of 0.0040 cm. Assuming that the data may be looked upon as a random sample from a normal distribution, construct a 95% confidence of the bearings?

(April 2005, Nov 2004)

3. What is the maximum error one can expect to make with probability 0.90 when using the mean of random sample of size 64 to estimate the mean of a population with variance is 2.56? **(Nov 2004)**

4. A sample of 10 cam shafts intended for use in gasoline engines has an average eccentricity of 1.02 and a standard deviation of 0.044 inch. Assuming the data may be treated a random sample from a normal population, determine a 95% confidence interval for the actual mean eccentricity of the cam shaft. **(Nov 2004)**

5. If we can assert with 95% that the maximum error is 0.05 and $p = 0.2$, find the sample size. **(Nov 2004)**

6. A random sample of size 100 has a standard deviation of 5. What can you say about the maximum error with 95% confidence? **(April 2005)**

7. Assuming that $\sigma = 20$, how large a random sample be taken to assert with probability 0.95 that the sample mean will not differ from the true mean by more than 3 points? **(April 2005, Nov 2005)**

8. A random sample of size 81 was taken whose variance is 20.25 and mean is 32, construct 98% confidence interval. **(Nov 2005)**

9. A random sample of size 64 is taken to estimate the mean annual income of 1000 families and the mean and standard deviation are ground to be Rs. 6,300 and Rs. 9.50 respectively. Find a 95% confidence interval for the population mean μ.

10. A cubical die is thrown 900 times and a throw of 3 or 4 is observed 3240 times. Find the extreme limits between which the probability of a throw of 3 or 4 lies.

11. A random sample of size 9 from a normal population gave $\bar{x} = 15.8$ and $s^2 = 10.9$. Obtain the 99% confidence interval for μ.

12. A sample of size 64 and mean 60 was taken from a population whose standard deviation is 10. Find the 95% confidence interval for the mean. **(April 2007)**

13. A random sample of 100 items has a mean of 80 and standard deviation of 20. Construct 99% confidence interval for the mean.

14. Maximum error = 0.7. The standard deviation of the population is 5. What is the sample size for 0.05 level?

15. The management of a manufacturing firm wishes to determine the average time required to complete a certain manual operation. There should be 0.95 confidence that the error of estimate will not exceed $\frac{1}{2}$ minute. Find the sample size to be required if the standard deviation is 2 minutes.

16. If the error of estimates will not exceed 0.49 with 0.95 probability and the sample size is 64. What is the standard deviation of the population from which the sample has been drawn.

17. What is the standard deviation of the population, if the maximum error of estimate with probability 0.99 is 0.15 and the size of the sample is 100.

ANSWERS

1. 267	**2.** $0.5032 < \mu < 0.5088$	**3.** 0.329
4. $1.047 < \mu < 0.993$	**5.** 246	**6.** 0.98
7. 171	**8.** $30.835 < \mu < 33.165$	**9.** Rs. $6297.75 \le \mu \le$ Rs. 6302.25
10. $6.1 \le \mu \le 19.39$	**12.** 57.55, 62.45	**14.** 196
15. 61	**16.** 2	**17.** 5

5.6 QUIZ

1. If $x_1, x_2, \ldots\ldots, x_n$ is a random sample from a normal population $N(\mu, 1)$ then $\frac{1}{n}\sum_{i=1}^{n} x_1^2$ is an unbiased estimator of..................

(a) $\frac{\mu^2}{n}$ (b) μ^2

(c) $\mu^2 + 1$ (d) none of them

2. The sample variance $s^2 = \frac{1}{n-1} \sum_{i=1}^{n} (x_i - \bar{x})^2$ is an unbiased estimator of

(a) $\frac{n-1}{n}\sigma^2$ (b) $\frac{n}{n-1}\sigma^2$

(c) $\frac{1}{n}\sigma^2$ (d) σ^2

3. The sample variance $s^2 = \frac{1}{n-1} \sum_{i=1}^{n} (x_i - \bar{x})^2$ is consistent estimator of

(a) σ^2 (b) $\frac{1}{n}\sigma^2$

(c) $\frac{n-1}{n}\sigma^2$ (d) none

4. A random sample of size $n = 100$ is taken from a population with $\sigma = 5.1$. If the sample mean is $\bar{x} = 21.6$ then the 95% confidence interval for the population mean μ is ....

(a) (21.0, 22.6) (b) (21.6, 22.6)

(c) (20.6, 22.6) (d) none

5. A sample of size 9 from a normal population gave $\bar{x} = 15.8$ and $s^2 = 10.9$. The 99% confidence interval for μ is......................

(a) (12.21, 19.39) (b) (6.1, 19.39)

(c) (12.21, 18.35) (d) none of these

6. A random sample of size 81 was taken whose variance is 20.25 and mean 32 the 98% confidence interval from mean is ....

(a) (30.835, 32.165) (b) (30.265, 33.165)

(c) (32.165, 33.165) (d) 30.835, 33.165)

7. Given $n = 12$, $\bar{x} = 66.3$, $\sigma = 8.4$, $\alpha = 0.05$ then the 95% confidence interval for the true mean μ is

(a) (61.0, 70.5) (b) (61.0, 72)

(c) (61.0, 71.6) (d) none

8. If $n = 100$, $p = \frac{51}{100}$, $\alpha = 0.01$ then the 99% confidence interval for p is

(a) (0.42, 0.68) (b) (0.40, 0.70)

(c) (0.41, 0.69) (d) none

9. Two random samples of sizes 40 and 50 whose standard deviations are 10 and 15. Variance of the difference between means of mean of sampling distribution is............

(*a*) 1 (*b*) 3

(*c*) 5 (*d*) 7

10. Two sample whose variances $\sigma_x^2 = 64$ and $\sigma_y^2 = 25$ and sizes are 100 and 64 then

$\sigma^2_{\bar{x}-\bar{y}}$ = ...

(*a*) 99 (*b*) 103

(*c*) 105 (*d*) 79

11. Two samples of sizes 200 and 300 and variances 100 and 144 are taken. The standard error for $\mu_1 - \mu_2$ is

(*a*) 0.99 (*b*) 0.79

(*c*) 0.59 (*d*) 0.69

12. If $n = 16$, $s = 2$, $\bar{x} = 18$. Then 95% confidence interval for mean is................

(*a*) (12.235, 15.625) (*b*) (14.12, 17.235)

(*c*) (16.935, 19.065) (*d*) (20.125, 23.235)

13. If $n = 169$. The variance of the population is 25 and $\bar{x} = 50$ then 99% confidence interval for μ is...................

(*a*) (42, 44) (*b*) (49, 51)

(*c*) (36, 38) (*d*) (24, 26)

14. If $n = 144$, $\sigma = 4$ and $\bar{x} = 150$ then 95% confidence interval for μ is................

(*a*) (139.72, 140.25) (*b*) (149.35, 150.65)

(*c*) (172.1, 182.12) (*d*) (179.1, 182.25) -

15. The sample size is = 25. The maximum error with 0.95 probability is 0.1. Then standard deviation of the sample is...................

(*a*) 1.5 (*b*) 0.5

(*c*) 2.5 (*d*) 3.5

16. The size of the sample is 16 and standard deviation of population is 3. Then maximum error with probability 0.95 is...................

(*a*) 1.6 (*b*) 2.6

(*c*) 1.2 (*d*) 3.6

17. If $n = 9$, $s = 0.15$. The maximum error with 0.99 probability is...................

(*a*) 0.168 (*b*) 0.272

(*c*) 0.324 (*d*) 0.468

18. If the maximum error with probability 0.99 is 0.86 and if the sample size is 144, the variance of the population is...............

(*a*) 16 (*b*) 25

(*c*) 36 (*d*) 30

19. If the maximum error with 95% confidence we can assert is 0.1 and the size of the sample is 144 then σ = ...

(*a*) 0.54 (*b*) 0.61

(*c*) 0.45 (*d*) 0.72

20. If the maximum error with 99% confidence is 0.06 and $\sigma = 1$ then size of the sample is................

(*a*) 1649 (*b*) 1849

(*c*) 1549 (*d*) 1249

21. A random sample of size *n* is taken from a population whose standard deviation is 2 and maximum error with 95% confidence is 0.04, then n = ...

(*a*) 9604 (*b*) 1256

(*c*) 7624 (*d*) 2925

22. A random sample of size *n* is taken from a population with standard deviation 2. The maximum error of estimate with probability 0.95 is 0.05 then n = ...

(*a*) 79 (*b*) 65

(*c*) 89 (*d*) 56

23. A sample of size *n* is taken a population whose variance is 9. The maximum error of estimate for μ with probability 0.95 is 0.5 then n = ...

(*a*) 139 (*b*) 239

(*c*) 169 (*d*) 196

24. If the maximum error with probability 0.95 = 1.2. Standard deviation of the population is 10. Then sample size is.......................

(*a*) 267 (*b*) 129

(*c*) 225 (*d*) 169

25. If the variance of a population is 144, from which a sample of size 81 is taken with 99% confidence, the maximum error is......................

(*a*) 6.44 (*b*) 2.44

(*c*) 3.44 (*d*) 4.5

ANSWERS

1. (*c*)	**2.** (*d*)	**3.** (*a*)	**4.** (*a*)	**5.** (*b*)	**6.** (*d*)	**7.** (*c*)	**8.** (*a*)	**9.** (*d*)	**10.** (*b*)
11. (*a*)	**12.** (*c*)	**13.** (*b*)	**14.** (*b*)	**15.** (*c*)	**16.** (*a*)	**17.** (*a*)	**18.** (*a*)	**19.** (*b*)	**20.** (*b*)
21. (*a*)	**22.** (*a*)	**23.** (*a*)	**24.** (*a*)	**25.** (*c*)					

Chapter 6

TEST OF HYPOTHESIS

6.0 TEST OF SIGNIFICANCE

A very important aspect of the sampling theory is the study of the tests of significance, which enable us to decide on the basis of the sample result, if

(*i*) The deviation between the observed sample statistic and the hypothetical parameter value, or

(*ii*) The deviation between two independent sample statistic and; is significant or might be attributed to chance or the fluctuations of sampling.

Since, for large n, almost all the distributions can be approximated very closely by a normal probability curve, we use the Normal Test of Significance for large samples.

Hypothesis. Any verifiable statement is called Hypothesis. It a hypothesis concerned with the behaviour is random variable then that hypothesis is called statistical hypothesis. It is two types.

(*i*) Simple hypothesis

(*ii*) Composite hypothesis

Simple hypothesis. If the statistical hypothesis specifies the population completely then it is called simple hypothesis.

Example:

1. Let $x_1, x_2, x_3, \ldots\ldots, x_n$ be a random sample drawn from an normal population with mean μ and variance σ^2. The hypothesis is that $\mu = \mu_0$ and $\sigma = \mu_0^2$ then it is called simple hypothesis.

2. Let a ample of size 'n' drawn from a normal population with mean μ and variance σ^2 is unity. The hypothesis is that $\mu = \mu_0$ then it is also called as simple hypothesis.

Composite hypothesis. If the statistical hypothesis does not specify the population completely then it is called composite hypothesis.

Example:

1. Let $x_1, x_2, x_3, \ldots\ldots, x_n$ be a sample of normal mean μ and variance σ^2. The hypothesis is that mean $\mu = \mu_0$ must be composite hypothesis because here variance is not specified completely.

Null hypothesis. The statistical hypothesis that is tested under the assumptions that it is true is called null hypothesis. If it is tested on the basis of sample observations and is liable to be rejected as well as depending upon the outcomes of the statistical test is applied. Null hypothesis is denoted by H_0.

For example, in case of a single statistic, H_0 will be that the sample statistic does not differ significantly from the hypothetical paramete value and in the case of two statistic, H_0 will be that the sample statistics do not differ significantly.

Alternative hypothesis. Any hypothesis which is complementary to the null hypothesis is called an alternative hypothesis. OR The hypothesis that is statistician formulate as an alternative to the hypothesis is called alternative hypothesis is always denoted by H_1. In testing of hypothesis the null hypothesis is rejected is implies the alternative hypothesis is accepted.

For example, if we want to test the null hypothesis that the population has a specified mean μ_0 *i.e.*, $H_0 : \mu = \mu_0$ then the alternative hypothesis could be :

(*i*) $H_1 : \mu \neq \mu_0$ (*ii*) $H_1 : \mu < \mu_0$ (*iii*) $H_1 : \mu > \mu_0$

The alternative hypothesis in (*i*) is known as a two-tailed alternative and the alternatives in (*ii*) and (*iii*) are known as left-tailed and right-tailed alternatives respectively.

Critical Region. Let $x_1, x_2, x_3, \ldots\ldots, x_n$ be the sample observations denoted O. All the values of O will be aggregate of a sample and they constitute a space called the sample space which is denoted S.

Since the sample values $x_1, x_2, \ldots\ldots, x_n$ can be taken as a point in n-dimensional space, we specify some region of the n-dimensional space and see whether this lies with in this region or out side this region. We divide the whole sample space S into two disjoint parts W and S-W say W′. The null hypothesis H_0 is rejected if the observed sample points falls in W and if it falls in W′, we reject the alternative hypothesis H_1 and accept H_0. The region of rejection of H_0 when H_0 is true then that region is called critical region. The complement of critical region is called acceptance region. The critical region is denoted by 'α'.

Example:

Suppose if the test is based on sample size "two" the total sample space S is divided into two disjoint subsets W and W′. If the sample points drawn in the space W then reject H_0 otherwise H_0 accepted.

The probability of rejecting H_0 when H_0 is true

$\therefore$ α = Size of critical region = Probability of {Sample points falls in W/H_0 is true}

= Probability {rejected region / H_0 is true}

= Probability $\{x \in W/H_0$ is true$\}$ where x is sample point in W

Errors in sampling. The main objective in sampling theory is to draw valid inferences about the population parameters on the basis of the sample results. In practice we decide to accept or reject the lot after examining a sample from it. As such we are liable to commit the following two types of errors.

Type-I error. The error of rejecting H_0 when H_0 is true is called type-I error. The probability of type-I error is denoted by α.

$\therefore$ Probability (reject H_0 when it is true)

$$= P\text{ (reject } H_0/H) = \alpha$$

Type-II error. Accept H_0 when it is wrong (*i.e.,*) accepting a wrong hypothesis.

$\therefore$ P(accept H_0 when it is wrong)

$$= P(\text{accept } H_0/H_1) = \beta$$

$$\alpha = \text{size of type I error}$$

$$\beta = \text{size of type II error.}$$

	Accept H_0	*Reject H_0*
H_0 is true	Correct decision	Type I error
H_0 is false	Probability = $1 - \alpha$	Probability = α
	Type II error	Correct decision
	Probability = β	Probability = $1 - \beta$

Level of significance:

The maximum probability of committing type I error is called the level of significance and is denoted by the Greek letter α.

Some times the level of significance is also called size of the test.

Level of significance measures the amount of risk or error associated in making decisions. A is generally taken as 0.05 or 0.01.

Level of significance $\alpha = 0.01$ is used for high precision and $\alpha = 0.05$ for moderate precision.

Level of significance is also expressed as percentage. Thus level of significance $\alpha = 5\%$ means there are 5 chances in 100 that null hypothesis is rejected when it is true or one is 95% confident that a right decision is made.

Power of test. Power of test is the probability of rejecting H_0 given that a specific alternative is true. The power of a test can be computed as $1 - \beta$. Often different types of tests are compared by contrasting power properties.

Standard error. The standard deviation of the sampling distribution of the statistics is known as standard error and is denoted by SE. The standard error are some well known large samples are given below where n is sample size and σ^2 is variance of population and P is population proportion and $Q = 1 - P$. n_1, n_2 represents the sizes of two independent random samples then

S.No.	*Statistics*	*Standard error*
1.	Sample Mean $\bar{x}$	$\frac{\sigma}{\sqrt{n}}$ or $\sqrt{\frac{\sigma^2}{n}}$
2.	Sample proportion p	$\sqrt{\frac{PQ}{n}}$
3.	Difference of two sample means $(\bar{x}_1 - \bar{x}_2)$	$\sqrt{\frac{\sigma_1^2}{n_1} + \frac{\sigma_2^2}{n_2}}$
4.	Difference of two sample proportion $(p_1 - p_2)$	$\sqrt{\frac{P_1 Q_1}{n_1} + \frac{P_2 Q_2}{n_2}}$

One tailed test. A test of any statistical hypothesis where the alternative hypothesis is expressed by the symbol < or > is called one tailed test. Since the entire critical region lies in one tail of the distribution of the test statistic. The critical region for all alternative hypothesis containing the symbol > lies entirely in the right tail of the distribution while the critical region for an alternative hypothesis containing the symbol < lies entirely in the left trail. The symbol indicates the deviation when the critical region lies.

Examples:

A test for testing the mean of a population $H_0 : \mu = \mu_0$ against the alternative hypothesis $H_1 : \mu < \mu_0$ or $H_1 : \mu > \mu_0$ is called one tail test. If alternative hypothesis $H_1 : \mu > \mu_0$ then that test is known as right tailed test. $H_1 : \mu < \mu_0$ then that test is known as left tailed test.

Two tailed test. A test of any statistical hypothesis when the alternative is written with a symbol ≠ is called a two tailed test, since the critical region is split into two equal parts.

Example:

A test of testing the mean of a population $H_0 : \mu = \mu_0$ against the alternative hypothesis $H_1 : \mu \neq \mu_0$ is known as two tailed test.

Procedure for testing of hypothesis. The various steps in testing of a statistical hypothesis are summarized below.

1. **Null Hypothesis.** Set up the null hypothesis H_0.
2. **Alternative Hypothesis.** Set up the alternative hypothesis H_1: This will enable us to decide whether we have to use a one tailed (right or left) test or two tailed test.
3. **Level of Significance.** Choose the appropriate level of significance (α) depending on the reliability of the estimates and permissible risk. This is to be decided before sample is drawn *i.e.,* α is fixed in advance.
4. **Test Statistic.** Compute the test statistic $Z = \dfrac{t - E(t)}{S.E(t)}$ under the null hypothesis.
5. **Conclusion.** Compare the computed value of Z in step 4. With the significant value Z_α at a given level of significance 'α'.

If $|Z| < Z_\alpha$ *i.e.,* if calculated value of Z is less than Z_α then accept the null hypothesis H_0. If $|Z| > Z_\alpha$ *i.e.,* if calculated value of Z is greater than Z_α then reject the null hypothesis H_0 or accept alternative hypothesis H_1.

Critical values:

Critical values (Z_α)	*Level of significance*		
	1%	5%	10%
Two tailed test	$\lvert Z_\alpha \rvert = 2.58$	1.96	1.645
Right tailed test	2.33	1.645	1.28
Left tailed test	–2.33	–1.645	–1.28

6.1 TEST OF SIGNIFICANCE OF A SINGLE MEAN(LARGE SAMPLE)

Let x_i (I = 1, 2, 3, ..., n) is a random sample of size n taken from a population with mean μ and variance σ^2 then the sample mean is normally distributed.

$H_0 : \mu = \mu_0$ the population has the specified mean value or there is no significant difference between the sample mean ($\bar{x}$) and population mean (μ).

$$\therefore \qquad Z = \frac{\bar{x} - E(\bar{x})}{S.E.(\bar{x})}$$

$$= \frac{\bar{x} - \mu}{\sigma / \sqrt{n}}$$

Note:

1. If the population standard deviation σ is unknown then we use its sample variance.

$$\therefore \qquad Z = \frac{\bar{x} - \mu}{s / \sqrt{n}}$$

$$S^2 = \frac{1}{n-1} \sum (x_i - \bar{x})^2$$

2. At α% level of confidence limits for μ are

$$\bar{x} - Z_\alpha \frac{\sigma}{\sqrt{n}} < \mu < \bar{x} + Z_\alpha . \frac{\sigma}{\sqrt{n}}.$$

6.2 SOLVED PROBLEMS

Problem 1: A sample of 900 members has a mean 3.4 cms and standard deviation 2.61 cms. Is the sample population of mean 3.25 cms and deviation 2.61 cm? If the population is normal and its mean is unknown find the 95% and 98% confidence limits for true mean.

Solution:

Null hypothesis $H_0 : \mu = 3.25$

Alternative hypothesis $H_1 : \mu \neq 3.25$ (two-tailed test)

Level of significance: The significant value of Z at 5% level of significance for two tailed test is $Z_\alpha = 1.96$

Test statistics $Z = \dfrac{\bar{x} - \mu}{\sigma / \sqrt{n}}$

Sample size $n = 900$

Mean $\bar{x} = 3.4$

Population mean $\mu = 3.25$, $\sigma = 2.61$

$$\therefore \qquad Z = \frac{3.4 - 3.25}{2.61/\sqrt{900}}$$

$$= \frac{0.15}{0.087} = 1.724.$$

Conclusion: Since Z calculated value is less than Z table value at given level of significance. Hence accept null hypothesis H_0.

$\therefore$ The sample is drawn from a population with mean 3.25 and standard deviation 2.61.

Suppose μ is unknown then 95% confidence limits are

$$\bar{x} - 1.96 \frac{\sigma}{\sqrt{n}} < \mu < \bar{x} - 1.96 \frac{\sigma}{\sqrt{n}}$$

$$3.4 - 0.171 < \mu < 3.4 + 0.171$$

$$3.229 < \mu < 3.571$$

98% confidence limits are

$$\bar{x} - 2.33 \frac{\sigma}{\sqrt{n}} < \mu < \bar{x} + 2.33 \frac{\sigma}{\sqrt{n}}$$

$$3.4 - 0.2027 < \mu < 3.6027$$

Problem 2: An oceanographer wants to check whether the average depth of the ocean in a certain region is 57.4 fathoms, as had previously been recorded. What can be conclude at the level of significance 0.05 if soundings take at 40 random locations in the given region yielded a mean of 59.1 fathoms with a standard deviation of 5.2 fathoms? **(May 2004, Nov 2003, May 2003)**

Solution:

Null hypothesis $H_0 : \mu = 57.4$

Alternative hypothesis $H_1 : \mu \neq 57.4$ (Two tailed test)

Level of significance. The significance value of Z at 5% level of significance for two tailed test is $Z_\alpha = 1.96$

Test statistics: $\bar{x} = \dfrac{\bar{x} - \mu}{s/\sqrt{n}}$

Sample size $n = 40$

Sample mean $\bar{x} = 59.1$

Sample standard deviation $s = 5.2$

Population mean $\mu = 57.4$

$$Z = \frac{59.1 - 57.4}{5.2/\sqrt{40}} = 2.068$$

Conclusion: Since Z calculated value is greater than Z table value at given level of significance. Hence reject the null hypothesis H_0.

Problem 3: A company manufacturing electric bulbs claims that the average life of its bulbs is 1600 hours. The average life and standard deviation of a random sample of 100 such bulbs were 1570 hours and 120 hours respectively. Should we accept the claim of the company?

Solution:

Null hypothesis $H_0 : \mu = 1600$

Alternative hypothesis $H_1 : \mu \neq 1600$ (Two tailed test)

Level of significance: The significance value of Z at 5% level of significance for two tailed test is $Z_\alpha = 1.96$

Test statistics: $Z = \dfrac{\bar{x}-\mu}{s/\sqrt{n}}$

Sample size $n = 100$

Sample mean $\bar{x} = 1570$

Sample standard deviation $s = 120$

Population mean $\mu = 1600$

$$Z = \frac{1570 - 1600}{120/\sqrt{100}} = -2.5$$

Conclusion: Since $|Z|$ calculated value greater than Z table value at 5% level of significance. Hence reject the null hypothesis H_0.

Problem 4: According to the norms established for a mechanical aptitude test persons who are 18 years old have an average height of 73.2 with standard deviation of 8.6. If 45 randomly selected persons of that age average 76.7. Test the null hypothesis $\mu = 73.2$ against the alternative hypothesis $\mu > 73.2$ at 0.01 level of significance. **(Nov 2004)**

Solution:

Null hypothesis $H_0 : \mu = 73.2$

Alternative hypothesis $H_1 : \mu > 73.2$ (Right tailed test)

Level of significance: The significance value of Z at 1% level of significance for right tailed test is $Z_\alpha = 2.33$

Test statistics: $Z = \dfrac{\bar{x}-\mu}{\sigma/\sqrt{n}}$

Sample size $n = 45$

Sample mean $\bar{x} = 76.7$

Population mean $\mu = 73.2$

Population standard deviation $\sigma = 8.6$

$$Z = \frac{76.7 - 73.2}{8.6/\sqrt{45}} = 2.73$$

Conclusion: Since Z calculated value is greater than Z table value at 1% level of significance. Hence reject the null hypothesis H_0.

Problem 5: It is claimed that a random sample of 100 tyres with a mean life of 15269 km is drawn from a population of tyres which has a mean life of 15200 km and a standard deviation of 1248 km. Test the validity of this claim. **(April 2007, May 2006)**

Solution:

Null hypothesis H_0: $\mu = 15200$

Alternative hypothesis $H_1 : \mu \neq 15200$ (Two tailed test)

Level of significance: Since the value of Z at 5% level of significance for two tailed test is $Z_\alpha = 1.96$

Test statistics: $$Z = \frac{\bar{x}-\mu}{\sigma/\sqrt{n}}$$

Sample size $n = 100$

Sample mean $\bar{x} = 15269$

Population mean $\mu = 15200$

Population standard deviation $\sigma = 1248$

$$Z = \frac{15269 - 15200}{1248/\sqrt{100}} = 0.55$$

Conclusion: Since Z calculated value less than Z table value at 5% level of significance. Hence accept the null hypothesis H_0.

Hence the claim is true.

Problem 6: A lady stenographer claims that she can take dictation at the rate of 118 words per minute. Can we reject her claim on the basis of 100 trials in which she demonstrates a mean of 116 words and a standard deviation of 15 words?

(Nov. 2007, Aug. 2007, May 2007, Nov. 2006)

Solution:

Null hypothesis H_0: $\mu = 118$ *i.e.*, she dictate

Alternate hypothesis $H_1 : \mu \neq 118$ *i.e.*, she cannot dictate (Two tailed test)

Level of significance: Since the value of Z at 5% level of significance for two tailed test is

$$Z_\alpha = 1.96$$

Test statistics: $$Z = \frac{\bar{x}-\mu}{\sigma/\sqrt{n}}$$

Sample size $n = 100$

Sample mean $\bar{x} = 116$

Population mean $\mu = 118$

Population standard deviation $\sigma = 15$

$$Z = \frac{116 - 118}{15/\sqrt{100}} = -1.33$$

Conclusion: Since $|Z|$ calculated value less than Z table value at 5% level of significance. Hence accept the null hypothesis H_0.

Problem 7: A sample of 64 students has a mean weight of 70 kgs. Can this be regarded as a sample from a population with mean weight 65 kgs and standard deviation 25 kgs?

(Nov 2006, May 2006)

Solution:

Null hypothesis H_0: $\mu = 65$

Alternate hypothesis H_1: $\mu \neq 65$ (Two tailed test)

Level of significance: Since the value of Z at 5% level of significance for two tailed test is

$$Z_\alpha = 1.96$$

Test statistics:

$$Z = \frac{\bar{x} - \mu}{\sigma / \sqrt{n}}$$

Sample size $n = 64$

Sample mean $\bar{x} = 70$

Population mean $\mu = 65$

Population standard deviation $\sigma = 25$

$$Z = \frac{70-65}{25/\sqrt{64}}$$

$$= 4.48.$$

Conclusion: Since Z calculated value greater than Z table value at 5% level of significance. Hence reject the null hypothesis H_0.

Problem 8: The mean and the standard deviation of a population are 11795 and 14054 respectively. If $n = 50$, find 95% confidence interval for the mean. **(May 2006, May 2005)**

Solution:

Mean of the population $m = 11795$

Standard deviation of population $\sigma = 14054$

Sample size $n = 50$

Significant value at 95% confidence is $Z_\alpha = 1.96$

Confidence interval for the mean $= \left[\bar{x} - Z_\alpha \frac{\sigma}{\sqrt{n}}, \bar{x} + Z_\alpha \cdot \frac{\sigma}{\sqrt{n}}\right]$

$$\left[\bar{x} - 1.96 \frac{\sigma}{\sqrt{n}}, \bar{x} + 1.96 . \frac{\sigma}{\sqrt{n}}\right]$$

$$[11795 - 1.96 \times \frac{14054}{\sqrt{50}},\ 11795 + 1.96 \times \frac{14054}{\sqrt{50}}]$$

$$[7899.43, 15690.57].$$

Problem 9: In a random sample of 60 workers the average time taken by them to get to work is 33.8 minutes with a standard deviation of 6.1 minutes, can we reject the null hypothesis $\mu = 32.6$ minutes in favour of alternative hypothesis $\mu > 32.6$ at $\alpha = 0.025$ level of significance? **(Nov 2005)**

Solution:

Null hypothesis H_0: $\mu = 32.6$

Alternative hypothesis H_1: $\mu > 32.6$ (Right tailed test)

Level of significance: Since the value of Z at $\alpha = 0.025$ level of significance is $Z_\alpha = 2.58$

Test statistics: $Z = \dfrac{\bar{x} - \mu}{\sigma/\sqrt{n}}$

Sample size $n = 60$

Sample mean $\bar{x} = 33.8$

Population mean $\mu = 32.6$

Population standard deviation $\sigma = 6.1$

$$Z = \frac{33.8 - 32.6}{6.1\sqrt{60}} = 1.5237.$$

Conclusion: Since Z calculated value less than Z table value at $\alpha = 0.025$ level of significance. Hence accept the null hypothesis H_0.

Problem 10: A sample of 100 iron bars is said to be drawn from a large number of bars. Whose lengths are normally distributed with mean 4 feet and standard deviation 0.6 ft. If the sample mean is 4.2 feet can the sample be regarded as a truly random sample? **(Nov 2007)**

Solution:

Null hypothesis $H_0 : \mu = 4$ *i.e.,* the sample can be regarded as a truly random sample

Alternative hypothesis $H_1 : \mu \neq 4$ *i.e.,* sample cannot be regarded as random sample

Level of significance: Since the value of Z at 5% level of significance is $Z_\alpha = 1.96$

Test statistics: $Z = \dfrac{\bar{x} - \mu}{\sigma/\sqrt{n}}$

Sample size $n = 100$

Sample mean $\bar{x} = 4.2$

Population mean $\mu = 4$

Population standard deviation $\sigma = 0.6$

$$Z = \frac{4.2 - 4}{0.6 / \sqrt{200}} = 3.3$$

Conclusion: Since Z calculated value greater than Z table value at 5% level of significance. Hence reject the null hypothesis H_0.

6.3 TEST OF SIGNIFICANCE OF TWO MEANS (LARGE SAMPLES)

Let $\bar{x}_1$ be the mean of a sample size n_1 drawn form a population with mean μ_1 and variance σ_1^2 and let $\bar{x}_2$ be the mean of an independent random sample of size n_2 drawn from another population with mean μ_2 and variance σ_2^2 since n_1 and n_2 are large samples.

$$\bar{x}_1 \sim N\left(\mu_1, \sigma_1^2\right) \bar{x}_2 \sim N\left(m_2, \sigma_2^2\right)$$

Also $\bar{x}_1 - \bar{x}_2$ being the difference of two independent normal variates is also a normal varaite we want to test the null hypothesis $\mu_1 - \mu_2 = \delta$, where δ is specifically constant.

In testing the significance for the difference of two means, we shall consider the alternative hypothesis

$$\mu_1 - \mu_2 < 0 , \mu_1 - \mu_2 > 0 , \mu_1 - \mu_2 \neq \delta$$

Test statistic:
$$z = \frac{(\bar{x}_1 - \bar{x}_2) - E(\bar{x}_1 - \bar{x}_2)}{S.E. (\bar{x}_1 - \bar{x}_2)}$$

Case *i*: Under the null hypothesis $H_0 : \mu_1 = \mu_2$ *i.e.*, there is no significance difference between two means.

$$E(\bar{x}_1 - \bar{x}_2) = E(\bar{x}_1) - E(\bar{x}_2)$$

$$= \mu_1 - \mu_2 = 0$$

$$\therefore \quad Z = \frac{(\bar{x}_1 - \bar{x}_2)}{\sqrt{\frac{\sigma_1^2}{n_1} + \frac{\sigma_2^2}{n_2}}}$$

Case *ii*: If the null hypothesis $H_0 : \mu_1 - \mu_2 = \delta$ or $\mu_1 - \mu_2 < \delta$ or $\mu_1 - \mu_2 \neq 0$

$$\therefore \quad Z = \frac{\bar{x}_1 - \bar{x}_2 - \delta}{\sqrt{\frac{\sigma_1^2}{n_1} + \frac{\sigma_2^2}{n_2}}}$$

Critical region for testing

For $\mu_1 - \mu_2 = \delta$

Alternative hypothesis	Reject Null hypothesis if
$\mu_1 - \mu_2 < \delta$	$Z < -Z_\alpha$
$\mu_1 - \mu_2 > \delta$	$Z > Z_\alpha$
$\mu_1 - \mu_2 \neq \delta$	$Z < -Z_\alpha$
	$Z > Z_\alpha$
For $\mu_1 - \mu_2 = 0$	
$\mu_1 < \mu_2$	$Z \leq -Z_\alpha$
$\mu_1 > \mu_2$	$Z > Z_\alpha$
$\mu_1 \neq \mu_2$	$Z < -Z_\alpha$
	$Z > Z_\alpha$

Note:

1. If $\sigma_1^2 = \sigma_2^2 = \sigma^2$ *i.e.*, the samples have been drawn from the population with common variance σ^2, then

$$Z = \frac{\bar{x}_1 - \bar{x}_2}{\sigma\sqrt{\frac{1}{n_1} + \frac{1}{n_2}}}$$

2. If σ^2 is unknown then its estimates based on the sample variances used

$$\sigma^2 = \frac{n_1 S_1^2 + n_2 S_2^2}{n_1 + n_2}$$

3. If $\sigma_1^2 \neq \sigma_2^2$ and σ_1 , σ_2 are unknown then they are estimate the sample value

$$Z = \frac{\bar{x}_1 - \bar{x}_2}{\sqrt{\frac{S_1^2}{n_1} + \frac{S_2^2}{n_2}}}$$

6.4 SOLVED PROBLEMS

Problem 1: Samples of students where drawn from two universities and from their weights in kgm and standard deviation are calculated. Make a large sample test to test the significance of the difference between the means

	Mean	*S.D.*	*Sample size*	
University A :	55	10	400	
University B :	57	15	100	**(May 2007, Nov 2005)**

Solution:

Null hypothesis H_1: $\mu_1 = \mu_2$

Alternative hypothesis H_1: $\mu_1 \neq \mu_2$ (two tail)

Level of significance: The value of Z at 5% level of significance for two tailed test is $Z_\alpha = 1.96$

Test statistics: $$Z = \frac{\bar{x}_1 - \bar{x}_2}{\sqrt{\frac{s_1^2}{n_1} + \frac{s_2^2}{n_2}}}$$

1st sample mean $\bar{x}_1 = 55$

2nd sample mean $\bar{x}_2 = 57$

Variance of 1st sample $s_1^2 = 100$

$s_2^2 = 225$

$n_1 = 400 \quad n_2 = 100$

$$Z = \frac{55 - 57}{\sqrt{\frac{100}{400} + \frac{225}{100}}} = -1.26$$

Conclusion: Since $|Z|$ calculated value less than Z table value at 5% level of significance.

$\therefore$ Accept null hypothesis H_0

$\therefore$ There is no significance difference between the means.

Problem 2: A sample of heights of 6400 English men has mean of 67.85 inches and standard deviation 2.56 inches. While a sample of heights of 1600 Australians has a mean of 68.55 inches and standard deviation 2.52 inches. Do the data indicate that Australians are on the average, taller than English men ?

Solution:

Null hypothesis H_0: $\mu_1 = \mu_2$

Alternative hypothesis H_1: $\mu_1 < \mu_2$ (left tail test)

Level of significance : The value of Z at 5% level of significance for left tail test is

$$Z_\alpha = -1.645$$

Test statistic :

$$Z = \frac{\bar{x}_1 - \bar{x}_2}{\sqrt{\frac{s_1^2}{n_1} + \frac{s_2^2}{n_2}}}$$

$$\bar{x}_1 = 67.85 \qquad \bar{x}_2 = 68.55$$

$$n_1 = 6400 \qquad n_2 = 1600$$

$$\sigma_1 = 2.56 \qquad \sigma_2 = 2.52$$

$$Z = \frac{67.85 - 68.55}{\sqrt{\frac{6.554}{6400} + \frac{6.35}{1600}}}$$

Conclusion : Since $|Z|$ calculated value greater than Z table value at 5% level of significance

$\therefore$ Reject null hypothesis H_0

$\therefore$ Australians are taller than English men.

Problem 3: The means of two large samples of 1000 and 2000 are 67.5 and 68 inches respectively. Can the samples be regarded as drawn from the same population of standard deviation 2.5 inches? At 5% level.

Solution:

Null hypothesis H_0: $\mu_1 = \mu_2$

Alternative hypothesis $H_1 = \mu_1 \neq \mu_2$ (two-tailed)

Level of significance : The value of Z at 5% level of significance for two tailed

Test is $\quad Z_\alpha = 1.96$

Test statistics:

$$Z = \frac{\bar{x}_1 - \bar{x}_2}{\sigma\sqrt{\frac{1}{n_1} + \frac{1}{n_2}}}$$

$$n_1 = 1000 \qquad n_2 = 2000$$

$$\bar{x}_1 = 67.5 \qquad \bar{x}_2 = 68$$

$$\sigma = 2.5$$

$$\therefore \qquad Z = \frac{67.5 - 68}{2.5\sqrt{\frac{1}{1000} + \frac{1}{2000}}} = -5.1$$

Conclusion: Since |Z| calculated value greater than Z table value at 5% level of significance difference between the means.

Problem 4: A sample of 100 electric light bulbs produced by manufacturer A showed a mean life time of 1190 hours and a standard deviation of 90 hours. A sample of 75 bulbs produced by manufacturer B showed a mean life time of 1230 hours with a standard deviation of 120 hours. Is then difference between the mean life times of the two brands at 0.05 level?

Solution:

Null hypothesis H_0: $\mu_1 = \mu_2$

Alternative hypothesis H_1: $\mu_1 \neq \mu_2$ (two tailed test)

Level of significance: The value of Z at 5% level of significance for two tailed test is $Z_\alpha = 1.96$

Test statistic:
$$= \frac{\bar{x}_1 - \bar{x}_2}{\sqrt{\frac{s_1^2}{n_1} + \frac{s_2^2}{n_2}}}$$

$$n_1 = 100 \qquad n_2 = 75$$

$$\bar{x}_1 = 1190 \qquad \bar{x}_2 = 1230$$

$$s_1 = 90 \qquad s_2 = 120$$

$$Z = \frac{1190 - 1230}{\sqrt{\frac{(90)^2}{100} + \frac{(120)^2}{75}}} = -2.42$$

Conclusion: Since 121 calculated value is greater than Z table value at 5% level of significance

∴ Reject null hypothesis H_0

∴ There is a significance difference the two means.

Problem 5: A random sample of 1000 men from Northern India gives their mean wage to be Rs. 30 per day with a standard deviation of Rs. 1.50. A sample of 1500 men from southern India gives a mean wage of Rs. 2. Discuss whether the mean rate of wages varies between the two regions.

Solution:

Null hypothesis $\mu_1 = \mu_2$

Alternative hypothesis $\mu_1 \neq \mu_2$ (two tail test)

Level of significance: Since the value of Z at 5% level of significance is $Z_\alpha = 1.96$

Test statistic: $$Z = \frac{\bar{x}_1 - \bar{x}_2}{\sqrt{\frac{\sigma_1^2}{n_1} + \frac{\sigma_2^2}{n_2}}}$$

$\bar{x}_1$ = mean of the 1st sample = Rs. 30

$\bar{x}_2$ = mean of the second sample = Rs. 32

σ_1^2 = variance of the first sample = 2.25

σ_2^2 = variance of the second sample = –4

n_1 = size of the first sample = 1000

n_2 = size of the second sample = 1500

$$Z = \frac{30 - 32}{\sqrt{\frac{2.25}{1000} + \frac{4}{1500}}} = -28.5$$

Conclusion: Since |Z| calculated value greater than Z table value at 5% level of significance. Hence reject the null hypothesis H_0.

Problem 6: A company claims that its light bulbs are superior to those of its main competitor. If a study showed that a sample of $n_1 = 40$ of its bulbs that a mean life time of 647 hrs of continuous use with a standard deviation of 27 hrs; while a sample of $n_2 = 40$ bulbs made by its main competitor had a mean life time of 638 hrs of continuous use with a standard deviation of 31 hrs. Does this substantiate the claim at the 0.05 level of significance.

Solution:

Null hypothesis $\mu_1 - \mu_2 = 0$

Alternative hypothesis $\mu_1 - \mu_2 > 0$

Level of significance: Since the value of Z at 5% level of significance is $Z_\alpha = 1.96$

Test significance: $$Z = \frac{\bar{x}_1 - \bar{x}_2}{\sqrt{\frac{s_1^2}{n_1} + \frac{s_2^2}{n_2}}}$$

$\bar{x}_1$ = mean of the 1st sample = 647

$\bar{x}_2$ = mean of the second sample = 638

s_1^2 = variance of the first sample = 729

s_2^2 = variance of the second sample = 961

n_1 = size of the first sample = 40

n_2 = size of the second sample = 40

$$Z = \frac{647 - 638}{\sqrt{\frac{729}{40} + \frac{961}{40}}} = 1.38$$

Conclusion: Since Z calculated value less than Z table value at 5% level of significance. Hence accept the null hypothesis H_0.

Therefore observed difference between the two samples is not significant.

Problem 7: The average income of persons was Rs.210 with standard deviation of Rs.10 in sample of 100 people of a city. For another sample of 150 persons, the average income was Rs.220 with standard deviation of Rs.12. The standard deviation of the people of the city was Rs.11. Test whether there is any significant difference between the average incomes of the localities.

Solution:

Null hypothesis $H_0 : \mu_1 = \mu_2$

Alternative hypothesis $H_1 : \mu_1 \neq \mu_2$ (two tailed test)

Level of significance: Since the value of Z at 5% level of significance for two tailed test is $Z_\alpha = 1.96$

Test statistics: $$Z = \frac{\bar{x}_1 - \bar{x}_2}{\sqrt{\frac{s_1^2}{n_1} + \frac{s_2^2}{n_2}}}$$

$\bar{x}_1$ = mean of the 1st sample = 210

$\bar{x}_2$ = mean of the second sample = 220

s_1^2 = variance of the first sample = 100

s_2^2 = variance of the second sample = 144

n_1 = size of the first sample = 100

n_2 = size of the second sample = 150

$$Z = \frac{210 - 220}{\sqrt{\frac{100}{100} + \frac{144}{150}}} = -7.142$$

Conclusion: Since $|Z|$ calculated value greater than Z table value at 5% level of significance. Hence reject the null hypothesis H_0.

∴ There is significant difference between the average incomes of the localities.

6.5 ESTIMATION OF PROPORTIONS

Test of significance of single proportion: If x is the number of successes in 'n' independent trials with constant probability P of success for each trial, then

$E(x) = nP$ $V(x) = nPQ$, where $Q = 1 - P$

Test statistics $$Z = \frac{x - E(x)}{\sqrt{V(x)}} = \frac{x - nP}{\sqrt{nPQ}}$$

Note:

1. Let in a sample size of n, x be the number of persons possessing the given attribute then the observed proportion of success is $\frac{x}{n} = p$

$$\therefore \quad E(p) = E\left(\frac{x}{n}\right) = \frac{1}{n} E(x) = \frac{1}{n} \cdot \times nP = \frac{P}{n}$$

$$V(p) = V\left(\frac{x}{n}\right) = \frac{1}{n^2} V(x) = \frac{1}{n^2} \times nPQ = \frac{PQ}{n}$$

$\therefore$ Test statistics $$Z = \frac{p - E(p)}{\sqrt{V(p)}} = \frac{p - P}{\sqrt{PQ/n}}$$

$\therefore$ Standard error of proportion $= \sqrt{\frac{PQ}{n}}$

2. α% confidence limits of observed proportion success are $P \pm Z_\alpha \sqrt{PQ/n}$
3. If P is not known then using sample proportion

   $\therefore$ The limits are $p \pm Z_\alpha \sqrt{pq/n}$.

6.6 SOLVED PROBLEMS

Problem 1: A die is thrown 9000 times and throw of 3 or 4 observed 3240 times. Show that the die cannot be regarded as an unbiased one and find the limits between which the probability of a throw of 3 or 4 lies.

Solution:

Null hypothesis H_0 : The die is an unbiased one

Alternative hypothesis $H_1 : P \neq \frac{1}{3}$ *i.e.* die is not unbiased one

Level of significance: The significant value of Z is 3

Test statistics: $$Z = \frac{x - nP}{\sqrt{nPQ}}$$

Sample of size $n = 9000$

Number of success $x = 3240$

Probability of getting 3 or 4 is $P = \frac{1}{3}$

$$\therefore \quad Z = \frac{3240 - 9000 \times 1/3}{\sqrt{9000 \times 1/3 \times 2/3}} = 5.36$$

Conclusion: Since Z calculated value is greater than Z table value at given level of significance.

$\therefore$ Reject the null hypothesis H_0

Hence die is not unbiased.

Limits: $p = \frac{x}{n} = \frac{3240}{9000} = 0.36 \Rightarrow q = 1 - p = 1 - 0.36 = 0.64$

Probable limits for observed proportion of successes are $P \pm Z_\alpha \sqrt{\frac{pq}{n}}$

$$= 0.36 \pm 0.015$$

$$= 0.375,\ 0.345.$$

Problem 2: A random sample of 500 pineapples was taken from a large consignment and 65 were found to be bad. Show that the standard error of the proportion of bad and in a sample of this size is 0.015 and deduce that the percentage of bad pineapples in the consignment almost certainly lies between 8.5 and 17.5.

Solution:

Given Number of pineapples $n = 500$

Number of bad pineapples $X = 65$

Proportion of bad pineapples $p = x/n = 0.13$

$$q = 0.87$$

Standard error of proportion $\sqrt{\frac{pq}{n}} = \sqrt{\frac{0.13 \times 0.87}{500}} = 0.015$

Probable limits for proportion of bad pineapples are $p \pm Z \sqrt{\frac{pq}{n}} = 0.85$ and 0.175

Percentage limits of bad pineapples are 8.5 and 17.5.

Problem 3: A random sample of 100 apples was taken and out of which 15 were found to be bad. What can we say that 95% confidence about the maximum error of proportions.

Solution:

Sample size $n = 100$

Number of bad apples $X = 15$

Proportion of bad apples $p = 0.015$

$$q = 0.85$$

$$Z_{\alpha/2} = 1.96$$

Maximum error $E = Z_{\alpha/2} \sqrt{\frac{pq}{n}} = 1.96 \sqrt{\frac{0.15 \times 0.85}{100}} = 0.07.$

Problem 4: In a random sample 400 industrial accident it was found that 231 were due to least partially to unsafe working conditions. Construct a 99% confidence interval and 95% confidence about the maximum error if we use the sample proportion to estimate the correspond true proportion. **(Feb 2007, May 2006)**

Solution:

$$n = 400$$
$$X = 231$$
$$p = X/n = 0.578$$
$$q = 1 - p = 0.422$$
$$Z_{\alpha/2} = 2.58$$

99% confidence interval for proportion

$$p - Z_{\alpha/2}\sqrt{\frac{pq}{n}} < P < p + Z_{\alpha/2}\sqrt{\frac{pq}{n}} = (0.514, 0.642)$$

Maximum error $\quad E = Z_{\alpha/2}\sqrt{\frac{pq}{n}}$

$$= 1.96\sqrt{\frac{0.578 \times 0.42}{400}} = 0.048.$$

Problem 5: If we can assert with 95% that the maximum error is 0.05 and P is given as 0.2. Find the size of the sample.

Solution:

Sample size $\quad = n = \left(\frac{Z_{\alpha/2}}{E}\right)^2 PQ$

$$E = 0.05$$
$$P = 0.2 \Rightarrow Q = 0.8$$
$$Z_{\alpha/2} = 1.96$$
$$n = \left(\frac{1.96}{0.05}\right)^2 (0.2)(0.8) = 244.$$

Problem 6: A coin is tossed 10,000 times and it turns up head 5195 times. Discuss whether the coin may be regarded as unbiased one.

Solution:

Null hypothesis H_0: coin is unbiased

Alternative hypothesis H_1: coin is biased one

Test statistic: $\quad Z = \frac{x - nP}{\sqrt{nPQ}}$

$$N = 10000$$

No. of successes $X = 5195$

Probability of getting head $p = 1/2$

$$q = 1/2$$

$$Z = \frac{5195 - 10000 \times \frac{1}{2}}{\sqrt{10000 \times \frac{1}{2} \times \frac{1}{2}}} = 3.9$$

Conclusion: Since Z calculated value is greater than 3.

∴ Reject null hypothesis H_0

Hence coin is biased one

Problem 7: A manufacturer claims that any 4% of his products are defective. A random sample of 500 was taken among which 100 were defective. Test the hypothesis at 0.05 levels.

(Nov. 2005)

Solution:

Null hypothesis H_0: P = 0.04

Alternative hypothesis H_1: P > 0.04 (right tailed test)

Level of significance: The value of Z at 5% level of significance for right tailed test is $Z_\alpha = 1.96$

Test statistic $Z = \frac{p - P}{\sqrt{PQ/n}}$

Sample size $n = 500$

Number of defective sample $X = 100$

Proportion of defective sample $P = X/n = 0.2$

$$P = 0.04 \qquad Q = 0.96$$

$$Z = \frac{0.2 - 0.04}{\sqrt{0.04 \times 0.96/500}} = 18.25$$

Conclusion: Since Z calculated value is greater than Z table value at 5% level

∴ Reject null hypothesis H_0.

Problem 8: A social worker believes that fewer than 25% of the couples in a certain area are used any form of birth control. A random sample of 120 couples was contacted. Twenty of them said that they have used. Test the belief of social worker at 0.05 level.

Solution:

Null hypothesis H_0: P = 0.25

Alternative hypothesis H_1: P < 0.25 (left tailed test)

Level of significance: The value of Z at 5% level of significance for left tail test is $Z_\alpha = -1.645$

Test statistic: $Z = \frac{p - P}{\sqrt{PQ/n}}$

Sample size $n = 120$

Number of couples who used birth control 20 = X

Proportion $p = X/n = 20/120 = 0.167$

$$P = 0.25 \qquad Q = 0.75$$

$$Z = \frac{0.167 - 0.25}{\sqrt{0.25 \times 0.75/120}} = -2.05$$

Conclusion: Since $|z|$ calculated value is greater than Z table value at 5% level. Reject the null hypothesis H_0.

Problem 9: In a large consignment of oranges a random sample of 64 oranges revealed that 14 oranges were bad. Is it reasonable to ensure that 20% of the oranges are bad?

(Nov 2007, Aug 2007, May 2007, Nov 2006)

Solution:

Null hypothesis $H_0 : p = 0.2$

Alternative hypothesis $H_1: p \neq 0.2$ (Two tailed test)

Level of significance: Since the value of Z at 5% level of significance for two tailed test is $Z_\alpha = 1.96$

Test statistics: $$Z = \frac{x - np}{\sqrt{npq}}$$

Let the number of bad oranges is $x = 14$

Sample of size $n = 64$

$$Z = \frac{14 - 64 \times 0.2}{\sqrt{64 \times 0.2 \times 0.8}} = 0.375$$

Conclusion: Since Z calculated value is less than Z table value at 5% level of significance. Hence accept the null hypothesis H_0.

Problem 10: In a random sample of 125 cola drinkers, 68 said they prefer thumsup to Pepsi. Test the null hypothesis P = 0.5 against the alternate hypothesis P > 0.5.

(Aug 2007, May 2007, May 2006)

Solution:

Null hypothesis $H_0 : P = 0.5$

Alternative hypothesis $H_1 : P > 0.5$ (Right tailed test)

Level of significance: Since the value of Z at 5% level of significance for right tailed test is $Z_\alpha = 1.96$

Test statistics: $$Z = \frac{p - P}{\sqrt{PQ/n}}$$

Number of people who prefer thums up to Pepsi is $x = 68$

Sample size $n = 125$

Sample proportion $p = \frac{x}{n} = \frac{68}{125} = 0.544$

Population proportion P = 0.5

$$\Rightarrow \quad Q = 1 - P$$

$$= 1 - 0.5 = 0.5$$

$$Z = \frac{0.544 - 0.5}{\sqrt{0.5 \times 0.5/125}} = 0.9838$$

Conclusion: Since Z calculated value less than Z table value at 5% level of significance. Hence accept the null hypothesis H_0.

Problem 11: Among 900 people in a state 90 are found to be chapatti eaters. Construct 99% confidence interval for the true proportion. **(Nov 2007, Feb 2007, Nov 2006, May 2005)**

Solution:

Sample size $n = 900$

Number of chapatti eaters $x = 90$

Sample proportion $p = \frac{x}{n} = \frac{90}{900} = 0.1$

$\Rightarrow$ $q = 1 - p$

$= 1 - 0.1 = 0.9$

The value of Z at 99% confidence level is $Z_\alpha = 2.575$

Confidence interval is,

$$p - Z_\alpha \sqrt{\frac{pq}{n}} < P < p + Z_\alpha \sqrt{\frac{pq}{n}}$$

$$Z_\alpha \sqrt{\frac{pq}{n}} = 2.575 \times \sqrt{\frac{0.1 \times 0.9}{900}} = 0.02575$$

$\therefore\ 0.1 - 0.02575 < P < 0.1 + 0.02575$

$0.07425 < P < 0.12575.$

Problem 12: A random sample of 1200 apples was taken from a large consignment and found that 10% of them are bad. The supplier claims that only 2% are bad. Test his claim at 95% level. **(Nov 2007, Feb 2007, Nov 2006, May 2005)**

Solution:

Null hypothesis H_0: $P = 0.02$

Alternative hypothesis H_1 : $P \neq 0.02$ (Two tailed test)

Level of significance: Since the value of Z at 5% level of significance for two tailed test is $Z_\alpha = 1.96$

Test statistics: $Z = \frac{x - nP}{\sqrt{nPQ}}$

Number of bad apples $x = 10$

Sample size $n = 1200$

$P = 0.02 \Rightarrow Q = 1 - P \Rightarrow 1 - 0.02 = 0.08$

$$Z = \frac{10 - 1200 \times 0.02}{\sqrt{1200 \times 0.02 \times 0.08}} = 2.886$$

Conclusion: Since $|Z|$ calculated value is greater than Z table value at given level of significance. Hence reject the null hypothesis H_0.

Problem 13: A manufacturer of electric bulbs claims that the percentage of defectives in his product doesn't exceed 6. A sample of 40 bulbs is found to contain 5 defectives. Would you consider the claim justified? **(Nov 2007, Feb 2007)**

Solution:

Null hypothesis H_0: percentage of defective bulbs

$$P = 6/100 = 0.06$$

Alternative hypothesis H_1: $P > 0.06$ (right tailed test)

Level of significance : Since the value of Z at 5% level of significance for right tailed test is $Z_\alpha = 1.645$

Test statistic: Let X be the discrete binomial random variable which is the number of defective bulbs with $P = 0.06$ and $n = 20$

Computation: $X = 5$

$$nP_0 = 20 \times 0.06 = 1.2$$

Since $X = 5 > nP_0 = 1.2$, then

$$\begin{aligned} P &= \text{Probability } (X \geq x \text{ when } P = P_0) \\ &= P(X \geq 5 \text{ with } P = 0.06) \\ &= 1 - P(X \leq 4) = 1 - \sum_{X=0}^{4} b(X; 20, 0.06) \\ &= 1 - \sum_{X=0}^{4} 20C_X (0.06)^X (0.94)^{20-X} \\ &= 1 - (0.29 + 0.37 + 0.22 + 0.08 + 0.02) \\ &= 1 - 0.98 \\ P &= 0.02 \end{aligned}$$

Conclusion: Since Z calculated value is less than Z table value at 0.05 level of significance. Hence accept H_0 *i.e.*, $P = 0.06$

Problem 14: In a study designed to investigate whether certain detonators used with explosives in coal mining meet the requirement that at least 90% will ignite the explosive when charged it is found that 174 of 200 detonators function properly. Find a 95% confidence interval for the true proportion? **(Feb 2007)**

Solution:

P – probability of success $= 174/200 = 0.87$

$$Q = 1 - P = 1 - 0.87 = 0.13$$

$$n = 200,\ Z_{\alpha/2} = 1.96 \text{ (for 95\%)}$$

Formula is given by

$$P - Z_{\alpha/2}\sqrt{PQ/n},\ P + Z_{\alpha/2}\sqrt{PQ/n}$$

$$= Z_{\alpha/2}\sqrt{PQ/n}$$

$$= 1.96\sqrt{0.87 \times 0.13/200} = 1.96 \times 0.0237 = 0.046452$$

$\therefore$ Confidence interval is $(0.87 - 0.046452,\ 0.87 + 0.046452)$

$$= (0.823548,\ 0.916452).$$

Problem 15: A die thrown 256 times an even digit turns up 150 times. Can we say that the die is unbiased. **(Nov 2005)**

Solution:

Null hypothesis : H_0: The die is unbiased

Alternative hypothesis : H_1: The die is biased

Level of significance: Since the value of Z at 0.05 level of significance for two type test is $Z_\alpha = 1.96$

Test statistics:

$$Z = \frac{X - np}{\sqrt{npq}}$$

$$X = \text{number of success} = 150$$

$$n = 256$$

$$P = \text{probability of getting an even digit}$$

$$(2 \text{ or } 4 \text{ or } 6) = 1/2$$

$$Q = 1 - P = 1/2$$

$$Z = \frac{150 - 256 \times \frac{1}{2}}{\sqrt{256 \times \frac{1}{2} \times \frac{1}{2}}} = 2.75$$

Conclusion: Since Z calculated value is greater than Z table value at 0.05 level. Hence reject the null hypothesis H_0. Therefore the die is biased.

6.7 TEST OF SIGNIFICANCE FOR DIFFERENCE OF PROPORTIONS

Let X_1, X_2 be the number of persons the given attribute in a random samples of sizes n_1 and n_2 taken from two populations respectively. Then sample proportions are given by

$$p_1 = \frac{X_1}{n_1} \qquad p_2 = \frac{X_2}{n_2}$$

If P_1 and P_2 are population proportional then

$$E(p_1) = P_1 \qquad E(p_2) = P_2$$

$$V(p_1) = \frac{P_1 Q_1}{n_1} \qquad V(p_2) = \frac{P_2 Q_2}{n_2}$$

$$\therefore \quad Z = \frac{(p_1 - p_2) - E(p_1 - p_2)}{\sqrt{V(p_1 - p_2)}}$$

Under the null hypothesis H_0: $P_1 = P_2$

i.e., there is no difference between proportions

$$\therefore \quad E(p_1 - p_2) = E(p_1) - E(p_2) = P_1 - P_2 = 0$$

$$V(p_1 - p_2) = V(p_1) + V(p_2) = \frac{P_1 Q_1}{n_1} + \frac{P_2 Q_2}{n_2} = PQ\left(\frac{1}{n_1} + \frac{1}{n_2}\right)$$

∴ Test statistic for difference of proportion

$$Z = \frac{p_1 - p_2}{\sqrt{PQ\left(\frac{1}{n_1} + \frac{1}{n_2}\right)}}$$

If population proportion P is not known

$$P = \frac{X_1 + X_2}{n_1 + n_2}$$

Confidence interval for difference of proportions

$$p_1 - p_2 \pm Z_{\alpha/2}\sqrt{PQ\left(\frac{1}{n_1} + \frac{1}{n_2}\right)}$$

Standard error $\quad p_1 - p_2 = \sqrt{PQ\left(\frac{1}{n_1} + \frac{1}{n_2}\right)}$

Maximum error $\quad = Z_{\alpha/2}\sqrt{PQ\left(\frac{1}{n_1} + \frac{1}{n_2}\right)}$

Note: Suppose of the population proportions P_1 and P_2 are given then the test for difference in population proportions

$$Z = \frac{p_1 - p_2}{\sqrt{\left(\frac{P_1 Q_1}{n_1} + \frac{P_2 Q_2}{n_2}\right)}}$$

6.8 SOLVED PROBLEMS

Problem 1: Random samples of 400 men and 600 women were asked whether they would like to have a flyover near their residence. 200 men and 325 women were in favour of the proposal. Test the hypothesis that proportions of men and women in favour of the proposal are same against that they are not at 5% level.

Solution:

Null hypothesis H_0: $P_1 = P_2$

i.e., There is no significance difference between the opinion of men and women

Alternative hypothesis H_1: $P_1 \neq P_2$ (two tailed test)

Level of significance: The value of Z at 5% level of significance for the tailed test is $Z_\alpha = 1.96$

Test statistic: $$Z = \frac{p_1 - p_2}{\sqrt{PQ\left(\frac{1}{n_1} + \frac{1}{n_2}\right)}}$$

Given $n_1 = 400 \quad n_2 = 600$

$X_1 = 200 \quad X_2 = 325$

$$p_1 = \frac{X_1}{n_1} = 0.5 \quad p_2 = \frac{X_2}{n_2} = 0.541$$

$$P = \frac{X_1 + X_2}{n_1 + n_2}$$

$$= \frac{200 + 325}{400 + 600} = 0.525$$

$$Q = 1 - P = 0.475$$

$\therefore$ $$Z = \frac{0.5 - 0.541}{\sqrt{0.525 \times 0.475\left(\frac{1}{400} + \frac{1}{600}\right)}} = -1.269$$

Conclusion: Since $|Z|$ calculated value is less than Z table value at 5% level

$\therefore$ Accept null hypothesis H_0.

Hence proportion of men and women favouring the proposal is same.

Problem 2: In a sample of 600 men form a certain large city 450 are found to be smokers. In one of 900 from another city 450 are smokers do the data indicate that the cities are significantly different with respect to prevalence of smoking among men?

Solution:

Null hypothesis H_0: $P_1 = P_2$

i.e., There is no significance difference in smoking between two cities.

Alternative hypothesis: H_1: $P_1 \neq P_2$ (two tailed test)

Level of significance: The value of Z at 5% level of significance for the tailed test is $Z_\alpha = 1.96$

Test statistic: $$Z = \frac{p_1 - p_2}{\sqrt{PQ\left(\frac{1}{n_1} + \frac{1}{n_2}\right)}}$$

Given

$n_1 = 600 \quad n_2 = 900$

$X_1 = 450 \quad X_2 = 450$

$$p_1 = \frac{X_1}{n_1} = 0.75 \qquad p_2 = \frac{X_2}{n_2} = 0.5$$

$$P = \frac{X_1 + X_2}{n_1 + n_2} = \frac{450 + 450}{600 + 900} = 0.6$$

$$Q = 1 - P = 0.4$$

$$\therefore \quad Z = \frac{0.75 - 0.5}{\sqrt{0.6 \times 0.4\left(\frac{1}{600} + \frac{1}{900}\right)}} = 9.689$$

Conclusion: Since Z calculated value is greater than Z table value at 5% level

∴ Reject null hypothesis H_0

Hence there is a difference in smoking between two cities.

Problem 3: A machine puts out 16 imperfect articles in a sample of 500. After a machine is overhauled it puts out 3 imperfect articles in a batch of 100. Has the machine improved?

Solution:

Null hypothesis H_0: $P_1 = P_2$

Alternative hypothesis H_1: $P_1 > P_2$ (right tailed test)

Level of significance: The value of Z at 5% level of significance for right tailed test is $Z_\alpha = 1.645$

Test statistic:

$$Z = \frac{p_1 - p_2}{\sqrt{PQ\left(\frac{1}{n_1} + \frac{1}{n_2}\right)}}$$

Given

$$n_1 = 500 \qquad n_2 = 100$$

$$X_1 = 16 \qquad X_2 = 3$$

$$p_1 = \frac{X_1}{n_1} = 0.032 \qquad p_2 = \frac{X_2}{n_2} = 0.03$$

$$P = \frac{X_1 + X_2}{n_1 + n_2} = \frac{16 + 3}{500 + 100} = 0.032$$

$$Q = 1 - P = 0.968$$

$$\therefore \quad Z = \frac{0.032 - 0.03}{\sqrt{0.032 \times 0.968\left(\frac{1}{500} + \frac{1}{100}\right)}} = 0.103$$

Conclusion: Since Z calculated value is less than Z table value at 5% level

∴ Accept null hypothesis H_0.

Problem 4: Before an increasing in excise duty on tea 800 persons out of a sample of 1000 persons were found to be tea drinkers. After an increase in excise duty 800 people were tea drinkers in a sample of 1200 people. Using standard error of proportion whether there is a significance decrease in the consumption of tea after the increase in excise duty.

Solution:

Null hypothesis H_0: $P_1 = P_2$

Alternative hypothesis $H_1 : P_1 > P_2$ (right tailed test)

Level of significance: The value of Z at 5% level of significance for right tailed test is $Z_\alpha = 1.645$

Test statistic:
$$Z = \frac{p_1 - p_2}{\sqrt{PQ\left(\frac{1}{n_1} + \frac{1}{n_2}\right)}}$$

Given

$$n_1 = 1000 \quad n_2 = 1200$$
$$X_1 = 800 \quad X_2 = 800$$
$$p_1 = \frac{X_1}{n_1} = 0.8$$
$$p_2 = \frac{X_2}{n_2} = 0.67$$
$$P = \frac{X_1 + X_2}{n_1 + n_2} = \frac{800 + 800}{1000 + 1200} = 0.73$$
$$Q = 1 - P = 0.27$$
$$\therefore \quad Z = \frac{0.8 - 0.67}{\sqrt{0.73 \times 0.27\left(\frac{1}{1000} + \frac{1}{1200}\right)}} = 6.84$$

Conclusion: Since Z calculated value is greater than Z table value at 5% level.

$\therefore$ Reject null hypothesis H_0.

Hence there is significant decrease in the consumption of tea after increase in excise duty.

Problem 5: A study shows that 16 of 200 tractors produced on one assembly line required extensive adjustments before they could be shipped, while the same was true for 14 of 400 tractors produced on another assembly line. At the 0.01 level of significance, does this support the claim that the second production line does superior work?

Solution:

Null hypothesis H_0: $P_1 = P_2$

Alternative hypothesis H_1: $P_1 < P_2$ (left tailed test)

Level of significance: The value of Z at 1% level of significance for left tailed test is $Z_\alpha = -2.33$

Test statistic:
$$Z = \frac{p_1 - p_2}{\sqrt{PQ\left(\frac{1}{n_1} + \frac{1}{n_2}\right)}}$$

Given

$$n_1 = 200 \quad n_2 = 400$$
$$X_1 = 16 \quad X_2 = 14$$
$$p_1 = \frac{X_1}{n_1} = 0.08 \qquad p_2 = \frac{X_2}{n_2} = 0.035$$
$$P = \frac{X_1 + X_2}{n_1 + n_2} = \frac{16 + 14}{200 + 400} = 0.05$$
$$Q = 1 - P = 0.95$$
$$\therefore \quad Z = \frac{0.08 - 0.03}{\sqrt{0.05 \times 0.95 \left(\frac{1}{200} + \frac{1}{400}\right)}} = 2.39$$

Conclusion: Since $|Z|$ calculated value is greater than Z table value at 1% level.

$\therefore$ Accept null hypothesis H_0.

Hence second production is not superior to first.

Problem 6: In a year there are 956 births in a town A of which 52.5% were males, while in towns A and B combined this proportion in a total of 1406 births was 0.496. Is there any significant difference in a proportion of male births in the two towns?

Solution:

Null hypothesis H_0: $P_1 = P_2$

Alternative hypothesis H_1: $P_1 \neq P_2$ (two tailed test)

Level of significance: The value of Z at 5% level of significance for two tailed test is $Z_\alpha = 1.96$

Test statistic:
$$Z = \frac{p_1 - p_2}{\sqrt{PQ\left(\frac{1}{n_1} + \frac{1}{n_2}\right)}}$$

Given $n_1 = 956$
$$n_1 + n_2 = 1406 \Rightarrow n_2 = 450$$
$$P = \frac{n_1 p_1 + n_2 p_2}{n_1 + n_2}$$
$$= \frac{956 \times 0.525 + 450 \times p_2}{956 + 450} = 0.496 \Rightarrow p_2 = 0.434$$
$$Q = 1 - P = 1 - 0.496 = 0.504$$
$$\therefore \quad Z = \frac{0.525 - 0.434}{\sqrt{0.496 \times 0.504 \left(\frac{1}{956} + \frac{1}{450}\right)}} = 0.368$$

Conclusion: Since Z calculated value is greater than Z table value at 5% level.

$\therefore$ Reject null hypothesis H_0.

Hence there is a significance difference in the proportion of male births in the towns A and B.

Problem 7: A manufacturer of electronic equipment subjects samples of two competing brands of transistors to an accelerated performance test. If 45 of 180 transistors of the first kind and 34 of 120 transistors of the second kind fail the test, what can we conclude at the level of significance $\alpha = 0.05$ about the difference between the corresponding sample proportions?

(May 2004, Nov 2003)

Solution:

Null hypothesis H_0: $P_1 = P_2$ *i.e.,* there is no difference.

Alternative hypothesis $H_1 : P_1 \neq P_2$ *i.e.,* there is a difference (Two tailed test)

Level of significance: Since the value of Z at 0.05 level of significance for two tailed test is $Z_\alpha = 1.96$

Test statistics:

$$Z = \frac{p_1 - p_2}{\sqrt{PQ\left(\frac{1}{n_1} + \frac{1}{n_2}\right)}}$$

Given $n_1 = 180$ $\quad n_2 = 120$

$X_1 = 45$ $\quad X_2 = 34$

$$p_1 = \frac{X_1}{n_1} = 0.25 \qquad p_2 = \frac{X_2}{n_2} = 0.283$$

$$P = \frac{X_1 + X_2}{n_1 + n_2}$$

$$= \frac{45 + 34}{180 + 120}$$

$$= 0.263$$

$$Q = 1 - P$$

$$= 0.737$$

$$\therefore \quad Z = \frac{0.25 - 0.283}{\sqrt{0.263 \times 0.737\left(\frac{1}{180} + \frac{1}{120}\right)}}$$

$$= -0.084$$

Conclusion: Since $|Z|$ calculated value is less than Z table value at given level of significance. Hence accept the null hypothesis H_0. *i.e.* there is no significance difference.

Problem 8: On the basis of their total scores, 200 candidates of a civil service examination are divided into two groups, the upper 30% and the remaining 70%. Consider the first question of the examination. Among the first group, 40 had the correct answer, whereas among the second group, 80 had the correct answer. On the basis of these results, can one conclude that the first question is no good at discriminating ability of the type being examined here?

(May 2004, Nov 2003)

Solution:

Null hypothesis H_0: $P_1 = P_2$

Alternative hypothesis H_1: $P_1 \neq P_2$ (Two tailed test)

Level of significance: Since the value of Z at 0.05 level of significance for two tailed test is $Z_\alpha = 1.96$

Test statistics:
$$Z = \frac{p_1 - p_2}{\sqrt{PQ\left(\frac{1}{n_1} + \frac{1}{n_2}\right)}}$$

Given
$$n_1 = 60 \qquad n_2 = 140$$
$$X_1 = 40 \qquad X_2 = 80$$
$$p_1 = \frac{X_1}{n_1} = 0.667 \qquad p_2 = \frac{X_2}{n_2} = 0.571$$
$$P = \frac{X_1 + X_2}{n_1 + n_2}$$
$$= \frac{40 + 80}{200} = 0.6$$
$$Q = 1 - P = 0.4$$

$$\therefore \qquad Z = \frac{0.667 - 0.571}{\sqrt{0.6 \times 0.4\left(\frac{1}{60} + \frac{1}{140}\right)}}$$
$$= 1.269$$

Conclusion: Since Z calculated value is less than Z table value at given level of significance. Hence accept the null hypothesis H_0.

6.9 EXERCISE

1. A manufacturer claims that only 4% of his products are defective. A random sample of 500 was taken among which 100 defective. Test the hypothesis at 0.05 levels. **(Aug 2007)**
2. A coin is tossed 512 times. Head turn up 244 times. Can you say that the coin is unbiased? **(Aug 2007, May 2007)**
3. A sample of size 64 and mean 60 was taken from a population whose standard deviation is 10. Find 95% confidence interval for the mean. **(Aug 2007, May 2007)**
4. A sample size of 36 was drawn and its mean was found to be 80. Test whether this sample could have come from a normal population with mean 90 and variance 225? **(May 2007)**

5. In a random sample of 160 workers exposed to a certain amount of radiation, 24 experienced some ill effects. Construct 995 confidence interval for the corresponding true percentage. **(May 2007, May 2006, Nov 2004)**

6. The performance of a computer is observed over a period of 2 years to check the claim that the probability is 0.20 that is drawn time will exceed 5 hours in any given week. Testing the null hypothesis $p = 0.20$ against the alternate hypothesis $p \neq 0.20$, what can we conclude at the level of significance $\alpha = 0.05$ if there were only 11 weeks in which the drawn time of the computer exceed 5 hours? **(May 2007, May 2006, Nov 2004)**

7. 400 articles from a factory are examined and 3% were found to be defective. Construct 95% confidence interval. **(Nov 2006)**

8. In a city 250 men out of 750 were found to be smokers. Does this information support the conclusion that the majority of men in this city are smokers? **(May 2006, May 2005)**

9. It is claimed that a random sample of 49 tyres is a mean life of 15200 km. This sample was drawn from a population whose mean is 15150 km and standard deviation of 1200 km. Test the significance at 0.05 level **(May 2006, Nov 2005)**

10. A sample of 400 items is taken from a population whose standard deviation is 10. The mean of the sample is 40. Test whether the sample has come from a population with mean 38. Also calculate 95% confidence interval for the proportion. **(Nov 2005)**

11. A random sample of size 81 was taken whose variance is 20.25 and means 32 construct 98% confidence interval. **(Nov 2005)**

12. A die is thrown 256 times an even digits turns up 150 times. Can we say that the die is unbiased? **(Nov 2005)**

13. If 80 patients are treated with antibiotic 59 got cured. Find a 99% confidence limits to the true proportion of curare? **(Nov 2004)**

14. In a random sample of 120 pumps includes 47 required repairs within the first 5 years, test the null hypothesis $p = 0.30$ against the alternative hypothesis $p > 0.30$ at 0.05 level of significance? **(Nov 2004)**

15. A die is thrown 1350 times 5 or 6 obtained 650 times test whether the die is unbiased. **(May 2007)**

16. Experiences had shown that 20% of a manufactured product is on the top quality. In one days production of 400 articles only 50 are of top quality. Test the hypothesis at 0.05 level. **(May 2005)**

17. A social worker believes that fewer than 25% of the couples in a certain area are ever used any form of birth control. A random sample of 120 couples was contacted. Twenty of them said that they have used. Test the belief if social worker at 0.05 level. **(Nov 2005)**

18. An ambulance service claims that it takes on the average less than 10 minutes to reach its destination in emergency calls. A sample of 36 calls has a mean of 11 minutes and the variance of 16 minutes. Test the significance at 0.05 level. **(Nov 2005)**

19. Tests performed with a random sample of 40 diesel engines produced by a large manufacturer show that they have a mean thermal efficiency of 31.45% with a standard deviation of 1.6% at the 0.01 level of significance. Test the null hypothesis $\mu = 32.5\%$ against the alternative hypothesis $\mu \neq 32.5\%$.

20. A sample of 900 members has a mean 3.4 cms and standard deviation 2.61 cms. Is this sample has been taken from a large population of mean 3.25 cms and standard deviation 2.61 cms. If the population is normal and its mean is unknown. Find the 95% and 98% confidence limits of the true mean?

21. If the mean breaking strength of copper wire is 505 lbs. with a standard deviation of 15 lbs. How large a sample must be in order to that there will be one chance in 100 that the mean breaking strength of the sample is less than 500?

22. An investigation of two kinds of photocopying equipment showed that 71 failures of the first kind of equipment took on the average 83.2 minutes to repair with a standard deviation of 19.3 minutes, while 75 failures of the second kind equipment took on the average 90.8 minutes to repair with a standard deviation of 21.4 minutes. Test the null hypothesis $\mu_1 - \mu_2 = 0$, the hypothesis that on the average it take an equal mount of time to repair either kind of equipment against the alternative hypothesis $\mu_1 - \mu_2 \neq 0$ at the level of significance $\alpha = 0.05$.

23. A random sample of 1000 men from northern India gives their mean wage to be Rs. 30 per day with a standard deviation of Rs. 1.50. A sample of 1500 men from southern India gives a mean wage of Rs. 32 per day with standard deviation of Rs. 2. Discuss whether the mean rate of wages varies between the two regions.

24. A biased coin was thrown 400 times and head resulted 240 times. Find the standard error of the observed proportion of heads and deduce that the probability of getting a head in a single throw of the coin lies almost certainly between 0.53 and 0.67.

25. In a city A 20% of a random sample of 900 school boys had a certain slight physical defect. In another city B 18.5% of a random sample of 1600 school boys had the same defect. Is the difference between the proportions significant at 0.05 level of significance.

26. In a sample of 100 ceramic pistons made for an experimental designed engine 18 cracked construct 95% confidence interval.

27. In a sample of 500 people in Maharashtra 300 are wheat eaters. What can you about the maximum error with 99% confidence.

28. If we can assert with 95% that the maximum error is 0.05 and p is given as 0.2. Find the size of the sample.

29. 20 people were attacked by a disease and only 18 survived. Will you reject the hypothesis that the survival rate if attacked by this disease is 85%, in favour of the hypothesis that is more at 5% level.

30. In a sample of 600 students of a certain college 400 are found to use ball pens. In another college from a sample of 900 students 450 were found to use ball pens. Test whether 2 colleges are significantly different with respect to the habit of using ball pens.

31. A sample of 81 iron bars whose mean length is 5 feet. Is drawn from a population whose mean is 7 ft. and standard deviation 3 ft?

32. It was found that the average income of 36 workers is 12,000 per annum taken was it taken from a population whose mean is Rs. 11,800 per annum and standard deviation is 800?

33. A random sample of 225 items is found to have a mean of 80 can it be regarded as taken from a population whose mean is 82 and standard deviation is 15? Test at 0.01 level.

34. The mean breaking strength of cubes supplied by a manufacturer is 2000 with a standard deviation of 200. A sample of 100 has a mean of 1950. Test the significance at 0.05 level.

35. The following data gives the means of two samples taken from a population. Test whether there is any significant difference between the two samples at 95% level. $\bar{x} = 60$, $\bar{y} = 59$, $n_1 = 100$, $n_2 = 200$ and $\sigma = 50$.

36. The following data was collected from coolies. Test whether the two means differ significantly at 99% level.

Samples	*No. of cases*	*Mean wages*	*Standard deviation of the wages*
Sample I	400	50	6
Sample II	900	49	6

37. Test whether the means of the samples are significantly different, given the following data

Samples	*Size*	*Mean*	*Standard deviation*
Sample I	100	110	12
Sample II	100	112	16

38. A normal population has a deviation of 0.2. Two samples of sizes 900 each are taken and found the mean as 15.10 and 15.30. Test the significant.

39. A random sample of 400 men from one stage gives the mean pay of Rs. 200 per day with a standard deviation of Rs. 10. Another random sample of 400 men has a mean pay of Rs. 190 per day with a standard deviation of Rs. 10. Construct 99% confidence interval for $\mu_1 - \mu_2$.

40. A random sample of 1200 apples was taken from a large consignment and found that 10% of then were bad. The supplier claims that only 2% are bad. Test his claim.

41. A sample of 400 men were taken from a city. 5% of them are found to be smokers. Find the standard error of proportions.

42. If the proportion P = 0.4 and maximum error with 95% confidence is 0.05. Find the size of the sample.

43. With 99% confidence, maximum error is 0.06 and the proportion $p = 0.5$. Find the sample size.

44. In a city among 900 people are smokers. Construct 99% upper limit for the true proportion.

45. In a sample of 600 students of a certain college, 400 are using ball pens. In another college among 900 students 450 are using ball pens. Test whether there is a significant difference between the proportions.

46. Random samples of 300 men and 500 women were asked whether they would like to have a fly over near their residence 150 men 200 women were in favour of the proposal. Find the 95% confidence interval for the true difference of proportion.

47. A machine puts out 20 imperfect articles in a sample of 500, after machine is overhauled it puts out 2 imperfect articles in a batch of 100. Find the 95% confidence interval for the difference of proportion.

ANSWERS

1. Z = 18.25	**2.** biased	**4.** Z = –12.67
5. {0.095, 0.205}	**6.** Z = 0.118	**7.** {0.954, 0.986}
8. Z = –9.128	**9.** Z = – 0.2916	**10.** Z= 4, {39.02, 40.98}
12. Z = 2.75	**13.** {0.6106, 0.8644}	**14.** Z= 2.19, **15.** biased
16. Z = –3.75	**17.** Z = –2.108	**18.** Z = 1.5
19. Z = – 3.33	**20.** Z = 1.73 limits 3.197, 3.603	**21.** Z = 2.33, size = 49
22. Z = 2.2	**23.** Z = –28.5	**24.** S.E. 0.024, limits 0.528, 0.672
25. Z = 0.96	**26.** SE = 0.038, interval 0.105, 0.255	
27. M.E. = 0.0568	**28.** Size of sample 244	**29.** Z = 0.625
30. Z = 6.49	**31.** Z = – 6	**32.** Z = 1.5
33. Z = 2	**34.** Z = 2.5	**35.** Z = 1.64
38. Z = 21.3	**39.** 8.26, 11.74	**40.** Z = 9.3
41. 0.011	**42.** 369	**43.** 463
44. 0.0188	**45.** Z = 6.5	**46.** 0.0784, 0.1216
47. Z = –0.123, 0.0523		

6.10 QUIZ

1. Given $n = 30$, $\bar{x} = 786$, $\sigma = 48$, $\mu = 800$ to test the hypothesis that $\bar{x} = 800$ against $\bar{x} \neq 800$. What is the test statistic used?

   (a) $Z = \dfrac{\bar{x} - \mu}{\sigma}$ (b) $Z = \dfrac{\bar{x} - \mu}{\sigma/\sqrt{n}}$

   (c) $t = \dfrac{\bar{x} - \mu}{\sigma}$ (d) $t = \dfrac{\bar{x} - \mu}{\sigma/\sqrt{n}}$

2. The critical value for right-tailed test at 1% level of significance is

   (a) 2.33 (b) 1.645

   (c) 1.28 (d) none

3. The critical value for two-tailed test at 5% level of significance is....................

   (a) 2.33 (b) 1.645

   (c) 1.28 (d) none

4. The critical value for two-tailed test at 10% level of significance

(*a*) 2.33 (*b*) 1.645

(*c*) 1.28 (*d*) none

5. What is type I error?

(*a*) accept H_0 if H_0 is true (*b*) accept H_0 if H_0 is false

(*c*) reject H_0 if H_0 is true (*d*) reject H_0 if H_0 is false

6. What is type II error?

(*a*) accept H_0 if H_0 is true (*b*) accept H_0 if H_0 is false

(*c*) reject H_0 if H_0 is true (*d*) reject H_0 if H_0 is false

7. A random sample of size 100 has SD of 5 then the maximum error with 95% confidence is

(*a*) E = 0.98 (*b*) 0.95

(*c*) 1.05 (*d*) 1.10

8. A research worker wishes to estimate the mean of a population by using sufficiently large sample. The probability is 0.95 that the sample mean will not differ from the true mean by more than 25% of the S.D. Then the maximum size of the sample is

(*a*) 25 (*b*) 36

(*c*) 47 (*d*) 61

9. If the maximum error of proportion with 95% confidence is 0.686 and P = 1/2 then the sample size is................

(*a*) 169 (*b*) 196

(*c*) 144 (*d*) 204

10. If $x_1 = 200$, $x_2 = 200$, $n_1 = 500$ and $n_2 = 400$ then the standard error of difference between the proportions is.....................

(*a*) 0.033 (*b*) 0.045

(*c*) 0.12 (*d*) 0.08

11. If $P_1 = 0.6$, $P_2 = 0.55$, $n_1 = 500$, $n_2 = 1000$ then Z =.................

(*a*) 2.3 (*b*) 1.9

(*c*) 1.7 (*d*) 3.2

12. If the mean of two samples of sizes 1000 and 2000 are 67.5 and 68 and the $\sigma = 2.5$ then Z = ...

(*a*) 5.1 (*b*) 4.7

(*c*) 5.9 (*d*) 3.5

13. If $\sigma_1 = 2.56$, $\sigma_2 = 2.52$, $n_1 = 6400$ and $n_2 = 1600$ then standard error of the means is ....

(*a*) 0.07 (*b*) 0.09

(*c*) 0.12 (*d*) 0.15

14. If standard deviation of samples of size $n_1 = 2000$ and $n_2 = 2000$ are σ and 2σ and standard error of the difference between the means is 0.15 then σ = ...

(a) 3 (b) 6

(c) 5 (d) 9

15. The standard deviations of two samples of sizes are 1000 and 1500 are 1.5 and 2 then the standard error of difference between the means is......................

(a) 0.07 (b) 0.05

(c) 0.12 (d) 0.22

16. If the sample sizes are 200 and 300 standard deviations are 4 and 4.2 then the standard error of difference between the means is...........................

(a) 0.372 (b) 0.22

(c) 0.512 (d) 0.625

17. If $\bar{x}$ = 485, $\bar{y}$ = 510, σ_1 = 50, σ_2 = 40, $n_1 = n_2$ = 200 then Z = ...

(a) –5.5 (b) 2.5

(c) 3.7 (d) 4.2

18. If 204 apples out of 300 were good. The maximum error with 95% confidence for the true proportion of good apples is........................

(a) 0.052 (b) 0.0812

(c) 0.09 (d) 0.12

19. If x = 84, n = 200 then maximum error of true proportions with 99% confidence interval is ....

(a) 0.09 (b) 0.15

(c) 0.05 (d) 0.22

20. If P = 0.4 maximum error with 99% confidence of true proportion is........................

(a) 1299 (b) 1150

(c) 1325 (d) 1000

21. If x = 69, n = 120 then maximum error with 95% confidence of true proportion is............

(a) 0.125 (b) 0.088

(c) 0.05 (d) 0.02

22. If P = 0.75 maximum error with 95% confidence is 0.06 then the sample size is............

(a) 185 (b) 250

(c) 200 (d) 150

23. If P = 0.5 and n = 200 then standard error of proportion is.....................

(a) 0.035 (b) 0.12

(c) 0.08 (d) 0.21

24. If P = 1/9 and n = 900 then the probable limits for the percentage of proportion is............

(a) (8, 13.25) (b) (6, 11)

(c) (4, 9.25) (d) (10, 15.25)

25. If the standard error of proportion is 0.035 then the maximum error with 95% confidence is..............

(*a*) 0.0686 (*b*) 0.0825
(*c*) 0.09 (*d*) 0.044

26. In a sample of 100 ceramic pistons made for an experimental designed engine 18 were cracked. Then the standard error of proportion is....................

(*a*) 0.05 (*b*) 0.038
(*c*) 0.062 (*d*) 0.08

27. In a sample of 100 ceramic pistons made for an experimental designed engine 18 were cracked, then 95% confidence interval of proportions is......................

(*a*) 0.105, 0.255 (*b*) 0.12, 0.24
(*c*) 0.1, 0.26 (*d*) 0.15, 0.21

28. A die is thrown 256 times. An even digit turns up 150 times then Z =.............

(*a*) 1.75 (*b*) 2.75
(*c*) 3.25 (*d*) 4.00

29. If P = 0.95 and the sample size is 200 then the standard error of proportion is..........

(*a*) 0.0154 (*b*) 0.0321
(*c*) 0.15 (*d*) 0.275

30. If P = 0.5 and the sample size is 750 then the standard error is..................

(*a*) 0.0183 (*b*) 0.18
(*c*) 0.0251 (*d*) 0.036

31. If the mean of a sample of size 100 is 12 and the standard deviation is 2. Then 95% confidence limits are................

(*a*) 11.04, 12.96 (*b*) 10.98, 13.02
(*c*) 11.608, 12.392 (*d*) 11.2, 12.8

32. If $\bar{x}$ = 116, μ = 120, variance = 225, sample size = 100 then the test statistic Z =............

(*a*) 2.1 (*b*) 3.12
(*c*) 2.67 (*d*) 3.72

33. If the mean of a sample of size 8 is 20 with a standard variation of 3 then 95% confidence interval is....................

(*a*) 9.347, 10.653 (*b*) 9.12, 10.88
(*c*) 8.72, 13.28 (*d*) 9.62, 10.38

34. If the mean of a sample of size 100 is 15 and standard deviation 2 then the 95% confidence interval is

(*a*) 14.608, 15.392 (*b*) 13.92, 16.08
(*c*) 14.5, 15.5 (*d*) 14.75, 15.25

35. A biased coin was thrown 400 times and head turned up 240 times. Then the standard error of proportion of head is......................

(*a*) 0.02 (*b*) 0.024
(*c*) 0.04 (*d*) 0.06

36. Given that on the average 5% of insured men of age 65 die within a year . If the sample size is 600 standard error corresponding to true proportion is..................

(*a*) 0.089 (*b*) 0.002
(*c*) .004 (*d*) 0.01

37. A random sample of 400 products contains 52 defective items. Standard error of proportion of ....

(*a*) 0.165 (*b*) 0.0375
(*c*) 0.05 (*d*) 0.0725

38. A random sample of 900 apples were taken among which 4% are bad with 0.95 probability the maximum error for true proportion is.....................

(*a*) 0.0065 (*b*) 0.001
(*c*) 0.002 (*d*) 0.005

39. 900 articles from a factory are examined and found to be 5% defective. With 0.99 probability, the maximum error corresponding to true proportion is...................

(*a*) 0.019 (*b*) 0.03
(*c*) 0.12 (*d*) 0.05

40. If true proportion $p = 1/3$, the maximum error with 0.95 probability is 0.02. The sample size is......................

(*a*) 2134 (*b*) 2324
(*c*) 1975 (*d*) 1725

41. If $p = 1/2$, maximum error with 95% confidence is 0.07, then sample size is..............

(*a*) 196 (*b*) 169
(*c*) 225 (*d*) 200

42. 100 among 600 articles are defective. If the maximum error with probability 0.99 is 0.02. The sample size is......................

(*a*) 4.7 (*b*) 5.7
(*c*) 6.7 (*d*) 7.7

43. We want to be at least 95% confidence that the error is at the most 0.06, then samples size is

(*a*) 215 (*b*) 267
(*c*) 313 (*d*) 425

44. We want to be at least 99% confidence that the error is at the most 0.03. Then sample size is

(*a*) 1849 (*b*) 2132
(*c*) 1627 (*d*) 1432

45. If the maximum error with 0.99 probability is 0.25 and sample size 400; then standard deviation of the population is.........................

(*a*) 1.23 (*b*) 2.23

(*c*) 1.93 (*d*) 3.12

46. Among 100 students in a class, 60 people use ball pens. With 95% confidence, the maximum error for true proportion is....................

(*a*) 0.096 (*b*) 0.02

(*c*) 0.03 (*d*) 0.05

47. A sample of 64 was taken and found that 15 are smokers. Standard error of proportions is.......................

(*a*) 0.0275 (*b*) 0.0425

(*c*) 0.0675 (*d*) 0.0825

48. In a sample survey conducted in a large city 95 out of 500 answered yes to a question with 95% confidence the maximum error for true proportion can be.......................

(*a*) 0.01 (*b*) 0.023

(*c*) 0.05 (*d*) 0.09

49. A random sample of 1000 people were taken and out of which 4% are chapatti eaters. The maximum error with probability 0.95 is.....................

(*a*) 0.012 (*b*) 0.025

(*c*) 0.04 (*d*) 0.06

50. A company manufacturing electric bulbs claims that the average life of its bulbs is 1600 hours. The average life and standard deviation of a random sample of 100 such bulbs were 1570 hours and 120 hours respectively. Then the test statistic Z is.....................

(*a*) 2.5 (*b*) 2.9

(*c*) 2.6 (*d*) none

ANSWERS

1. (*b*)	**2.** (*a*)	**3.** (*d*)	**4.** (*b*)	**5.** (*c*)	**6.** (*b*)	**7.** (*a*)	**8.** (*d*)	**9.** (*d*)	**10.** (*a*)
11. (*b*)	**12.** (*a*)	**13.** (*a*)	**14.** (*a*)	**15.** (*d*)	**16.** (*a*)	**17.** (*a*)	**18.** (*a*)	**19.** (*a*)	**20.** (*a*)
21. (*b*)	**22.** (*c*)	**23.** (*a*)	**24.** (*a*)	**25.** (*a*)	**26.** (*b*)	**27.** (*b*)	**28.** (*b*)	**29.** (*a*)	**30.** (*a*)
31. (*c*)	**32.** (*c*)	**33.** (*a*)	**34.** (*a*)	**35.** (*b*)	**36.** (*a*)	**37.** (*a*)	**38.** (*a*)	**39.** (*a*)	**40.** (*a*)
41. (*a*)	**42.** (*a*)	**43.** (*b*)	**44.** (*a*)	**45.** (*c*)	**46.** (*a*)	**47.** (*a*)	**48.** (*b*)	**49.** (*a*)	**50.** (*a*)

Chapter 7

TEST OF SIGNIFICANCE

7.0 STUDENTS 't' DISTRIBUTION

Let x_i ($i = 1, 2, \ldots n$) be a random sample of size n from a normal population with mean μ and variance σ^2 then student's t is defined by the statistic:

$$t = \frac{\bar{x} - \mu}{S/\sqrt{n}} \qquad \ldots(1)$$

where $\bar{x} = \frac{1}{n}\sum_{i=1}^{n} x_i$, is the sample mean and $S^2 = \frac{1}{n-1}\sum_{i=1}^{n}(x_i - \bar{x})^2$, ...(2)

is an unbiased estimate of the population variance σ^2, and it follows student's t-distribution with $v = (n - 1)$ degree freedom with probability density function.

Applications of t-Distribution

The t-distribution has a wide number of applications in statistics, some of which are enumerated below.

(*i*) To test if the sample mean ($\bar{x}$) differs significantly from the hypothetical value μ of the population mean.

(*ii*) To test the significance of the difference between two sample means.

(*iii*) To test the significance of an observed sample correlation coefficient and sample regression coefficient.

(*iv*) To test the significance of observed partial correlation coefficient.

In the following sections we will discuss these applications in detail, one by one.

7.1 t-TEST FOR SINGLE MEAN

Suppose we want to test:

(*i*) If a random sample x_i ($i = 1, 2, \ldots, n$) of size n has been drawn from a normal population with a specified mean, say μ_0, or (*ii*) If the sample mean differs significantly from the hypothetical value μ_0 of the population mean. Under the null hypothesis, H_0

(*a*) The sample has been drawn from the population with mean μ_0 or

(*b*) There is no significant difference between the sample mean $\bar{x}$ and the population mean μ_0,

The statistic
$$t = \frac{\bar{x} - \mu}{S/\sqrt{n}} \quad ...(1)$$

where
$$\bar{x} = \frac{1}{n}\sum_{i=1}^{n} x_i \text{ and } S^2 = \frac{1}{n-1}\sum_{i=1}^{n}(x_i - \bar{x})^2 \quad ...(2)$$

Follows student's t-distribution with $(n - 1)$ degree freedom.

We now compare the calculated value of t with the tabulated value at certain level of significance. If calculated $|t| >$ tabulated t, null hypothesis is rejected and if calculated $|t| <$ tabulated t, H_0 may be accepted at the level of significance adopted.

Confidence or Fiducial limits for μ: If $t_{0.05}$ is the tabulated value of t for $v = (n - 1)$ degree freedom at 5% level of significance, *i.e.*, $P(|t| > t_{0.05}) = 0.05 \Rightarrow P(|t| \le t_{0.05}) = 0.95$, the 95% confidence limits for μ are given by

$$|t| \le t_{0.05}, \text{ i.e., } \left|\frac{\bar{x} - \mu}{S/\sqrt{n}}\right| \le t_{0.05}$$

$$\Rightarrow \bar{x} - t_{0.05} \cdot \frac{s}{\sqrt{n}} \le \mu \le \bar{x} + t_{0.05} \frac{s}{\sqrt{n}}$$

Thus, 95% confidence limits for μ are : $\bar{x} \pm t_{0.05} \cdot \left(\frac{s}{\sqrt{n}}\right)$ ...(3)

Similarly, 99% confidence limits for μ are $x \pm t_{0.01} \cdot \left(\frac{s}{\sqrt{n}}\right)$ ...(4)

where $t_{0.01}$ is the tabulated value of t for $v = (n - 1)$ degree of freedom at 1% level of significance.

7.2 SOLVED PROBLEMS

Problem 1: According to the norms established for a mechanical aptitude test, persons who are 18 years old have an average height of 73.2 with a standard deviation 8.6. If 4 randomly selected persons of that age averaged 76.7, test the null hypothesis $\mu = 73.2$ against the alternate hypothesis $\mu > 73.2$ at the 0.01 level of significance?

Solution:

Null hypothesis H_0: $\mu = 73.2$

Alternative hypothesis H_1: $\mu > 73.2$ (Right tailed test)

Level of significance: $\alpha = 0.01$ level of significance for right tailed test with 3 degree freedom is $t_\alpha = 2.33$

Test statistics: Here $\bar{x}$ = Mean of the sample = 76.7

Sample size $n = 4$

$$\sigma = 8.6$$

$$t = \frac{\bar{x} - \mu}{\sigma/\sqrt{n}}$$

$$t = \frac{76.7 - 73.2}{8.6/\sqrt{4}} = 0.8139$$

Conclusion: Since t calculated value is less than t table value at given level of significance. Hence accept the null hypothesis H_0.

Problem 2: A random sample from a company's very extensive files shows that the orders for a certain kind of machinery were filed, respectively in 10, 12, 19, 14, 15, 18, 11 and 13 days. Use the level of significance $\alpha = 0.01$ to test the claim that on the average such orders are filed in 10.5 days. Assume normality.

Solution:

Null hypothesis H_0: $\mu = 10.5$

Alternative hypothesis H_1: $\mu > 10.5$ (Right tailed test)

Level of significance: At $\alpha = 0.01$ level of significance for right tailed test with 7 degree freedom is $t_\alpha = 2.998$

Test statistics: $$t = \frac{\bar{x} - \mu}{S\sqrt{n}}$$

$$\bar{x} = \frac{10 + 12 + 19 + 14 + 15 + 11 + 13}{8} = 14$$

$$S^2 = \frac{\sum (x_i - \bar{x})^2}{n - 1}$$

$$= \frac{(10 - 14)^2 + (12 - 14)^2 + (19 - 14)^2 + \dots\dots + (13 - 14)^2}{7}$$

$$= 10.28$$

$$S = 3.206$$

$$t = \frac{14 - 10.5}{3.206/\sqrt{8}} = 3.08$$

Conclusion: Since t calculated value is greater than t table value at given level of significance. Hence reject the null hypothesis H_0.

Problem 3: A sample of 10 cam shafts intended for use in gasoline engines has on average eccentricity of 1.02 and a standard deviation of 0.44 inches. The data may be treated a random sample from a normal population. Determine a 95% confidence interval for the actual mean eccentricity of the cam shaft.

Solution:

The confidence interval is $\bar{x} \pm t_{\alpha/2} \dfrac{S}{\sqrt{n}}$

$$n = \text{sample size } 10$$
$$S = \text{standard deviation} = 0.044 \text{ inches.}$$
$$t_{\alpha/2} \cdot \frac{S}{\sqrt{n}} = 2.26 \times \frac{0.044}{\sqrt{n}} = 0.031$$
$$\bar{x} + t_{\alpha/2} \frac{S}{\sqrt{n}} = 1.02 + 0.031 = 1.051$$
$$\bar{x} - t_{\alpha/2} \frac{S}{\sqrt{n}} = 1.02 - 0.031$$
$$= 0.989$$

$\therefore$ Confidence interval $= (0.989, 1.051)$

Problem 4: The mean weekly sales of soap bars in departmental stores was 146.3 bars per store. After an advertising campaign the mean weekly sales in 22 stores for a typical week increased to 153.7 and showed a standard deviation of 17.2. Was the advertising campaign successful?

Solution:

Null Hypothesis H_0: The advertising campaign is not successful, *i.e.*, H_0: $\mu = 146.3$

Alternative Hypothesis H_1: $\mu > 146.3$ (Right-tail)

Level of significance: At 5% level significance for right tail test with 21 degree freedom is

$$t_\alpha = 1.72$$

Test statistics:

We are given;

$$n = 22, \quad \bar{x} = 153.7, \quad S = 17.2$$

Under Null Hypothesis H_0,

$$t = \frac{\bar{x} - \mu}{S/\sqrt{n-1}}$$
$$= \frac{153.7 - 146.3}{17.2/\sqrt{21}}$$
$$= 9.03$$

Conclusion: Since t calculated value is greater than t table value at 5% level of significance. Hence reject the null hypothesis H_0.

Problem 5: A random sample of 10 boys had the following I.Q.'s : 70, 120, 110, 101, 88, 83, 95, 98, 107, 100. Do these data support the assumption of a population mean I.Q. of 100? Find a reasonable range in which most of the mean I.Q. values of samples of 10 boys lie. **(Jan 2005)**

Solution:

Null hypothesis, H_0: The data are consistent with the assumption of a mean I.Q. of 100 in the population *i.e.*, $\mu = 100$.

Alternative hypothesis H_1: $\mu \neq 100$. (Two tailed test)

Level of significance: At 5% level of significance for two tailed test with 9 degree freedom is $t_\alpha = 2.262$

Test statistic:

Under H_0, the test statistic is :

$$t = \frac{\bar{x} - \mu}{S/\sqrt{n}}$$

where $\bar{x}$ and S^2 are to be computed from the sample values of I.Q.'s

Calculations for sample mean and standard deviation

x	$(x - \bar{x})$	$(x - \bar{x})^2$
70	–27.2	739.84
120	22.8	519.84
110	12.8	163.84
101	3.8	14.44
88	–9.2	84.64
83	–14.2	201.64
95	–2.2	4.84
98	0.8	0.64
107	9.8	96.04
100	2.8	7.84
Total 972		1833.60

Here $$n = 10, \quad \bar{x} = \frac{972}{10} = 97.2$$

and $$S^2 = \frac{1833.60}{9} = 203.73$$

$\therefore$ $$|t| = \frac{|97.2 - 100|}{\sqrt{203.73/10}}$$

$$= \frac{2.82}{\sqrt{20.37}} = 0.62$$

Conclusion: Since calculated t value is less than t table value for two tailed test with 9 degree freedom. Hence H_0 may be accepted at 5% level of significance and we may conclude that the data are consistent with the assumption of mean I.Q. of 100 in the population.

The 95% confidence limits within which the mean I.Q. values of samples of 10 boys will lie are given by:

$$\bar{x} \pm t_{0.05} \frac{s}{\sqrt{n}} = 97.2 \pm 2.262 \times 4.514 = 97.2 \pm 10.21 = 107.41 \text{ and } 86.99$$

Hence the required 95% confidence interval is [86.99, 107.41].

Problem 6: Five measurements of the tar content of a certain kind of cigarette yielded 14.5, 14.2, 14.4, 14.3 and 14.6 mg per cigarette. Show that the difference between the mean of this sample $\bar{x} = 14.4$ and the average tar claimed by the manufacturer $\mu = 14.0$ is significant at $\alpha = 0.05$.

(April 2003)

Solution:

Null hypothesis H_0: $\mu = 14.0$

Alternative hypothesis H_1: $\mu \neq 14$

Level of significance: At $\alpha = 0.05$ level of significance for two tailed test with 4 degree freedom is $t_\alpha = 2.78$

Test statistics: $t = \dfrac{\bar{x} - \mu}{S/\sqrt{n}}$

$\bar{x}$ = mean of the sample

$$\bar{x} = \frac{14.5 + 14.2 + 14.4 + 14.3 + 14.6}{5} = \frac{72}{5} = 14.4$$

$$S^2 = \text{variance of the sample} = \frac{\sum(x_i - \bar{x})^2}{n-1}$$

$$= \frac{(14.5-14.4)^2 + (14.2-14.4)^2 + (14.4-14.4)^2 + (14.3-14.4)^2 + (14.6-14.4)^2}{4}$$

$$= \frac{0.10}{4} = 0.025$$

μ = mean of the population = 14.0

n = sample size = 5

$$t = \frac{14.4 - 14.0}{\sqrt{0.025/5}}$$

$$= \frac{0.4}{\sqrt{0.05}}$$

$$= 5.5$$

Conclusion: Since t calculated value greater than t table value at 0.05 levels.

Hence the null hypothesis H_0 is rejected.

H_1 is accepted *i.e.*, the average tar claimed by the manufacturer $\mu = 14.0$ is significant $\mu \neq 14$.

Problem 7: A random samples of 10 bags of pesticides are taken whose weights are 50, 49, 52, 44, 45, 48, 46, 45, 49,45 (in kgs). Test whether the average packing can be taken to be 50 kgs.

Solution:

Null Hypothesis H_0: $\mu = 50$

Alternative hypothesis H_1: $\mu \neq 50$ (Two tailed test)

Level of significance: At 0.05 level of significance for two tailed test with 9 degree freedom is $t_\alpha = 2.26$

Test statistic:
$$t = \frac{\bar{x}-\mu}{S/\sqrt{n}}$$

$$\bar{x} = \frac{50 + 49 + 51 + 44 + 45+48+46+45+49+45}{10} = \frac{472}{10} = 47.2$$

x	$(x - \bar{x})$	$(x - \bar{x})^2$
50	2.8	7.84
49	1.8	3.24
51	3.8	14.44
44	–3.2	10.24
45	–2.2	4.84
48	0.8	0.64
46	–1.2	1.44
45	–2.2	4.84
49	1.8	3.24
46	–1.2	1.44

Total = 52.20

$$S^2 = \frac{\sum(x_i - \bar{x})^2}{n-1} = \frac{52.20}{10-1} = 5.8$$

$\bar{x}$ mean of the sample = 47.2

μ = mean of the population = 50

n = sample size 10

$$t = \frac{47.2-50}{\sqrt{5.8/10}} = -3.6$$

Conclusion: Since $|t|$ calculated value is greater than t table value at given level.

∴ Hence H_0 should be rejected.

∴ The average packing cannot be taken as 50 kgs.

Problem 8: A process for making certain bearing is under control if the diameters of the bearing have the mean of 0.5000 cm. What can we say about this process if a sample of 10 of these bearings has a mean diameter of 0.5060 cm and a standard deviation of 0.0040 cm?

Solution:

Since the sample size n = 10 is small (n < 30) use t-statistics $t = \frac{\bar{x}-\mu}{S/\sqrt{n}}$

Here $\bar{x}$ = mean of sample = 0.5060

S = standard deviation of sample = 0.00400

n = sample of size = 10

μ = mean of population = 0.5000

Null hypothesis H_0: μ = 0.5000

Alternative hypothesis H_1: $\mu \neq 0.5000$ (Two tailed test)

Level of significance: At α = 0.05 level of significance for two tailed test with 9 degrees of freedom is t_α = 3.250

Test statistics:

$$t = \frac{\bar{x}-\mu}{S/\sqrt{n}}$$

$$= \frac{0.5060 - 0.5000}{0.0040/\sqrt{10}}$$

$$= 4.7434$$

Conclusion: Since t calculated value greater than t table value at given level. Hence reject null hypothesis H_0.

Problem 9: A random sample of 6 steel beams has a mean compressive strength of 58,392 psi (pounds per square inch) with a standard deviation of 648 psi. Use this information and level of significance α = 0.05 to test whether the true average compressive strength of the steel from which this sample came is 58,000 psi(8).

Solution:

Sample size n = 6

Sample mean $\bar{x}$ = 58.392

Sample standard deviation S = 648

Population mean μ = 58000

Null hypothesis H_0: $\bar{x} = \mu$

Alternative hypothesis H_1: $\bar{x} \neq \mu$ (Two tailed test)

Level of significance: At α = 0.05 level for two tailed test with 5 degree freedom is 2.015

Test statistics:

$$t = \frac{\bar{x}-\mu}{S/\sqrt{n}}$$

$$= \frac{58392 - 58000}{648/\sqrt{6}} = 1.482$$

Conclusion: Since t calculated value is greater than t table at given level.

Hence accept the null hypothesis H_0.

Hence true average compressive strength of the steel from which this beam is cut is 58,000.

Problem 10: A random sample of size 25 from a normal population has the mean $\bar{x} = 47.5$ and the standard deviation S = 8.4. Does this information support or refute the claim that the mean of the population is $\mu = 42.1$?

Solution:

Sample size $n = 25$

Sample mean $\bar{x} = 47.5$

Sample standard deviation $S = 8.4$

Population mean $\mu = 42.1$

Null hypothesis H_0: $\mu = 42.1$

Alternative hypothesis H_1: $\mu \neq 42.1$ (Two tailed test)

Level of significance: At $\alpha = 0.05$ level of significance for two tailed test with 24 degree freedom is $t_\alpha = 1.711$

Test statistics: $$t = \frac{\bar{x}-\mu}{S/\sqrt{n}} = \frac{47.5-42.1}{8.4/\sqrt{25}} = 3.21$$

Conclusion: Since t calculated value greater than t table value at given level. Hence reject the null hypothesis H_0.

7.3 *t*-TEST FOR DIFFERENCE OF MEANS

Suppose we want to test if two independent samples x_i $(i = 1, 2, ..., n_1)$ and y_j, $(j = 1, 2, ..., n_2)$ of sizes n_1 and n_2 have been drawn from two normal populations with means μ_x and μ_y respectively.

Under the null hypothesis (H_0) that the samples have been drawn from the normal populations with means μ_x and μ_y and under the assumption that the population variance are equal, *i.e.*, $\sigma_x^2 = \sigma_y^2 = \sigma^2$ (say) the statistic

$$t = \frac{(\bar{x}-\bar{y})-(\mu_x-\mu_y)}{S\sqrt{\left(\frac{1}{n_1}+\frac{1}{n_2}\right)}} \quad ...(1)$$

where $$\bar{x} = \frac{1}{n_1}\sum_{i=1}^{n_1} X_i, \quad \bar{y} = \frac{1}{n_2}\sum_{j=1}^{n_2} y_j$$

$$S^2 = \frac{1}{n_1+n_2-2}\left[\sum_i (x_i-\bar{x})^2 + (y_1-\bar{y})^2\right] \quad ...(2)$$

is an unbiased estimate of the common population variance σ^2, follows students t-distribution with $(n_1 + n_2 - 2)$ degree freedom.

Under null hypothesis H_0: $\mu_1 = \mu_2$ test statistics t is

$$t = \frac{(\bar{x}-\bar{y})}{S\sqrt{\left(\frac{1}{n_1}+\frac{1}{n_2}\right)}}.$$

7.4 SOLVED PROBLEMS

Problem 1: Two horses A and B were tested according to the time (in seconds) to run a particular track with the following results.

Horse A:	28	30	32	33	33	29	34
Horse B:	29	30	30	24	27	29	–

Test whether the two horses have the same running capacity. **(April 2006)**

Solution:

Null hypothesis H_0: $\mu_1 = \mu_2$

Alternative Hypothesis H_1: $\mu_1 \neq \mu_2$ (Two tailed test)

Level of significance: At $\alpha = 0.05$ level of significance for two tailed test with 11 degree freedom is $t_\alpha = 2.2$

Test statistics:
$$t = \frac{(\bar{x}-\bar{y})-(\mu_x - \mu_y)}{S\sqrt{\left(\frac{1}{n_1}+\frac{1}{n_2}\right)}}$$

$\bar{x}$ = mean of 1st sample

$\bar{y}$ = mean of 2nd sample

$$S = \frac{1}{n_1+n_2-2}\left[\sum_i (x_i-\bar{x})^2 + (y_1-\bar{y})^2\right]$$

n_1 = First sample size = 7

$$\bar{x} = \frac{219}{7} = 31.3$$

n_2 = second sample size = 6

$$\bar{y} = \frac{169}{6} = 28.17$$

x	y	$x-\bar{x}$	$(x-\bar{x})^2$	$(y-\bar{y})$	$(y-\bar{y})^2$
28	29	–3.3	10.89	0.83	0.6889
30	30	–1.3	1.69	1.83	3.3489
32	30	0.7	0.49	1.83	3.3489
33	24	17	2.89	–4.17	17.389
33	27	1.7	2.87	–1.17	1.3689
29	29	–2.3	5.29	0.83	0.6889
34		2.7	7.29		
219	169		31.41		26.8335

$$S^2 = \frac{31.41 + 26.8335}{11}$$

$$= 5.29$$

$$S = 2.3$$

$$t = \frac{31.41 + 28.17}{2.3\sqrt{6^{-1} + 7^{-1}}}$$

$$= \frac{3.13}{1.28} = 2.44$$

Conclusion: Since t calculated value is greater than t table value at given level. Hence reject the null hypothesis H_0.

$\therefore$ Both the horses do not have the same running capacity.

Problem 2: To examine the hypothesis that the husbands are more intelligent that the wives, an investigator took a sample of 10 couples and administered then a test which measures the IQ as follows:

Husbands:	117	105	97	105	123	109	86	78	103	107
Wives:	106	98	87	104	116	95	90	69	108	85

Test the hypothesis with a reasonable test at level of significance of 0.05.

(Aug 2007, April 2007, Nov 2005, Nov 2004)

Solution :

Null hypothesis H_0: $\mu_1 - \mu_2 = 0$; no difference in I.Q.

Alternative hypothesis H_1: $\mu_1 - \mu_2 > 0$, husbands are more intelligent than the wives

(Right tailed test)

Level of significance: At $\alpha = 0.05$ level of significance for right tailed test with 18 degree freedom is $t_\alpha = 1.734$.

Test statistics: $$t = \frac{(\bar{x} - \bar{y}) - (\mu_x - \mu_y)}{S\sqrt{\left(\frac{1}{n_1} + \frac{1}{n_2}\right)}}$$

$$\bar{x}_1 = \frac{117 + 105 + 97 + 105 + 123 + 109 + 86 + 78 + 103 + 107}{10}$$

$$= \frac{1030}{10} = 103$$

$$\bar{y}_1 = \frac{106 + 98 + 87 + 104 + 116 + 95 + 90 + 69 + 108 + 85}{10}$$

$$= 95.8$$

$$S_1^2 = \frac{(117 - 103)^2 + (105 - 103)^2 + \ldots.. + (78 - 103)^2 + (103 - 103)^2 + (107 - 103)^2}{9}$$

$$S_1^2 = 178.44$$

$$S_1 = 13.358$$

$$S_2^2 = \frac{(106-95.8)^2+(98-95.8)^2+(87-95.8)^2+.....+(108-95.8)^2+(85-95.8)^2}{9}$$

$$= 186.622$$

$$S_2 = 13.66$$

$$n_1 = 10, \quad n_2 = 10$$

$$S_p = \sqrt{\frac{S_1^2(n_1-1)+S_2^2(n_2-1)}{n_1+n_2-2}} = \sqrt{\frac{178.44(9)+186.622(9)}{18}}$$

$$= \sqrt{\frac{3285.558}{18}} = 13.5$$

$$t = \frac{(103-95.8)-0}{13.5\sqrt{\frac{1}{10}+\frac{1}{10}}} = \frac{7.2}{6.0373}$$

$$= 1.19256$$

Conclusion: Since t calculated value less than t table at given level. Hence accept the Null Hypothesis H_0. We can conclude that there is no difference in I.Q.'s.

Problem 3: Two independent samples of sound 7 items respectively had the following values.

Sample I:	11	11	13	11	15	9	12	14
Sample II:	9	11	10	13	9	8	10	–

Is the difference between the means of samples significant? **(Feb 2007, April 2005)**

Solution:

Null hypothesis H_0: $\mu_1 - \mu_2 = 0$ *i.e.,* no difference between the means

Alternative hypothesis H_1: $\mu_1 - \mu_2 \neq 0$ (Two tailed test)

Level of significance: At $\alpha = 0.05$ level of significance for two tailed test with 13 degree freedom is $t_{\alpha/2} = 2.160$

Test statistics:

$$t = \frac{(\bar{x}-\bar{y})-(\mu_x-\mu_y)}{S\sqrt{\left(\frac{1}{n_1}+\frac{1}{n_2}\right)}}$$

$$\bar{x} = \frac{11+11+13+11+15+9+12+14}{8}$$

$$= \frac{96}{8} = 12$$

x_i	y_i	$(x_i - \bar{x})$	$(x_i - \bar{x})^2$	$(y_i - \bar{y})$	$(y_i - \bar{y})^2$
11	9	–1	1	–1	1
11	11	–1	1	1	1
13	10	1	1	0	0
11	13	–1	1	3	9
15	9	3	9	–1	1
9	8	–3	9	–2	4
12	10	0	0	0	0
14	–	2	4		
96	70		26		16

$$\bar{y} = \frac{9+11+10+13+9+8+10}{7} = \frac{70}{7} = 10$$

$$S^2 = \frac{1}{n_1 + n_2 - 2}\left[\sum_i (x_i - \bar{x})^2 + \sum (y_1 - \bar{y})^2\right]$$

$$= \frac{26 + 16}{13} = 3.23$$

$$S = 1.79$$

$$t = \frac{(12-10)-0}{1.79\sqrt{\frac{1}{8}+\frac{1}{7}}} = 2.15$$

Conclusion: Since t calculated value less than t table value at given level. Hence accept the Null Hypothesis H_0.

Problem 4: Ten soldiers participated in a shooting competition in the first week. After intensive training they participated in the competition in the second week. Their scores before and after training are given as follows.

Scores before:	67	24	57	55	63	54	56	68	33	43
Scores after:	70	38	58	58	56	67	68	75	42	38

Do the data indicate that the soldiers have been benefited by the training? **(April 2005)**

Solution:

Null hypothesis: H_0: $\mu_1 - \mu_2 = 0$ *i.e.,* there is no benefit of the training

Alternative hypothesis H_1: $\mu_1 - \mu_2 < 0$ (Left tailed test)

Level of significance: At $\alpha = 0.05$ level of significance for left tailed test with 18 degree freedom is $t_\alpha = -1.734$.

Test statistics:
$$t = \frac{(\bar{x}-\bar{y}) - (\mu_x - \mu_y)}{S\sqrt{\left(\frac{1}{n_1} + \frac{1}{n_2}\right)}}$$

$$\bar{x} = \frac{67+24+57+55+63+54+56+68+33+43}{10}$$

$$= \frac{520}{10}$$

$$= 52$$

Similarly,

$$\bar{y} = \frac{570}{10}$$

$$= 57$$

$$S^2 = \frac{1}{n_1 + n_2 - 2}\left[\sum_i (x_i - \bar{x})^2 + \sum (y_1 - \bar{y})^2\right]$$

x_i	y_i	$(x_i - \bar{x})$	$(x_i - \bar{x})^2$	$(y_i - \bar{y})$	$(y_i - \bar{y})^2$
67	70	15	225	13	169
24		–28	784	–19	361
57	38	5	25	1	1
55	58	3	9	1	1
63	58	11	121	–1	1
54	56	2	4	10	100
56	67	4	16	11	121
68	68	16	256	18	324
33	75	–19	361	–15	225
43	42	–9	81	–19	361
	38				
520	570		1882		1664

$$S^2 = \frac{1882 + 1664}{18} = \frac{3546}{18} = 197$$

$$S^2 = 197$$

$$S = 14.03$$

$$t = \frac{(52-57)-0}{14.03.\sqrt{\frac{1}{10}+\frac{1}{10}}} = 0.796$$

Conclusion: Since t calculated value less than t table value at given level. Hence accept the null hypothesis H_0.

Accept the claim that soldiers have been benefited by the training.

Problem 5: To compare two kinds of bumper guards, 6 of each kind were mounted on a car and then the car has run into a concrete wall. The following are the costs of repairs:

Guard 1:	107	148	123	165	102	119
Guard 2:	134	115	112	151	133	129

Use the 0.01 level of significance to test whether the difference between two sample means is significant?

Solution:

Null hypothesis: $H_0: \mu_1 = \mu_2$

Alternative hypothesis: $H_1: \mu_1 \neq \mu_2$ (Two tailed test)

Level of significance: At $\alpha = 0.01$ level of significance for two tailed test with 10 degree freedom is $t_{\alpha/2} = 3.169$

Test statistics:

$$t = \frac{\bar{x} - \bar{y}}{S\sqrt{\left(\frac{1}{n_1} + \frac{1}{n_2}\right)}}$$

$$\bar{x} = \frac{107 + 148 + 123 + 165 + 102 + 119}{6}$$

$$= \frac{764}{6} = 127.33$$

$$\bar{y} = \frac{134 + 115 + 112 + 151 + 133 + 129}{6} = 129$$

x_i	y_i	$(x_i - \bar{x})$	$(x_i - \bar{x})^2$	$(y_i - \bar{y})$	$(y_i - \bar{y})^2$
107	134	–20.3	413.3	5	25
148	115	20.67	427.24	–14	196
123	112	–4.33	18.74	–17	289
165	151	37.67	1419.02	22	484
102	133	–25.33	641.6	4	16
119	129	–8.33	69.3889	0	0
764	774		2989.28		1010

$$S^2 = \frac{1}{n_1 + n_2 - 2}\left[\sum_i (x_i - \bar{x})^2 + \sum (y_1 - \bar{y})^2\right]$$

$$S^2 = \frac{2989.28 + 1010}{6 + 6 - 2} = \frac{3999.28}{10}$$

$$= 399.928$$

$$S = 19.998$$

$$t = \frac{(127.33 - 129)}{19.998\sqrt{\frac{1}{6} + \frac{1}{6}}} = -0.1446$$

Conclusion: Since t calculated value less than t table value at given level. Hence accept the null hypothesis H_0.

Problem 6: Measurements of the fat content of two kinds of ice cream, brand A and Brand B, yielded the following sample data:

Brand A(%)	13.5	14	13.6	12.9	13
Brand B(%)	12.9	13	12.4	13.5	12.7

Test the null hypothesis $\mu_1 = \mu_2$ (where μ_1 and μ_2 are the respective true average fat contests of the two kinds of ice cream) against the alternative hypothesis μ_1 and μ_2 at the level of significance $\alpha = 0.05$. **(Nov 2007, May 2005, Nov 2003)**

Solution:

Null hypothesis H_0: $\mu_1 = \mu_2$

Alternative hypothesis H_1: $\mu_1 \neq \mu_2$ (Two tailed test)

Level of significance: At 0.05 level of significance for two tailed test with 8 degree freedom is $t_{\alpha/2} = 2.306$

Test statistics:

$$t = \frac{\bar{x} - \bar{y}}{\sqrt{n_1 s_1^2 + n_2 s_2^2}}$$

$\bar{x}$ = mean of brand A = 13.4

$\bar{y}$ = mean of brand B = 12.9

S_1 = standard deviation brand A = 0.406

S_2 = standard deviation brand B = 0.453

$n_1 = 5,\ n_2 = 5$

$$t = \frac{13.4 - 12.9}{\sqrt{5(0.406)^2 + 5(0.453)^2}}$$

$$= \frac{0.5}{1.36} = 0.368.$$

Conclusion: Since t calculated value less than t table value at given level. Hence accept the null hypothesis H_0.

Problem 7: Samples of two types of electric lights bulbs were tested for length of life and following data were obtained

	Type I	*Type II*
Sample size	8	7
Sample mean	1,234	1,036
Sample standard deviation	36	40

Is the difference in the means sufficient to warrant that Type-I is superior to Type-II regarding length of life?

Solution:

Null hypothesis H_0: $\mu_1 = \mu_2$

Alternative hypothesis H_1: : $\mu_1 > \mu_2$ (Right tailed test)

Level of significance: At 0.05 level of significance for right tailed test with 13 degree freedom is $t_\alpha = 1.77$.

Test statistics:

$$t = \frac{(\bar{x} - \bar{y})}{S\sqrt{\frac{1}{n_1} + \frac{1}{n_2}}}$$

Sample mean $\bar{x} = 1{,}234$

Sample mean $\bar{y} = 1{,}036$

S_1 = standard deviation = 36

S_2 = standard deviation = 40

$n_1 = 8$, $n_2 = 7$

$$S^2 = \frac{n_1 s_1^2 + n_2 s_2^2}{n_1 + n_2 - 2} = \frac{8 \times 1296 + 7 \times 1600}{8 + 7 - 2} = 1659.08$$

$$t = \frac{1234 - 1036}{\sqrt{1659.08\left(\frac{1}{8} + \frac{1}{7}\right)}} = 9.39$$

Conclusion: Since t calculated value greater than t table value at given level. Hence reject the null hypothesis H_0.

Problem 8: The following are the average weekly loss of worker hours due to accidents in 10 industrial plants before and after a certain safety program was put into operation.

Before:	45	73	46	124	33	57	83	34	26	17
After:	36	60	44	119	35	51	77	29	24	11

Test whether the safety programme is effective in reducing the number of accidents at the level of significance of 0.05. **(Nov 2007, Aug 2007, April 2007, May 2005, Nov 2004)**

Solution:

$$n_1 = n_2 = 10$$

$$\bar{x} = \frac{538}{10} = 53.8$$

$$S_1^2 = \frac{1}{n_1 - 1}[(45-53.8)^2 + (73-53.8)^2 + (46-53.8)^2 + (124-53.8)^2$$
$$+ (33-53.8)^2 + (57-53.8)^2 + (83-53.8)^2 + (34-53.8)^2$$
$$+ (26-53.8)^2 + (17-53.8)^2]$$

$$= \frac{1}{9} [(-8.8)^2 + (19.2)^2 + (-7.8)^2 + (70.2)^2 + (20.8)^2 + (3.2)^2 + (29.2)^2 + (-19.8)^2 + (-27.8)^2 + (-36.8)^2]$$

$$= \frac{9251.6}{9} = 1027.96$$

$$S^2_2 = \frac{1}{9} [(-12.6)^2 + (11.4)^2 + (-4.6)^2 + (70.4)^2 + (-13.6)^2 + (2.4)^2 + (28.4)^2 + (-19.6)^2 + (-24.6)^2 + (-37.6)^2]$$

$$= \frac{8665.6}{9} = 962.84$$

Null hypothesis: H_0: $\bar{x} = \bar{y}$

Alternative hypothesis: H_1: $\bar{x} > \bar{y}$ (Right tailed test)

Level of significance: At 0.05 level the value of significant for right tailed test with 18 degree freedom is 2.101

Test statistics :

$$= \frac{(\bar{x} - \bar{y})}{\sqrt{\left(\frac{s_1^2}{n_1} + \frac{s_2^2}{n_2}\right)}} = \frac{53.8 - 48.6}{\sqrt{\left(\frac{1027.96}{10} + \frac{962.84}{10}\right)}}$$

$$= \frac{5.2}{14.11} = 0.369.$$

Conclusion: Since calculated t value is less than t table value at given lecel. Hence accept the null hypothesis H_0

∴ There is no improvement in the number of accidents after adopting the safety program.

7.5 F-DISTRIBUTION

Suppose we want to test

(*i*) Whether two independent samples $x_i (i = 1, 2, \ldots\ldots, n_1)$ and y_j, $(j = 1, 2, \ldots, n_2)$ have been drawn from the normal populations with the same variance σ^2 (say) or

(*ii*) Whether the two independent estimates of the population variance are homogeneous or not.

Under the null hypothesis (H_0) that (*i*) $\sigma_1^2 = \sigma_2^2 = \sigma^2$, *i.e.*, the population variances are equal, or (*ii*) Two independent estimates of the population variance are homogeneous, the statistic F is given by:

$$F = \frac{S_1^2}{S_2^2} \quad \ldots(1)$$

where

$$S_1^2 = \frac{1}{n_1 - 1}\sum(x_i - \bar{x})^2 \qquad S_2^2 = \frac{1}{n_2 - 1}\sum(y_i - \bar{y})^2 \quad \ldots(2)$$

are unbiased estimates of the common population variance σ^2 obtained from two independent samples and it follows Snedecor's F-distribution with (v_1, v_2) degree freedom, where $v_1 = n_1 - 1$ and $v_2 = n_2 - 1$.

7.6 SOLVED PROBLEMS

Problem 1: There are two firms A and B which produce rivets. Both the firms produce rivets whose mean diameters are same but their standard deviations may differ. A sample of 22 rivets produced by the firm A showed a S.D. of 2.9 mm while a sample of size 16 rivets from the form B showed a S.D. of 3.8 mm. Test whether both the firms have the same variability.

Solution:

Data: $n_1 = 22$ $\quad n_2 = 16$

Sample S.D. $s_1 = 2.9$ $\quad s_2 = 3.8$

H_0 : $\sigma_1^2 = \sigma_2^2$

H_1: $\sigma_1^2 = \sigma_2^2$

$$S_1^2 = \frac{n_1\, s_1^2}{n_1 - 1} = \frac{22\,(2.9)^2}{21} = 8.810$$

$$S_2^2 = \frac{n_2\, s_2^2}{n_2 - 1} = \frac{16\,(3.8)^2}{15} = 15.403$$

Since $S_2^2 > S_1^2$

$$\therefore \quad F = \frac{S_1^2}{S_2^2} = \frac{15.403}{8.810} = 1.748$$

This is a F distribution with $(v_2, v_1) = (15, 21)$ degree freedom

The table value of F (15, 21) at 5% level of significance and conclude that the variability is same.

Problem 2: Two random samples of sizes 8 and 11 drawn from two normal populations are characterized as follows:

Sample	*Size*	*Sum*	*Sum of squares*
1	8	9.6	61.52
2	11	16.5	73.26

Test whether the two population have the same variance use $\alpha = 0.05$.

Solution:

Null Hypothesis H_0: $\sigma_1^2 = \sigma_2^2$

$n_1 = 8$ $\quad n_2 = 11$

$\Sigma x_i = 9.6$ $\quad \Sigma y_i = 16.5$

$\Sigma x_i^2 = 61.52$ $\quad \Sigma y_i^2 = 73.26$

$$\bar{x} = \frac{\sum x_i}{n_1} = \frac{9.6}{8} = 1.2$$

$$\bar{y} = \frac{\sum x_i}{n_2} = \frac{16.5}{11} = 1.5$$

$$S_1^2 = \frac{\sum x_i^2}{n_1} - (\bar{x})^2 = \frac{61.52}{8} - (1.2)^2 = 6.25$$

$$\therefore \quad S_1^2 = \frac{n_1 s_1^2}{n_1 - 1} = \frac{8(6.25)}{7} = 7.14$$

$$S_2^2 = \frac{\sum y_i^2}{n_2} - (\bar{y})^2 = \frac{73.26}{11} - (1.5)^2 = 4.41$$

$$S_2^2 = \frac{n_2 s_2^2}{n_2 - 1} = \frac{11(4.41)}{10} = 4.851$$

Since $S_2^2 > S_1^2$

$$\therefore \quad F = \frac{7.14}{4.851} = 1.472$$

The table value $F_{0.05}(7, 10) = 3.64$

Since $F_{cal} < F_{tab}$ $\therefore$ We accept H_0 at 5% level of significance and conclude that both the samples have the same variances.

Problem 3: The time taken by workers in performing a job by two methods A and B is given below.

A:	20	16	26	27	23	22	
B:	27	33	42	35	32	34	38

Do the data show that the variances of the time distribution do not differ significantly?

Solution: Null Hypothesis $H_0 : \sigma_1^2 = \sigma_2^2$

x_i	x_i^2
20	400
16	256
26	676
27	729
23	529
22	484
$\Sigma x_i = 134$	$\Sigma x_i^2 = 3074$

y_i	y_i^2
27	729
33	1089
42	1764
35	1225
32	1225
34	1024
38	1444
$\Sigma y_i = 241$	$\Sigma y_i^2 = 8500$

$$n_1 = 6 \qquad n_2 = 7$$

$$\bar{x} = \frac{\sum x_i}{n_1} = \frac{134}{6} = 22.33$$

$$\bar{y} = \frac{\sum y_i}{n_2} = \frac{241}{7} = 34.43$$

$$s_1^2 = \frac{\sum x_i^2}{n_1} - (\bar{x})^2 = \frac{3074}{6} - (22.33)^2 = 13.70$$

$$\therefore \quad s_1^2 = \frac{n_1 s_1^2}{n_1 - 1} = \frac{6(13.70)}{5} = 16.44$$

$$s_2^2 = \frac{\sum y_i^2}{n_2} - (\bar{y})^2 = \frac{8500}{7} - (34.43)^2 = 28.86$$

$$s_2^2 = \frac{n_2 s_2^2}{n_2 - 1} = \frac{7(28.86)}{6} = 33.67$$

Since $s_2^2 > s_1^2$

$$\therefore \quad F = \frac{s_1^2}{s_2^2} = \frac{33.67}{16.44} = 2.05$$

Table value $F_{0.5}(5, 6) = 4.95$

Since $F_{cal} < F_{tab}$

$\therefore$ We accept H_0 at 5% of level of significance and conclude that the difference in the variance is not significant.

Problem 4: Five measurements of the output of two units have given the following results. (in kilograms of material per one hour of operation). Assuming that both samples have been obtained from normal populations test at 10% significance level it two populations have the same variance.

Unit A	14.1	10.1	14.7	13.7	14.0
Unit B	14.0	14.5	13.7	12.7	14.1

Solution:

x_i	$x_i - \bar{x}$	$(x_i - \bar{x})^2$	y_i	$y_i - \bar{y}$	$(y_i - \bar{y})^2$
14.1	0.8	0.64	14.0	0.2	0.04
10.1	–3.2	10.24	14.5	0.7	0.49
14.7	1.4	1.96	13.7	–0.1	0.01
13.7	0.4	0.16	12.7	–1.1	1.21
14.0	0.7	0.49	14.1	0.3	0.09
66.6		13.49	69.0		1.84

$$\bar{x} = \frac{66.6}{5} = 13.3,$$

$$\bar{y} = \frac{69}{5} = 13.8$$

Null hypothesis $H_0 : s_1^2 = s_2^2$
Alternative Hypothesis H_1: $s_1^2 \neq s_2^2$
Level of significance: $\alpha = 0.10$

Test statistics: $$F = \frac{s_1^2}{s_2^2},\ n_1 = n_2 = 5$$

$$s_1^2 = \sum\left(\frac{(x_i - \bar{x})^2}{n_1 - 1}\right) = \frac{13.49}{4} = 3.37$$

$$s_2^2 = \sum\left(\frac{(y_i - \bar{y})^2}{n_2 - 1}\right) = \frac{1.84}{4} = 0.46$$

$$\therefore \quad F = \frac{3.37}{0.46} = 7.3$$

Conclusion: $F_{0.95}(4, 4) = 6.39 \quad F > F_\alpha$

∴ Null hypothesis is rejected.

∴ There is significant difference between the variances.

***Problem** 5:* The daily wages in rupees of skilled workers in two cities are as follows:

City	*Size of sample*	*S.D. of wages in the sample*
City A	16	25
City B	13	32

Test at 5% level the equality of variances of the wage distribution in the two cities.

Solution:

Null hypothesis $H_0 : s_1^2 = s_2^2$

Alternative hypothesis H_1: $s_1^2 \neq s_2^2$

Level of significance: $\alpha = 0.05$

Test statistics : $F = \frac{s_1^2}{s_2^2}$

where $s_1^2 = \sum\left(\frac{(x_i - \bar{x})^2}{n_1 - 1}\right)$ and $s_2^2 = \sum\left(\frac{(y_i - \bar{y})^2}{n_2 - 1}\right)$

Standard deviation of 1st sample is = 25

$$\sigma_1^2 = (25)^2 = \sum\left(\frac{(x_i - \bar{x})^2}{n_1}\right)$$

$$s_1^2 = \frac{\sigma_1^2 . n_1}{n_1 - 1} = \frac{16 \times (25)^2}{15} = 666.67$$

$$s_2^2 = \frac{\sigma_2^2 . n_2}{n_2 - 1} = \frac{(25)^2 \times 0.13}{12} = 1109.33$$

$\therefore$ $F = \frac{1109.33}{666.67}$ (taking the greater value in the numerator)

$$= 1.66$$

Conclusion: $F_{0.05}(12, 15) = 2.48 \; F < F_\alpha$

$\therefore$ Null hypothesis is accepted.

$\therefore$ There is no significant difference between the variances.

7.7 CHI-SQUARE (χ^2) TESTS

Chi-square test for independences: Chi-square criteria based on observed and expected cell frequencies play an important role in many problems dealing with the analysis of count data. In this section we shall use such a criterion to analyze contingency tables, which are also known as $r \times c$ tables.

In $r \times c$ contingency table we find the association between two attributes A and B under consideration. If the hypothesis is rejected then the two attributes are considered to be associated, otherwise they are treated as independent. For example to see whether a given medicine 'A' is effective in curing a certain disease 'B' or whether there is any association between the intelligence of father and son, the $r \times c$ tables are used.

An $r \times c$ contingency tables has c columns representing the different categories $A_1, A_2, \ldots, A_c$ of one attribute, r rows representing the different categories $B_1, B_2, \ldots, B_r$ of another attribute and O_{ij} is the observed cell frequency for the cell belonging to the ith row and jth column schematically.

	A_1	A_2	....	A_c	Totals
B_1	$(A_1 B_1)$	$(A_2 B_1)$	....	$(A_c B_1)$	(B_1)
B_2	$(A_1 B_2)$	$(A_2 B_2)$	....	$(A_c B_2)$	(B_2)
.	.	.	.	.	.
.	.	.	.	.	.
.	.	.	.	.	.
B_r	$(A_1 B_r)$	$(A_2 B_r)$	....	$(A_c B_r)$	(B_r)
Totals	(A_1)	(A_2)	....	(A_c)	N

Expected frequencies (e_{ij}) are calculated as follows :

$$e_{(A_1B_1)} = \frac{(A_1)(B_1)}{N}$$

$$e_{(A_2B_1)} = \frac{(A_2)(B_1)}{N}$$

$$\vdots$$

$$e_{(A_cB_r)} = \frac{(A_c)(B_r)}{N}$$

$$\chi^2 = \sum_{i=1}^{r}\sum_{j=1}^{c}\frac{\left(O_{ij} - e_{ij}\right)^2}{e_{ij}}$$ follow χ^2 distribution with $(r-1)(c-1)$ degree freedom.

7.8 SOLVED PROBLEMS

Problem 1: In an experiment on immunization of cattle from tuberculosis the following results were obtained:

	Affected	*Not affected*
Inoculated	12	26
Not inoculated	16	6

Analyze and discuss the effect of vaccine in controlling susceptibility to tuberculosis.

Solution:

Null Hypothesis H_0: Vaccine is not effective in controlling susceptibility to tuberculosis (they are independent).

Alternative Hypothesis H_1 : They are associated.

Level of significance: If $\alpha = 0.05$ then with $(2-1)(2-1) = 1$ degree freedom is $\chi^2_{0.05,1} = 3.84$

Test statistics:

			Total
	12	26	38
	16	6	22
Total	28	32	60

Observed Frequency O	*Expected Frequency E*	*(O – E)*	*(O – E)²*	*(O – E)²/E*
12	17.7	–5.7	32.49	1.8355
26	20.3	5.7	32.49	1.6004
16	10.3	5.7	32.49	3.1543
6	11.7	–5.7	32.49	2.7769

Total: 9.3671

$O_{11} = 12 \qquad e_{11} = \frac{38 \times 28}{60} = 17.7$

$O_{12} = 26 \qquad e_{12} = \frac{38 \times 32}{60} = 20.3$

$O_{21} = 16 \qquad e_{21} = \frac{22 \times 28}{60} = 10.3$

$O_{22} = 6 \qquad e_{22} = \frac{22 \times 32}{60} = 11.7$

$$\therefore \qquad \chi^2 = \sum_{i=1}^{r}\sum_{j=1}^{c}\frac{(O_{ij} - e_{ij})^2}{e_{ij}} = 9.3671$$

Conclusion: Since calculated $\chi^2 = 9.3671$ is greater than $\chi^2_{0.05,1} = 3.84$ we reject H_0. Hence we conclude that vaccine is effective in controlling susceptibility to tuberculosis.

Problem 2: In an antimalarial campaign in a certain area quinine was administered to 812 persons out of a total population of 3248. The numbers of fever cases are shown below.

	Fever	*No fever*
Quinine	20	792
No quinine	220	2216

Discuss the usefulness of quinine in checking malaria

Solution:

Null Hypothesis H_0 : Quinine is not effective in checking malaria.

Alternative Hypothesis H_1: Quinine is effective in checking malaria

Level of significance: $\alpha = 0.05$, with (2–1) (2–1) = 1 degree freedom is $\chi^2_{0.05,1} = 3.84$

			Total
	20	792	812
	220	2216	2436
Total	240	3008	3248

Test statistics:

$O_{11} = 20$ $\quad e_{11} = \dfrac{812 \times 240}{3248} = 60$

$O_{12} = 792$ $\quad e_{12} = \dfrac{812 \times 3008}{3248} = 752$

$O_{21} = 220$ $\quad e_{21} = \dfrac{2436 \times 240}{3248} = 180$

$O_{22} = 2216$ $\quad e_{22} = \dfrac{2436 \times 3008}{3248} = 2256$

Observed Frequency O	*Expected Frequency E*	*(O – E)*	*(O – E)²*	*(O – E)² /E*
20	60	–40	1600	26.67
792	752	40	1600	2.13
220	180	40	1600	8.89
2216	2256	–40	1600	0.71

Total : 38.40

$$\therefore \quad \chi^2 = \sum_{i=1}^{r}\sum_{j=1}^{c}\frac{(O_{ij}-e_{ij})^2}{e_{ij}}$$

$$= 38.40$$

Conclusion: Since calculated $\chi^2 = 38.40$ is greater than $\chi^2_{0.05,1} = 3.84$, we reject H_0. Hence we conclude that quinine is effective in checking malaria.

Problem 3: The following data represents a person's ability in mathematics and his interest in statistics:

Ability in Mathematics

		Low	*Average*	*High*
	Low	63	42	15
Interest in	Average	58	61	31
statistics	High	14	47	29

Use the 0.01 level of significance and test for independence between person's ability in mathematics and interest in statistics.

Solution:

Null Hypothesis H_0 : Ability in mathematics and interest in statistics are independent.

Alternate Hypothesis H_1 : These two attributes are not independent.

Level of significance: $\alpha = 0.01$,

with (3–1)(3–1) = 4 degree freedom is $\chi^2_{0.01,4} = 13.277$

Test statistics:

				Total
	63	42	15	120
	58	61	31	150
	14	47	29	90
Total	135	150	75	360

$O_{11} = 63$ $\quad e_{11} = \dfrac{120 \times 135}{360} = 45$

$O_{12} = 42$ $\quad e_{12} = \dfrac{120 \times 150}{360} = 50$

$O_{13} = 15$ $\quad e_{13} = \dfrac{120 \times 75}{360} = 25$

$O_{21} = 58$ $\quad e_{21} = \dfrac{150 \times 135}{360} = 56$

$O_{22} = 61$ $\quad e_{22} = \dfrac{150 \times 150}{360} = 62$

$O_{23} = 31$ $\quad e_{23} = \dfrac{150 \times 75}{360} = 31$

$O_{31} = 14$ $\quad e_{31} = \dfrac{90 \times 135}{360} = 34$

$O_{32} = 47$ $\quad e_{32} = \dfrac{90 \times 150}{360} = 38$

$O_{33} = 29$ $\quad e_{33} = \dfrac{90 \times 75}{360} = 19$

Observed Frequency O	*Expected Frequency E*	$(O - E)$	$(O - E)^2$	$(O - E)^2/E$
63	45	18	324	7.2
42	50	–8	64	1.28
15	25	–10	100	4
58	56	2	4	0.0714
61	62	–1	1	0.0161
31	31	0	0	0
14	34	–20	400	11.7647
47	38	9	81	2.1315
29	19	10	100	5.2631

Total: 31.73

$$\therefore \qquad \chi^2 = \sum_{i=1}^{r}\sum_{j=1}^{c}\frac{\left(O_{ij}-e_{ij}\right)^2}{e_{ij}} = 31.73$$

Conclusion: Since calculated $\chi^2 = 31.73$ is greater than χ^2 table value we reject H_0.

Hence we conclude that ability in mathematics and interest in statistics are not independent.

Problem 4: Four methods are under development for making discs of a superconducting material. Fifty discs are made by each method and they are checked for superconductivity when cooled with liquid.

Super-	1st *method*	2nd *method*	3rd *method*	4th *method*
conductors	31	42	22	25
Failures	19	8	28	25

Test the significant difference between the proportions of superconductors at 0.05 levels.

(Nov 2006, April 2006)

Solution:

Null Hypothesis H_0: $p_1 = p_2 = p_3 = p_4$

Alternative hypothesis H_1: p_1, p_2, p_3 and p_4 are not all equal

Level of significance: At $\alpha = 0.05$ level of significant with 3 degree freedom is $\chi^2_{0.05} = 7.815$

Test statistics:

31	42	22	25	120
19	8	28	25	80
50	50	50	50	200

$O_{11} = 31 \qquad e_{11} = \dfrac{120 \times 50}{200} = 30$

$O_{12} = 42 \qquad e_{12} = \dfrac{120 \times 50}{200} = 30$

$O_{13} = 22 \qquad e_{13} = \dfrac{120 \times 50}{200} = 30$

$O_{14} = 25 \qquad e_{14} = \dfrac{120 \times 50}{200} = 30$

$O_{21} = 19 \qquad e_{21} = \dfrac{80 \times 50}{200} = 20$

$O_{22} = 8 \qquad e_{22} = \dfrac{80 \times 50}{200} = 20$

$O_{23} = 28 \qquad e_{23} = \dfrac{80 \times 50}{200} = 20$

$O_{24} = 25 \qquad e_{24} = \dfrac{80 \times 50}{200} = 20$

Observed Frequency O	Expected Frequency E	(O – E)	$(O - E)^2$	$(O - E)^2/E$
31	30	1	1	0.033
42	30	12	144	4.8
22	30	8	64	2.133
25	30	5	25	0.833
19	20	1	1	0.05
8	20	12	144	7.2
28	20	8	64	3.2
25	20	5	25	1.25

Total: 19.50

$$\therefore \quad \chi^2 = \sum_{i=1}^{r}\sum_{j=1}^{c}\frac{\left(O_{ij} - e_{ij}\right)^2}{e_{ij}} = 19.50$$

Conclusion: Since χ^2 calculated value is greater than χ^2 table value at given level. Hence rejected the null hypothesis H_0.

∴ There is a significant difference between the proportions.

Problem 5: From the following data find whether there is any significant liking in the habit of taking soft drinks among the categories of employees

Employees

Soft drinks	*Clerks*	*Teachers*	*Officers*
Pepsi	10	25	65
Thumsup	15	30	65
Fanta	50	60	30

(Nov 2006, April 2006)

Solution:

Null Hypothesis H_0 : There is a significant liking in the habit of taking soft drinks.

Alternative hypothesis H_1: There is no significant liking.

Level of signature: At $\alpha = 0.05$ level of significant value with 4 degree freedom is $\chi^2_{0.05} = 9.488$

Test statistics:

				Total
	10	25	65	100
	15	30	65	110
	50	60	30	140
Total	75	115	160	350

$O_{11} = 10 \qquad e_{11} = \dfrac{100 \times 75}{350} = 21.42$

$O_{12} = 25 \qquad e_{12} = \dfrac{100 \times 115}{350} = 32.9$

$O_{13} = 65 \qquad e_{13} = \frac{100 \times 160}{350} = 45.71$

$O_{21} = 15 \qquad e_{21} = \frac{100 \times 75}{350} = 23.57$

$O_{22} = 30 \qquad e_{22} = \frac{110 \times 115}{350} = 36.14$

$O_{23} = 65 \qquad e_{23} = \frac{110 \times 160}{350} = 50.28$

$O_{31} = 50 \qquad e_{31} = \frac{140 \times 75}{350} = 30$

$O_{32} = 60 \qquad e_{32} = \frac{140 \times 115}{350} = 46$

$O_{33} = 30 \qquad e_{33} = \frac{140 \times 60}{350} = 64$

Observed Frequency O	*Expected Frequency E*	*(O – E)*	*(O – E)²*	*(O – E)²/E*
10	21.4	–11.4	129.96	6.07
25	32.9	–7.9	62.41	1.9
65	45.7	19.3	372.49	8.15
15	23.6	–8.6	73.96	3.1
30	36.1	–6.1	37.21	1.03
65	50.3	14.7	216.09	4.29
50	30	20	400	13.3
60	46	14	196	4.26
30	64	–34	1156	18.06

Total : 60.16

$$\therefore \qquad \chi^2 = \sum_{i=1}^{r}\sum_{j=1}^{c} \frac{\left(O_{ij} - e_{ij}\right)^2}{e_{ij}} = 60.16$$

Conclusion: Since χ^2 calculated value is greater than χ^2 table value at given level. Hence reject the null hypothesis H_0.

Hence conclude that the habit of taking soft drink depends on categories of employees.

Problem 6: The following data gives the fields of interest and attitude to religion:

	Arts and commerce	*Science and engineering*	*Total*
Conformist	109	51	160
Non-conformist	23	17	40
Total	132	68	200

Examine whether the field of interest and attitude to religion are associated.

(Apr 2007, Feb 2007, Nov 2006, Nov 2005, May 2005)

Solution:

Let p_1, p_2 be the proportion of peoples whose interest and attitude to religion is given in the table

Null Hypothesis: $H_0 = p_1 = p_2$ (*i.e.,* associated)

Alternative hypothesis: H_1: $p_1 \neq p_2$ (*i.e.* not associated)

Level of significance: At $\alpha = 0.05$ level of significance with 1 degree freedom is $\chi^2_{0.05} = 3.841$

Test statistics:

$$O_{11} = 109 \qquad e_{11} = \frac{160 \times 132}{200} = 105.6$$

$$O_{21} = 23 \qquad e_{21} = \frac{40 \times 132}{200} = 26.4$$

$$O_{12} = 51 \qquad e_{12} = \frac{160 \times 68}{200} = 54.4$$

$$O_{22} = 17 \qquad e_{22} = \frac{40 \times 68}{200} = 13.6$$

Observed Frequency O	*Expected Frequency E*	*(O – E)*	*(O – E)²*	*(O – E)²/E*
109	105.6	3.4	11.56	0.109
51	54.4	–3.4	11.56	0.212
23	26.4	–3.4	11.56	0.43
17	13.6	3.4	11.56	0.85

Total: 1.6098

$$\therefore \quad \chi^2 = \sum_{i=1}^{r}\sum_{j=1}^{c}\frac{\left(O_{ij} - e_{ij}\right)^2}{e_{ij}}$$

$$= 1.6098$$

Conclusion: Since χ^2 calculated value less than χ^2 table value at given level. Hence accept the null hypothesis H_0.

Hence, both interest and attitude are associated to religion.

7.9 CHI-SQUARE TEST OF GOODNESS OF FIT

Chi-square test is used to check the significance of the discrepancy between theory and experiment. It is given by Karl Pearson in 1900 and is known as chi-square test of goodness of fit. It enables us to find the deviation of the experiment from theory is just by chance or is it really due to the inadequacy of the theory to fit the observed data.

If $O_i = [i = 1, 2, ..., n]$ is a set of observed (experimental) frequencies and $E_i[i = 1, 2, ..., n]$ is the corresponding set of expected (theoretical) frequencies then chi-square test is given by,

$$\chi^2 = \sum_{i=1}^{n}\frac{(O_i - E_i)^2}{E_i} \sim \chi^2_{(n-1)} \text{ degree freedom where } \Sigma O_i = \Sigma E_i$$

This is an approximate test for large values of n. For the validity of χ^2 –test between theory and experiment;

The following conditions must be satisfied:

1. The sample observations should be independent of each other.
2. Constraints on the cell frequencies if any, should be linear *e.g.,* $\Sigma O_i = \Sigma E_i$
3. The total frequency N, should be reasonably large, say greater than 50.
4. It does not make any assumptions regarding the parent population from which the observations are taken. Hence it is non parametric or distribution free test.
5. This test is wholly dependent on degrees of freedom.
6. The expected frequency of any item or cell must not be less than 5. If it is less than 5, then the frequencies of adjacent items should be pooled together in order to make it 5 or more than 5.

7.10 SOLVED PROBLEMS

Problem 1: A pair of dice is thrown 360 times and the frequency of each sum is indicated below

Sum	2	3	4	5	6	7	8	9	10	11	12
Frequency	8	24	35	37	44	65	51	42	26	14	14

Would you say that the dice are fair on the basis of the chi-square test at 0.05 level of significance?

Solution:

Null hypothesis H_0: The dice are fair

Alternative hypothesis H_1: The dice are not fair

Level of significance: $\alpha = 0.05$

Test statistics: $\chi^2 = \sum_{i=1}^{n} \frac{(O_i - E_i)^2}{E_i}$ O_i = Observed frequencies, E_i = Expected frequencies

The probabilities of getting a sum 2, 3, 4, 5, 6, 7, 8, 9, 10, 11, 12 are

X	2	3	4	5	6	7	8	9	10	11	12
P(X)	1/36	2/36	3/36	4/36	5/36	6/36	5/36	4/36	3/36	2/36	1/36

Sum	O_i	$E_i = 360 \times P(X)$	$O_i - E_i$	E_i
2	8	360 × 1/36 = 10	–2	0.4
3	24	360 × 2/36 = 20	4	0.8
4	35	360 × 3/36 = 30	5	0.833
5	37	360 × 4/36 = 40	–3	0.225
6	44	360 × 5/36 = 50	–6	0.72
7	65	360 × 6/36 = 60	5	0.417
8	51	360 × 5/36 = 50	1	0.02
9	42	360 × 4/36 = 40	2	0.1
10	26	360 × 3/36 = 30	–4	0.53
11	14	360 × 2/36 = 20	–6	1.8
12	14	360 × 1/36 = 10	4	1.6

$$\text{Sum} = 7.445$$

$$\therefore \quad \sum_{i=2}^{n} \frac{(O_i - E_i)^2}{E_i} = 7.445$$

$$\chi^2_{\infty} = 18.3 \text{ at 10 degree of freedom}$$

Conclusion: Since χ^2 calculated value is less than χ^2 table value at given level of significance. Accept the null hypothesis that the dice is fair since

$$\chi^2 = 7.445 < \chi^2_{\infty} = 18.3$$

Problem 2: The following is the distribution of the Hourly number of trucks arriving at a company's warehouse

Trucks arriving per hour	0	1	2	3	4	5	6	7	8
Frequency	52	151	130	102	45	12	5	1	2

Fit a Poisson distribution, and using it as parameter λ, fit a Poisson distribution. Test for goodness of fit at the 0.05 level of significance. **(Aug 2007, Feb 2007, Nov 2005)**

Solution:

x_i	f_i	f_iX_i
0	52	0
1	151	151
2	130	260
3	102	306
4	45	180
5	12	60
6	5	30
7	1	7
8	2	16
	Σf_i = 500N	$\Sigma f_i x_i$ = 1010

$$\text{Mean} = \lambda = \frac{\sum f_i x_i}{\sum f_i}$$

$$= \frac{1010}{500} = 2.02$$

The expected frequencies are given by $f(x) = \text{N. P}(X = x)$

$$\therefore \quad f(x) = 500.\ P(X = x)$$

$$f(0) = 500.\ \frac{e^{-\lambda}\lambda^0}{0!} = 500\ e^{-2.02} = 66$$

$$f(1) = 500.\ \frac{e^{-\lambda}\lambda^1}{1!} = 134$$

$$f(2) = 500 \frac{e^{-\lambda} \lambda^2}{2!} = 135$$

$$f(3) = 500 \frac{e^{-\lambda} \lambda^3}{3!} = 91$$

$$f(4) = 500. \frac{e^{-\lambda} \lambda^4}{4!} = 46$$

$$f(5) = 500. \frac{e^{-\lambda} \lambda^5}{5!} = 18$$

$$f(6) = 500 \frac{e^{-\lambda} \lambda^6}{6!} = 6$$

$$f(7) = 500 \frac{e^{-\lambda} \lambda^7}{7!} = 2$$

$$f(8) = 2$$

Observed Frequency O_i	*Expected Frequency* E_i	(O_i-E_i)	$(O_i-E_i)^2$	$(O_i-E_i)^2/E_i$
52	66	–14	196	2.96
151	134	17	289	2.16
130	135	–5	25	0.19
102	91	11	121	1.32
45	46	–1	1	0.02
12	18	–6	36	2.00
5	6	–1	1	0.17
1	2	–1	1	0.50
2	2	0	0	0.00

H_0: Poisson distribution is a good fit

$\therefore$ $\chi^2_{cal} = 9.32$ $\chi^2_{0.05,\,8} = 15.507$

Since $\chi^2_{cal} < \chi^2_{0.05,\,8}$

$\therefore$ Poisson distribution is a good fit.

Problem 3: Fit a Poisson distribution to the data and test the goodness of fit at 5% level of significance

X:	0	1	2	3	4	5	6	7	8	9	10	11	12	13	14	15	16
Y:	0	20	43	53	86	70	54	37	18	10	5	2	2	0	0	0	0

Solution:

X_i	f_i	f_iX_i	X_i^2	$f_iX_i^2$
0	0	0	0	0
1	20	20	1	20
2	43	86	4	172
3	53	159	9	477
4	86	344	16	1376
5	70	350	25	1750
6	54	324	36	1944
7	37	259	49	1813
8	18	144	64	1152
9	10	90	81	810
10	5	50	100	500
11	2	22	121	242
12	2	24	144	288
13	0	0	169	0
14	0	0	196	0
15	0	0	225	0
16	0	0	256	0
	$\Sigma f_i = 400$	$\Sigma f_iX_i = 1872$		$\Sigma f_iX_i^2 = 10544$

$$n = \Sigma f_i = 400$$

$$\bar{X} = \frac{\sum f_iX_i}{\sum f_i} = \frac{1872}{400} = 4.68$$

$$\sigma^2 = \frac{\sum f_iX_i^2}{\sum f_i} - (\bar{X})^2$$

$$= \frac{10544}{400} - 21.9024 = 4.4576$$

$$\therefore \quad \sigma^2 = 4.4576$$

$$\therefore \quad \lambda = \sigma^2 = \bar{X} = 4$$

$$E_1 = N\ P(X = 0) = 400\ \frac{e^{-\lambda}\lambda^0}{0!} = 400\ e^{-4} = 400\ (0.0183) = 0$$

$$E_2 = N\ P(X = 1) = 400\ \frac{e^{-\lambda}\lambda^1}{1!} = 400\ e^{-4}\ (4) = 29$$

$$E_3 = N\ P(X = 2) = 400\ \frac{e^{-\lambda}\lambda^2}{2!} = 400\ \frac{e^{-\lambda}(16)}{2!} = 58$$

$$E_4 = N\ P(X = 3) = 400\ \frac{e^{-\lambda}\lambda^3}{3!} = 400\ \frac{e^{-\lambda}(64)}{3!} = 78$$

$$E_5 = N\,P(X = 4) = 400\,\frac{e^{-\lambda}\lambda^4}{4!} = 400\,\frac{e^{-\lambda}(256)}{4!} = 78$$

$$E_6 = N\,P(X = 5) = 400\,\frac{e^{-\lambda}\lambda^5}{5!} = 400\,\frac{e^{-\lambda}4^5}{5!} = 62$$

$$E_7 = NP(X = 6) = 400\,\frac{e^{-\lambda}\lambda^6}{6!} = 400\,\frac{e^{-\lambda}4^6}{6!} = 42$$

$$E_8 = N\,P(X = 7) = 400\,\frac{e^{-\lambda}\lambda^7}{7!} = 400\,\frac{e^{-\lambda}4^7}{7!} = 24$$

$$E_9 = N\,P(X = 8) = 400\,\frac{e^{-\lambda}\lambda^8}{8!} = 400\,\frac{e^{-\lambda}4^8}{8!} = 12$$

$$E_{10} = NP(X = 9) = 400\,\frac{e^{-\lambda}\lambda^9}{9!} = 400\,\frac{e^{-\lambda}4^9}{9!} = 5$$

$$E_{11} = NP(X = 10) = 400\,\frac{e^{-\lambda}\lambda^{10}}{10!} = 400\,\frac{e^{-\lambda}4^{10}}{10!}\,2$$

$$E_{12} = NP(X = 11) = 400\,\frac{e^{-\lambda}\lambda^{11}}{11!} = 400\,\frac{e^{-\lambda}4^{11}}{11!} = 1$$

Observed Frequency O_i	*Expected Frequency E_i*	$(O_i - E_i)$	$(O_i - E_i)^2$	$(O_i - E_i)^2/E_i$
0	0	0	0	0
20	29	–9	81	2.79
43	58	–15	225	3.88
53	78	–25	625	8.01
86	78	8	64	0.82
70	62	8	64	1.03
54	42	12	144	3.42
37	24	13	169	7.04
18	12	6	36	3.00
10	5	5	25	5.00
5	2	3	9	4.50
4	1	3	9	9.00

$$\chi^2_{0.05,11} = 19.675$$

Since $\chi^2_{cal} < \chi^2_{0.05,11}$

∴ There is a difference between theory and experiment.

7.11 EXERCISE

1. The measurements of the output of two units have given the following results. Assuming that both samples have been level whether the two populations have the same variance.

Unit A:	14.1	10.1	14.7	13.7	14
Unit B:	14.0	14.5	13.7	12.7	14.1

(Nov 2007)

2. The following are the samples of skills. Test the significant difference between the means at 0.05 level.

Sample I:	74.1	77.7	74.4	74	73.8	
Sample II:	70.8	74.9	74.2	70.4	69.2	72.2

(Nov 2007)

3. It has been suggested that an average college teacher in AP spends less than 10 hours in a week on his own academic schedule. The figures for the time spent during a week are given below for 12 teachers.

7.1	13.1	7.8	3.6	8.4	4.9
9.6	3.4	0.1	7.2	20.3	11.1

Is the claim justified with the level of significance of 0.05? **(Nov 2007)**

4. Test of the fidelity and the selectivity of 190 radio receivers produced the results shown in the following table. **(Nov 2007, April 2007)**

		Fidelity		
		Low	Avg	High
Selectivity	Low	6	12	32
	Avg	33	61	18
	High	13	15	0

5. In an investigation on the machine performance the following results were obtained

	No. of units inspected	*No. of defectives*
Machine 1:	375	17
Machine 2:	450	22

Test whether there is any significant performance of two machines at $\alpha = 0.05$.

(Aug 2007, April 2007, Nov 2004)

6. The following is the distribution of hourly number of trucks arriving at a company's warehouse.

No. of trucks:	0	1	2	3	4	5	6	7	8
Frequency:	52	151	130	102	45	12	5	1	2

Find the mean of the distribution and using it as parameter λ fit a Poisson distribution. Test for goodness of fit at the 0.05 level of significance. **(Aug 2007, Feb 2007, Nov 2005)**

7. Suppose during 400 five minute intervals the air traffic control of an airport received 0, 1, 2, ..., 12, messages and observed frequencies are given in the following table.

X:	0	1	2	3	4	5	6	7	8	9	10	11	12
f:	4	15	47	76	68	74	46	39	15	9	5	2	0

Test whether the Poisson distribution with $\lambda = 4.5$ provides a good fit to the above data.

(April 2007)

8. The following is the distribution of the daily number failures reported in a western city on 300 days.

No. of power failures:	0	1	2	3	4	5	6	7	8	9
No. of days:	9	43	64	62	42	36	22	14	6	2

Test the goodness of fit of Poisson distribution at 0.05 level. **(April 2007)**

9. The following data gives the fields of interest and attitude to region.

	Arts and commerce	*Science and Engg.*	*Total*
Conformist	109	51	160
Non-conformist	23	17	40
Total	132	68	200

Examine whether the field of interest and attitude to region are associated.

(April 2007, Feb 2007, Nov 2006, 2005, May 2005)

10. A die is thrown 60 times with the following results.

Face:	1	2	3	4	5	6
Frequency:	8	7	12	8	14	11

Test the goodness of fit at 5% level. **(Feb 2007)**

11. Two independent samples of sound 7 items respectively had the following values.

Sample I:	11	11	13	11	15	9	12	14
Sample II:	9	11	10	13	9	8	10	–

Is the difference between the means of samples significant? **(Feb 2007, May 2005)**

12. From the following data find whether there is any significant liking in the habit of taking soft drinks among the categories of employees.

	Employees		
Soft drinks	*Clerks*	*Teachers*	*Officers*
Pepsi	10	25	65
Thums up	15	30	65
Fantha	50	60	30

13. Four methods are under development for making discs of a superconducting material. Fifty discs are made by each method and they are checked for superconductivity when cooled with liquid.

	1st method	*2nd method*	*3rd method*	*4th method*
Superconductors	31	42	22	25
Failures	19	8	28	25

Test the significant difference between the proportions of superconductors at 0.05 level.

14. A pair of dice are thrown 360 times and the frequency of each sum is indicated as follows.

Sum:	2	3	4	5	6	7	8	9	10	11	12
Frequency:	8	24	35	37	44	65	51	42	26	14	14

Would you say that the dice are fair as the basis of the chi-square test at 0.05 level of significance

15. Two horses A and B were tested according to the time (in seconds) to run a particular track with the following results.

Horse A:	28	30	32	33	33	29	34
Horse B:	29	30	30	24	27	29	–

Test whether the two horses have the same running capacity.

16. A survey of 400 families with 5 children each revealed the following distribution. Is this result with the hypothesis that male and female births are equally probable?

No. of boys:	0	1	2	3	4	5
No. of families:	25	60	110	113	62	30

17. Five coins are tossed 320 times the number of heads observed is given below. Examine whether the coin is unbiased.

No. of heads:	0	1	2	3	4	5
Frequency:	15	45	85	95	60	20

18. Fit a Poisson distribution and test the goodness of fit for the following data

x:	0	1	2	3	4
$f(x)$:	109	65	22	3	1

19. The heights of 10 students of a college are 70, 62, 67, 71, 68, 70, 64, 66, 64, 68 inches. It is reasonable to believe that the average height is greater than 64 inches. Test the null hypothesis at 0.05 level.

20. 11 school boys were given a test in painting. They are given a month's further tuition and a second test of equal difficulty was held at the end of the month. Do the marks give evidence that the students have benefited by extra coaching.

First test marks:	25	23	19	22	21	19	22	21	25	18	20
Second test marks:	26	22	22	19	23	21	24	24	25	22	18

21. Two working designs are under consideration for adoption in a plant. A time and motion study shows that 12 workers using design A have an average assembly time of 300 seconds with a S.D. of 12 seconds and that of 15 workers using design B have an average assembly time of 335 seconds and S.D. of 15 seconds. Is there a significant difference in the average assembly time between two working designs?

22. Two random samples gave the following results

	Sample size	*Sample mean*	*Sum of squares of deviation from the mean*
I:	10	15	90
II:	12	14	108

Test whether the samples could have come from the same normal population (*i*) using F test, (*ii*) using t-test.

ANSWERS

16. 41.78 **17.** 12.5 **18.** 0.738, good fit

7.12 QUIZ

1. If there are $x = 3$ failures among 3000 parts used continuously for a month one sided 95% confidence interval is ....

   (a) $p < 0.0026$ (b) $p < 0.01$

   (c) $p < 0.05$ (d) $p < 0.005$

2. If there are 5 defective items among 4000, one sided 99% confidence interval for proportion is ....

   (a) 0.0033 (b) 0.001

   (c) 0.05 (d) 0.09

3. If a coin is tossed 300 times and head turns up 250 times. Then standard error of P = ...

   (a) 1.25 (b) 3.25

   (c) 5.65 (d) 8.65

4. A die is thrown 954 times getting 1 or 2 is success. Then standard error of P =.............

   (a) 7.8 (b) 12.6

   (c) 11.2 (d) 14.6

5. A die is thrown 1024 times. Getting an even digit is a success. Then standard error of true proportion is...............

   (a) 6 (b) 8

   (c) 12 (d) 16

6. Random samples of 200 men and 100 women were asked whether they want bus stop in their colony. 50 men and 80 women said yes. Then standard error of proportion is ....

   (a) 16.12 (b) 18.2

   (c) 15.4 (d) 17.9

7. In a locality 50 persons were randomly selected and enquired about educational achievements

   The results are

	School	*College*	*Total*
Male	15	15	30
Female	5	15	20
Total	20	30	50

   Expected frequency e_{21} = ...

   (a) 10 (b) 8

   (c) 12 (d) 18

8. The table given below shows the data obtained during an epidemic of cholera. The data is given as follows.

35	40	75
65	10	75
100	50	150

e_{11} = ...

(a) 25 (b) 50

(c) 75 (d) 40

9. To test the goodness of fit Poisson distribution was fitted. If the $\mu = 0.5$ and expected frequency when $x = 1$ = is 118, then $f(2)$ = ...

(a) 19.5 (b) 29.5

(c) 17.5 (d) 25.6

10. If standard deviations of two samples, whose samples sizes were 9 and 10 are 18 and 24, then F = ...

(a) 1.2 (b) 1.4

(c) 1.36 (d) 1.8

11. The variance of two samples were 25 and 225 sample sizes are 9 and 15 then standard error of $\mu_1 - \mu_2$ is..................

(a) 1.2 (b) 3.6

(c) 4.2 (d) 6.4

12. If expected frequencies E_i and $(O_i - E_i)^2$ are given by

$(O_i-E_i)^2$:	100	25	4	100	4	25
E_i:	20	20	20	20	20	20

Then χ^2 =

(a) 15.2 (b) 12.9

(c) 10.2 (d) 9.8

13. If expected frequencies E_i and $(O_i - E_i)^2$ are given by

$(O_i - E_i)^2$:	64	25	36	49
E_i:	18	30	30	18

Then χ^2 =

(a) 7.2 (b) 9.5

(c) 8.3 (d) 10.25

14. If $s_1^2 = 666.67$, $s_2^2 = 1109.33$, then F = ...

(a) 0.91 (b) 0.82

(c) 1.66 (d) 1.1

15. If $\bar{x} = 13.4$, $\bar{y} = 12.9$, $s = 0.43$ and $n_1 = n_2 = 5$, then t =................

(a) 1.85 (b) 2.2

(c) 0.92 (d) 3.1

16. If $\bar{x} = 31$, $\bar{y} = 28$, $s = 2.13$, $n_1 = 6$ and $n_2 = 7$ then t = ...

(a) 2.53 (b) 1.74

(c) 3.12 (d) 0.92

17. If $s_1^2 = 2.5$, $s_2^2 = 2.8$, $n_1 = 5$ and $n_2 = 6$ then standard error is..................

(a) 0.985 (b) 1.2

(c) 0.77 (d) 1.72

18. If $\bar{x} = 70$, $\bar{y} = 74$, $s = 4.59$, $n = 5$, then t = ...

(a) 1.94 (b) 1.72

(c) 2.41 (d) 3.25

19. If $\bar{x} = 14.9$, $\mu = 10.5$, $s = 10.3$, $n = 8$, then t = ...

(a) 4.8 (b) 2.4

(c) 1.79 (d) 4.1

20. If $\bar{x} = 0.25$, $\mu = 0$, $s = 2.5$ and $n = 8$, then t = ...

(a) 0.283 (b) 0.45

(c) 0.62 (d) 0.85

21. The 99% confidence interval for μ for the above problem is..................

(a) 11.9864, 19.6136 (b) 13.1781, 19.6136

(c) 13.1781, 18.4219 (d) none

22. $t = \dfrac{x - \mu}{S/\sqrt{n}}$s is t-variate with v degree freedom then v = ...

(a) n (b) $n - 1$

(c) $n + 1$ (d) none

23. If $n = 25$ and the maximum error with probability 0.95 is 0.1 then the sample variance is

(a) 6 (b) 6.25

(c) 7 (d) 7.25

24. The maximum error with probability 0.99 when the sample size is 9 and then sample standard deviation is ....

(a) 0.161 (b) 0.163

(c) 0.164 (d) 0.168

25. A machinist is making engine parts with axle diameter of 0.700 cm. A random sample of 10 parts shows a mean diameter of 0.742 cm with a standard deviation of 0.040 cm then the value of t is

(a) 3.0 (b) 3.1

(c) 3.15 (d) none

26. The value of t for the following data is

Type	*Sample size*	*Sample mean*	*Sample variance*
A	4	52	42
B	9	42	56

(*a*) 2.56 (*b*) 2.40

(*c*) 2.31 (*d*) none

27. If $n_1 = 10$, $\Sigma(x_i - \bar{x})^2 = 120$, $n_2 = 12\Sigma(y_i - \bar{y})^2 = 314$ then the value of f is

(*a*) 2.96 (*b*) 1.645

(*c*) 3.91 (*d*) none

28. If a Poisson distribution is fitted to the following data

x:	0	1	2	3	4	Total
f:	211	90	19	5	0	325

then the parameters λ is

(*a*) 0.44 (*b*) 0.48

(*c*) 0.66 (*d*) none

29. If a Binomial distribution is fitted to the following data

x:	0	1	2	3	4	Total
f:	32	178	290	236	64	800

taking $n = 4$ then the expected value of $f(3)$ under the hypothesis that male and female births are equally probable is

(*a*) 50 (*b*) 200

(*c*) 300 (*d*) none

30. The value of χ^2 for the above distribution is

(*a*) 18.63 (*b*) 20.63

(*c*) 19.63 (*d*) none

	A	*B*	*C*	*D*	Total
High HDL:	53	80	68	57	258
Not high HDL:	67	120	82	73	342
Total:	120	200	150	130	600

31. The value of e_{11} is......................

(*a*) 51.6 (*b*) 52

(*c*) 57.6 (*d*) none

32. The value of e_{22} is ....

(*a*) 114.8 (*b*) 115

(*c*) 114 (*d*) none

33. The value of e_{23} is ....

(*a*) 85.5 (*b*) 87.2

(*c*) 83.3 (*d*) none

A random sample of 500 pineapples was taken from a large consignment and 65 of them were found to be bad then

34. The standard error of the population of bad ones in a sample is

(*a*) 0.02 (*b*) 0.015

(*c*) 0.03 (*d*) none of these

35. The interval for the percentage of bad pineapples is..................

(*a*) 8.5, 17.5 (*b*) 18, 175

(*c*) 17, 35 (*d*) none

36. A machine puts out 16 imperfect articles in a sample of 500. After the machine is overhauled it puts out 3 imperfect articles in a batch of 100. In order to test the improvement of the machine, Z was calculated then Z =

(*a*) 0.123 (*b*) 0.191

(*c*) 0.103 (*d*) none

37. A random sample of 64 mean showed 15 smokers. The standard error of proportions is

(*a*) 0.035 (*b*) 0.0275

(*c*) 0.0391 (*d*) none

38. 900 bolts from a factory were examined and it was found that 5% of them were defective. The maximum error due to true proportion is

(*a*) 0.19 (*b*) 0.21

(*c*) 0.32 (*d*) 0.35

39. If n = 400 and the maximum error with probability 0.99 is 0.25 then the standard deviation of the population is

(*a*) 1.23 (*b*) 1.53

(*c*) 1.93 (*d*) 1.73

ANSWERS

1. (*a*)	**2.** (*a*)	**3.** (*d*)	**4.** (*d*)	**5.** (*d*)	**6.** (*b*)	**7.** (*b*)	**8.** (*b*)	**9.** (*a*)	**10.** (*b*)
11. (*c*)	**12.** (*b*)	**13.** (*c*)	**14.** (*c*)	**15.** (*a*)	**16.** (*a*)	**17.** (*a*)	**18.** (*a*)	**19.** (*a*)	**20.** (*a*)
21. (*a*)	**22.** (*b*)	**23.** (*b*)	**24.** (*d*)	**25.** (*c*)	**26.** (*b*)	**27.** (*d*)	**28.** (*a*)	**29.** (*d*)	**30.** (*c*)
31. (*d*)	**32.** (*c*)	**33.** (*d*)	**34.** (*b*)	**35.** (*a*)	**36.** (*c*)	**37.** (*b*)	**38.** (*a*)	**39.** (*c*)	

Chapter **8**

QUEUEING THEORY

8.0 INTRODUCTION

In everyday life, it is seen that a number of people arrive eat a cinema ticket window. If the people arrive "too frequently" they will have to wait for getting their tickets or some times do without it. Under such circumstances, the only alternative is to form a queue, called the waiting line, in order to maintain a proper discipline. Occasionally, it also happens that the person issuing tickets will have to wait, until additional people arrive. Here the arriving people are called the customers and the person issuing the tickets is called a server.

Another example is represented by letters arriving at a typist's desk. Again, the letters represent the customers and the typist represents the server. A third example is illustrated by a machine breakdown situation. A broken machine represents a customer calling for the service of a repairman.

By the term 'customer' we mean the arriving unit that requires some service to be performed. The customer may be of persons machines, vehicles, parts , etc. Queues (waiting line) stands for a number of customers waiting to be serviced. The queue does not include the customer being serviced. The process are system that performs the services to the customer is termed by service channel or service facility.

The subject of queueing is not directly concerned with optimization (maximum or minimization). Rather, it attempts to explore, understand and compare various queueing situations and thus indirectly achieves optimization approximately.

8.1 QUEUEING SYSTEM

The mechanism of a queueing process is very simple, customers arrive at a service counter and are attended to by one or more of the servers. As soon as a customer is served, it departs form the system. Thus a queueing system can be described as consisting of customers arriving for service, waiting for service if it is not immediate, and leaving the system after being served.

The general frame work of queueing system is shown below:

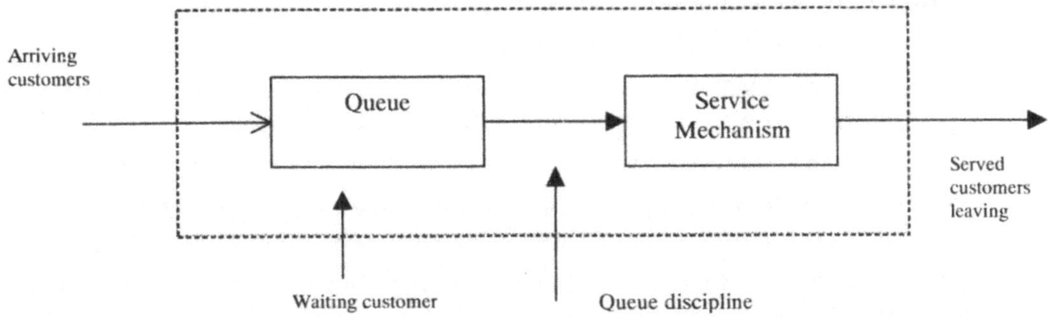

8.2 ELEMENTS OF A QUEUEING SYSTEM

The basic elements of a queueing system are as follows:

(*i*) *The input (or arrival pattern):* The input describes the way in which the customers arrive and join the system. Generally, the customers arrive in a more or less random fashion which is not worth making the prediction. Thus, the arrival pattern can best be described in terms of probabilities and consequently the probability distribution for inter-arrival times (the time between two successive arrivals) or the distribution of number of customers arriving in unit time must be defined.

(*ii*) *The service mechanism (or service pattern):* It is specified when it is known how many customers can be served at a time, what the statistical distribution of service time is and when service is available. It is true in most situations that service time is a random variable with the same distribution for all arrivals, but cases occur where there are clearly two or more classes of customers (*e.g.,* machines waiting repair) each with a different service time distribution. Service time may be constant or a random variable. Distribution of service time which are important in practice are 'negative exponential distribution' and the related 'Erlang (Gamma) distribution'.

(*iii*) *The queue discipline:* The queue discipline is the rule determining the formation of the queue, the manner of the customer's behaviour while waiting, and the manner in which they are chosen for service. The simplest discipline is 'first come, first served', according to which the customers are served in the order of their arrival. For example, such type of queue discipline is observed at a ration shop, at cinema ticket windows, at railway stations, etc. If the order is reversed, we have the 'last come, first served' discipline, as in the case of a big godown the items which come last are taken out first. An extremely difficult queue discipline to handle might be 'service in random order' or 'might is right'. Properties of a queueing system which are concerned with waiting times, in general, depend on queue discipline. For example, the variance of waiting time will be much greater with the queue discipline 'first come, last served' than with 'first come, first served'. Although mean waiting time will remain unaffected.

The following notations are used for describing the nature of service discipline.

FIFO – First In, First Out or **FCFS** ⟶ First Come, First Service discipline

LIFO – Last In, First Out or **FILO** ⟶ First In Last Out .

SIRO – Service in Random Order

(*iv*) *Customer's behaviour:* The customers generally behave in four ways

(*a*) *Balking*–A customer may leave the queue because the queue is too long and he has no time to wait or there is not sufficient waiting space.

(*b*) *Reneging*–This occurs when a waiting customer leaves the queue due to impatience.

(*c*) *Priorities*–In certain applications some customers are served before others regardless of their order of arrival. These customers have priority over others.

(*d*) *Jockeying*–Customers may jockey from one waiting line to another. It may be seen that this occurs in the super market.

(*v*) *Size of a population:* The collection of potential customers may be very large or of a moderate size. In a railway booking counter the total number of potential passengers is so large that although theoretically finite it can be regarded as infinity for all practical purposes. The assumption of infinite population is very convenient for analyzing a queueing model. However, this assumption is not valid where the customer group is represented by few machines in workshop that require operator facility from time to time. If the population size is finite then the analysis of queueing model becomes more involved.

(*vi*) *Maximum length of a queue:* Some times only a finite number of customers are allowed to stay in the system although the total number of customers in the population may or may not be finite. For example, a doctor may have appointments with k patients in a day. If the number of patients asking for appointment exceeds k, they are not allowed to join the queue. Thus, although the size of the population is infinite, the maximum number permissible in the system is k.

8.3 OPERATING CHARACTERISTICS OF QUEUEING SYSTEM

Some of the operational characteristics of queueing system, that are of a general interest for the evaluation of the performance of an existing queueing system and to design a new system are as follows :

(*a*) Expected number of customers in the system denoted by E(n) or L is the average number of customers in the system, both waiting and in service. Here, n stands for the number of customers in the queueing system.

(*b*) Expected number of customers in the queue denoted by E(m) or L_q is the average number of customers waiting in the queue. Here $m = n - 1$, *i.e.,* Excluding the customer being served.

(*c*) Expected waiting time in the system denoted by E(v) or W is the average total time spent by a customer in the system. It is generally taken to be the waiting time plus servicing time.

(*d*) Expected waiting time in queue denoted by E(w) or W_q is the average time spent by a customer in the queue before the commencement of his service.

(*e*) The server utilization facto (or busy period) denoted by $P(= \lambda/\mu)$ is the proportion of time that a server actually spends with the customers. Here λ stands for the average number of customers arriving per unit of time and μ stands for the average number of customers completing service per unit of time.

The server utilization factor is also known as traffic intensity.

8.4 PROBABILITY DISTRIBUTION IN QUEUEING SYSTEMS

It is assumed that customers joining the queueing system arrive in a random manner and follow a Poisson distribution or equivalently the inter-arrival times obey exponential distribution. In most of the cases, service times are also assumed to be exponentially distributed. It implies that the probability of service completion in any short time period is constant and independent of the length of time that the service has been in progress.

In this section the arrival and service distribution for Poisson queues are derived. The basic assumptions governing this type of queues are stated below:

Axiom 1: The number of arrivals in non-overlapping intervals are statistically independent, that is the process has independent increments.

Axiom 2: The probability of more than one arrival between time t and time $t + \Delta t$ is $0(\Delta t)$; that is the probability of two or more arrivals during small time interval Δt is negligible. Thus $P_0(\Delta t) + P_1(\Delta t) + 0(\Delta t) = 1$.

Axiom 3: The probability that an arrival occurs between time t and time $t + \Delta t$ is equal to $\lambda \Delta t + 0(\Delta t)$.

Thus $P_1(\Delta t) = \lambda \Delta t + 0(\Delta t)$,

where λ is constant independent of the total number of arrivals upto time t, Δt is an incremental element, and $0(\Delta t)$ represents the terms such that $\lim_{\Delta t \to 0} \frac{\sigma(\Delta t)}{\Delta t} = 0$.

8.5 DISTRIBUTION OF ARRIVALS (PURE BIRTH PROCESS)

The model in which only arrivals are counted and no departure takes place are called pure birth models. Stated in terms of queueing, birth-death processes usually arise when an additional customer increases the arrival (referred as birth) in the system and decreases by departure (referred as death) of a serviced customer from the system.

Let $P_n(t)$ denote the probability of n arrivals in a time interval of length t (both waiting and in service), where $n \geq 0$ is an integer. Then $P_n(t + \Delta t)$ being the probability of n arrivals in a time interval of length $t + \Delta t$ (making use of axiom 1) is as follows.

$P_n(t + \Delta t)$ = P{ n arrivals in time t and zero arrival in time Δt} + P{ $n - 1$ arrivals in time t and one arrival in time Δt} + P{$n - 1$ arrivals in time t and one arrival in time Δt} + P {$n - 2$ arrivals in time t and two arrivals in time Δt} + ...+ P{ no arrival in time t and n arrivals in time Δt}, for $n \geq 1$.

Making use of axiom 2 and axiom 3, this difference equation reduces to

$$P_n(t + \Delta t) = P_n(t)\, P_0(\Delta t) + P_{n-1}(t)\, P_1(\Delta t) + 0(\Delta t)$$

$$= P_n(t)\,[1 - \lambda \Delta t - 0(\Delta t)] + P_{n-1}(t)\,\{\lambda \Delta t + 0(\Delta t)\} + 0(\Delta t)$$

where the last term , $0(\Delta t)$ represents the terms

P[$n - k$ arrivals in time t and k arrivals in time Δt] $2 \le k \le n$

The equation given above can be re-written as

$$P_n(t + \Delta t) - P_n(t) = -\lambda \Delta t.\, P_n(t) + \lambda \Delta t.\, P_{n-1}(t) + 0(\Delta t)$$

Dividing it by Δt on both sides and then taking the limit as $\Delta t \to 0$, the equation reduces to

$$\frac{d}{dt} P_n(t) = -\lambda P_n(t) + \lambda P_{n-1}(t),\ n \ge 1 \qquad ...(1)$$

For the case when $n = 0$

$$P_0(t + \Delta t) = P_0(t)\, P_0(\Delta t) = P_0(t)[1 - \lambda \Delta t - 0(\Delta t)]$$

Rearranging the terms and then dividing on both sides by Δt, taking the limit as $\Delta t \to 0$, we have

$$\frac{d}{dt} P_0(t) = -\lambda P_0(t) \qquad ...(2)$$

To solve the $n + 1$ differential-difference equations given in (1) and (2),we make use of the generating function.

$$\phi(z, t) = \sum_{n=0}^{\infty} P_n(t).z^n \text{, in the unit circle } |z| \le 1.$$

Now multiplying the differential-difference equations given in (2) and (1) by $z^0, z^1, z^2, \ldots\ldots, z^n$ respectively and then taking summation over n from 0 to ∞, we get

$$\sum_{n=0}^{\infty} \frac{d}{dt} P_n(t) z^n = -\lambda \phi(z, t) + \lambda z \phi(z, t)$$

This can also be written as

$$\frac{d}{dt} \phi(z,t) = \lambda (z - 1) \phi(z, t).$$

An obvious solution of this differential equation is

$$\phi(z, t) = Ce^{-\lambda(z-1)t}, \text{ where C is an arbitrary constant.}$$

To determine the value of C, we use the initial condition that there is no arrival by time $t = 0$ and this gives

$$\phi(z, 0) = P_0(0) + \sum_{n=0}^{\infty} P_n(0).z^n = 1$$

Because $P_n(0) = 0$ for $n \ge 1$

Therefore, $C = 1$

Hence $\phi(z, t) = e^{\lambda(z-1)t}$ ...(3)

Now, $\frac{d}{dz}\phi(z, t)\big|_{z=0} = P_1(t)$, $\frac{d^2}{dz^2}\phi(z, t)\big|_{z=0} = 2!\, P_2(t)$

$$\frac{d^n}{dz^n}\phi(z, t)\Big|_{z=0} = n!\, P_n(t)$$

Therefore, using the value of $\phi(z, t)$ as given in equation (3), we get

$$P_0(t) = e^{-\lambda t},\ P_1(t) = (\lambda t)\, e^{-\lambda t}$$

$$P_2(t) = \frac{1}{2!}(\lambda t)^2 e^{-\lambda t} \qquad P_n(t) = \frac{1}{n!}(\lambda t)^n e^{-\lambda t}$$

The general formula, therefore, is

$$P_n(t) = \frac{1}{n!}(\lambda t)^n e^{-\lambda t}, \text{ for } n \geq 0$$

Which is the well known Poisson probability law with mean λt. Thus the random variable defined as the number of arrivals to a system in time t, has the Poisson distribution with a mean of λt arrivals, or a mean arrival rate of λ.

8.6 DISTRIBUTION OF INTER-ARRIVAL TIMES (EXPONENTIAL PROCESS)

Inter-arrival times are defined as the time intervals between two successive arrivals. Here, we shall show that if the arrival process follows the Poisson distribution an associated random variable defined as the time between successive arrivals (inter- arrival time) follows the exponential distribution $f(t) = \lambda e^{-\lambda t}$ and *vice-versa.*

Let the random variable T be the time between successive arrivals ; then

$P(T > t) = P$ (no arrival in time t) $= P_0(t) = e^{-\lambda t}$.

The cumulative distribution function of T denoted by $F(t)$ is given by

$$F(t) = P(T \leq t) = 1 - P(T > t)$$
$$= 1 - P_0(t) = 1 - e^{-\lambda t} \qquad t > 0$$

The density function $f(t)$ for inter-arrival times, therefore is

$$f(t) = \frac{d}{dt} F(t) = \lambda e^{-\lambda t} \qquad t > 0$$

The expected (or mean) inter-arrival time is given by

$$E(t) = \int_0^{\infty} t.f(t)dt = \int_0^{\infty} \lambda te^{-\lambda t}\, dt = 1/\lambda$$

where λ is the mean arrival rate.

Thus, T has the exponential distribution with mean $1/\lambda$, we would intuitively expect that if the mean arrival rate is λ, then the mean time between arrival is $1/\lambda$. Conversely, we can also show that if the inter-arrival times are independent and have the same exponential distribution then the arrival rate follows the Poisson distribution.

8.7 DISTRIBUTION OF DEPARTURES (PURE DEATH PROCESS)

The model in which only departures are counted and no other arrivals allowed are called pure death models. The queueing system starts with N customers at time $t = 0$, where $N \geq 1$. Departures occur at the rate of μ customers per unit time. To develop the differential-difference equations for the probability of n customers remaining after 't' time units $P_n(t)$, we make use of similar assumptions as was done for arrivals. Let the three axioms, given at the beginning of this section, be changed by using the word service instead of arrival of arrival and condition the probability statements by requiring the system to be non-empty.

Let us define

$\mu\Delta t$ = probability that a customer in service at time t will complete service during time Δt.

For small time interval $\Delta t > 0$, $\mu\,\Delta t$ gives probability of one departure during Δt. Using the same arguments as in pure birth process case, the differential-difference equations for this can easily be obtained.

$$P_n(t + \Delta t) = P_n(t)\ \{1 - \mu\,\Delta t + 0(\Delta t)\} + P_{n+1}(t).\ \{\mu\Delta t + 0(\Delta t)\},\ 1 \le n \le N - 1$$
$$P_0(t + \Delta t) = P_0(t) + P_1(t)\ \{\mu\,\Delta t + 0(\Delta t)\}\ n = 0$$
$$P_N(t + \Delta t) = P_N(t).\ \{1-\mu\Delta t + 0(\Delta t)\}.\ n = N$$

Re-arranging the above equations, dividing them by Δt on both sides and then taking the limits as $\Delta t \to 0$, we get

$$\frac{d}{dt} P_n(t) = -\mu\, P_n(t) + \mu\, P_{n+1}(t)\ ; \quad 0 \le n \le N-1$$
$$\frac{d}{dt} P_0(t) = \mu\, P_1(t) \qquad n = 0,\ t \ge 0$$
$$\frac{d}{dt} P_N(t) = -\mu P_N(t);$$

The solution of these equations with initial conditions:

$P_n(0) = \begin{cases} 1; & n = N \ne 0 \\ 0; & n \ne N \end{cases}$ can easily be obtained as earlier. The general solution to the above equation so obtained is

$$P_n(t) = \frac{(\mu t)^{N-n}\, e^{-\mu t}}{(N-n)!};\ 1 \le n \le N \text{ and } P_0(t) = 1 - \sum_{n=1}^{N} P_n(t)$$

which is known as a truncated Poisson law.

8.8 DISTRIBUTION OF SERVICE TIMES

Making similar assumptions as done above for arrivals, one could utilize the same type of process to describe the service pattern. Let the three axioms be changed by using the word service instead of arrival and condition the probability statements by requiring the system to be non-empty. Then we can easily show that, the time t to complete the service on a customer follows the exponential distribution

$$s(t) = \mu e^{-\lambda t},\ t > 0$$

where μ is the mean service rate for a particular service channel. This shows that service times follows exponential distribution with mean $1/\mu$. The number, n of potential services in time T will follow the Poisson distribution given by

$$\phi(n) = P[n \text{ service in time T, if serving is going on throughout T}]$$
$$= \frac{(\mu T)^N}{n!}\, e^{-\mu T}$$

Consequently, we can also show that

P[no service in Δt] = $1 - \mu\,\Delta t + 0(\Delta t)$ and P[one service in Δt] = $\mu\Delta t + 0(\Delta t)$.

8.9 POISSON QUEUEING SYSTEMS

Queues that follow the Poisson arrivals (exponential inter-arrival time) and Poisson services (exponential service time) are called Poisson queues. In this section, we shall study a number of Poisson queues with different characteristics.

8.9 (A) MODEL I {(M/M/1): (∞/FIFO)}

This model deals with a queueing system having single service channel. Poisson input, exponential service and there is no limit on the system capacity while the customers are served on a "first in, first out" basis.

The solution procedure of this queueing model can be summarized in the following three steps:

Step 1: Construction of differential-difference equations. Let $P_n(t)$ be the probability that there are n customers in the system at time t. The probability that the system has n customers at time $(t + \Delta t)$ can be expressed as the sum of the joint probabilities of the four mutually exclusive and collectively exhaustive events as follows:

$$\begin{aligned} P_n(t + \Delta t) = {} & P_n(t).\ P\,[\text{ no arrival in } \Delta t]\ .P\,[\text{no service completion in } \Delta t] \\ & + P_n(t).P[\text{one arrival in } \Delta t]\ .P\,[\text{one service completed in } \Delta t] \\ & + P_{n+1}(t).P[\text{no arrival in } \Delta t]\ .P\,[\text{one service completed in } \Delta t] \\ & + P_{n-1}(t).P[\text{one arrival in } \Delta t]\ .P\,[\text{no service completion in } \Delta t] \end{aligned}$$

This is re-written as:

$$\begin{aligned} P_n(t + \Delta t) = {} & P_n(t)\ [1+\lambda\Delta t + 0(\Delta t)]\ [1-\mu\Delta t + 0(\Delta t)] + P_n(t)\ [\lambda\Delta t][\mu\Delta t] + P_n + 1(t) \\ & [1-\lambda\Delta t + 0(\Delta t)]\ [\mu t\Delta t + 0(\Delta t)] + P_{n-1}(t)\ [\lambda\Delta t + 0(\Delta t)]\ [1- \mu\Delta t + 0(\Delta t)] \end{aligned}$$

or
$$P_n(t + \Delta t)-P_n(t) = -(\lambda + \mu)\,\Delta t\, P_n(t) + \mu\Delta t\, P_{n+1}(t) + \lambda\Delta t\, P_{n-1}(t) + 0(\Delta t)$$

Since Δt is very small, terms involving $(\Delta t)^2$ can be neglected. Dividing the above equations by Δt on both sides and then taking limit as $\Delta t \to 0$, we get

$$\frac{d}{dt} P_n(t) = -(\lambda + \mu)\, P_n(t) + \mu\, P_{n+1}(t) + \lambda\, P_{n-1}(t);\ n \geq 1$$

Similarly, if there is no customer in the system at time $(t + \Delta t)$, there will be no service completion during Δt. Thus for $n = 0$ and $t \geq 0$, we have only two probabilities instead of four. The resulting equation is

$$P_0(t + \Delta t) = P_0(t)\ \{1 - \lambda\Delta t + 0(\Delta t) + P_1(t)\ \{\mu\Delta t + 0(\Delta t)\}\ \{1-\lambda\Delta t + 0(\Delta t)\}$$

or
$$P_0(t + \Delta t) - P_0(t) = -\lambda\Delta t\, P_0(t) + \mu\Delta t\, P_1(t) + 0(\Delta t)$$

Dividing both sides of this equation by Δt and then taking limit as $\Delta t \to 0$, we get

$$\frac{d}{dt} P_0(t) = -\lambda\, P_0(t) + \mu\, P_1(t);\ n = 0$$

Step 2: Deriving the steady-state difference equations. In the steady state, $P_n(t)$ is independent of time t and $\lambda < \mu$ when $t \to \infty$

Consequently the differential-difference equations obtained in step 1 reduce to

$$0 = -(\lambda + \mu)P_n + \mu P_{n+1} + \lambda P_{n-1};\ n \geq 1$$

And

$$0 = -\lambda P_n + \mu P_1;\ n = 0$$

These constitute the steady-state difference equations. For the solution of the above difference equations there exist three methods, namely, the iterative method, use of generating functions and the use of linear operators. Out of these three the first one is the most straight forward and therefore the solution of the above equations will be obtained here by using the iterative method.

Using iteratively, the difference-equations yield

$$P_1 = \frac{\lambda}{\mu} P_0,\ P_2 = \frac{\lambda + u}{\mu} P_1 - \frac{\lambda}{\mu} P_0 = \left(\frac{\lambda}{\mu}\right)^2 P_0$$

$$P_3 = \frac{\lambda + u}{\mu} P_2 - \frac{\lambda}{\mu} P_1 = \left(\frac{\lambda}{\mu}\right)^3 P_0$$

and in general

$$P_n = \left(\frac{\lambda}{\mu}\right)^n P_0$$

Now,

$$P_{n+1} = \frac{\lambda + u}{\mu} P_n - \frac{\lambda}{\mu} P_{n-1},\ n \geq 1,\ n \geq 1$$

Substituting the values of P_n and P_{n-1} the equation yields

$$P_{n+1} = \frac{\lambda + u}{\mu}\left(\frac{\lambda}{\mu}\right)^n P_0 - \frac{\lambda}{\mu}\left(\frac{\lambda}{\mu}\right)^{n-1} P_0 = \left(\frac{\lambda}{\mu}\right)^{n+1} P_0$$

Thus by the principle of mathematical induction, the general formulae for P_n, is valid for $n \geq 0$.

To obtain the value of P_0, we make use of the boundary condition $\sum_{n=0}^{\infty} P_n = 1$

$$\therefore\ 1 = \sum_{n=0}^{\infty}\left(\frac{\lambda}{\mu}\right)^n P_0 = P_0 \sum_{n=0}^{\infty}\left(\frac{\lambda}{\mu}\right)^n;\ \text{since } P_n = P_0 = P_0 \frac{1}{1 - \lambda/\mu},\ \text{since } \lambda/\mu < 1$$

This gives $P_0 = 1 - \lambda/\mu$

Hence, the steady-state solution is

$P_n = (\lambda/\mu)^n (1-\lambda/\mu) = \rho^n(1 + \rho);\ \rho = \lambda/\mu < 1$ and $n \geq 0$

This expression gives us the probability distribution of queue length.

Characteristics of Model I

(*i*) Probability of queue size being greater than or equal to n, the number of customers is given by

$$P(\geq n) = \sum_{k=n}^{\infty} P_k = \sum_{k=n}^{\infty}(1-\rho)\rho^k = (1-\rho)\,\rho^n \sum_{k=n}^{\infty}\rho^{k-n} = (1-\rho)\rho^n \sum_{k-n=0}^{\infty}\rho^{k-n}$$

$$= \frac{(1-\rho)\rho^n}{1-\rho} = \rho^n$$

(*ii*) Average number of customers in the system is given by

$$E(n) = \sum_{n=0}^{\infty} nP_n = \sum_{n=0}^{\infty} n(1-\rho)\rho^n = (1-\rho)\sum_{n=0}^{\infty} n\rho^n = \rho(1-\rho)\sum_{n=0}^{\infty} n\rho^{n-1}$$

$$= \rho(1-\rho)\sum_{n=0}^{\infty}\frac{d}{d\rho}\rho^n = \rho\,(1-\rho)\frac{d}{d\rho}\sum_{n=0}^{\infty}\rho^n,\ \text{ since } \rho < 1$$

$$= \rho\,(1-\rho)\frac{1}{(1-\rho)^2} = \frac{\rho}{(1-\rho)} = \frac{\lambda}{(\mu-\lambda)}$$

(*iii*) Average queue length is given by

$E(m) = \sum_{\mu=0}^{\infty} mP_n$, where $m = n - 1$ being the number of customers in the queue, excluding the customer, which is in service.

$$\therefore \quad E(m) = \sum_{n=0}^{\infty}(n-1)\,P_n = \sum_{n=0}^{\infty} nP_n - \sum_{n=0}^{\infty} P_n = \sum_{n=0}^{\infty} nP_n - \left[\sum_{n=0}^{\infty} P_n - P_0\right]$$

$$\frac{\rho}{(1-\rho)} - [1-(1-\rho)] = \frac{\rho}{(1-\rho)} - \rho$$

$$= \frac{\rho^2}{(1-\rho)} = \frac{\lambda^2}{\mu(\mu-\lambda)}$$

(*iv*) Average length of non-empty queue is given by

$$E(m \backslash m > 0) = \frac{E(m)}{p(m>0)} = \frac{\lambda^2}{\mu(\mu-\lambda)}\cdot\frac{1}{(\lambda/\mu)^2} = \frac{\mu}{\mu-\lambda}$$

Since $\quad P(m > 0) = P(n > 1) = \sum_{n=0}^{\infty} P_n - P_0 - P_1 = \left(\frac{\lambda}{\mu}\right)^2$

(*v*) The fluctuation (variance) of queue length is given by

$$V(n) = \sum_{n=0}^{\infty}[n - E(n)]^2\,P_n = \sum_{n=0}^{\infty}\left[n^2\,P_n - [E(n)]\right]^2$$

Using some algebraic transformations and the value of P_n the result reduces to

$$V(n) = (1-\rho)\frac{\rho+\rho^2}{(1-\rho)^3} - \left[\frac{\rho}{(1-\rho)}\right]^2 = \left[\frac{\rho}{(1-\rho)^2}\right] = \frac{\lambda\mu}{(\mu-\lambda)^2}$$

Waiting time distribution for Model I. The waiting time of a customer in the system is for the most part, a continuous random variable except that there is a non-zero probability that the delay will be zero, that is a customer entering service immediately upon arrival. Therefore, if we denote the time spent in the queue by w and if $\psi_w(t)$ denotes its cumulative probability distribution then from the complete randomness of the Poisson distribution, we have

$$\psi_m(0) = P(w = 0) = P(\text{no customer in the system upon arrival})$$
$$= P_0 = (1 - \rho).$$

It is now required to find $\psi_w(t)$ for $t > 0$.

Let there be n customers in the system upon arrival, then in order for a customer to go into service at a time between 0 and t, all the n customers must have been served by time t. Let s_1, s_2,.........., s_n denote service times of n customers respectively, then

$$w = \sum_{i=1}^{n} s_i, (n \geq 1), \text{ and } w = 0 \quad (n = 0)$$

The distribution function of waiting time, w, for a customer who has to wait is given by

$$P(w \leq t) = P\left[\sum_{i=1}^{n} s_i, \leq t\right]; n \geq 1 \text{ and } t > 0$$

Since the service time for each customer is independent and identically distributed, therefore its probability density function is given $\mu e^{-\mu t}$ $(t > 0)$, where μ is the mean service rate. Thus

$$\psi_n(t) = \sum_{n=1}^{\infty} P_n \times P\ (n\text{–1 customers are served at time } t) \times P(1 \text{ customer is served in time } \Delta t)$$

$$= \sum_{n=1}^{\infty}\left(1 - \frac{\lambda}{\mu}\right)\left(\frac{\lambda}{\mu}\right)^n \frac{(\mu t)^{n-1} e^{-\mu t}}{(n-1)!} . \mu \Delta t$$

The expression for $\psi_m(t)$, therefore, can be written as

$$\psi_w(t) = P(w \leq 1) = \sum_{n=1}^{\infty} P_n \int_0^1 \psi_n(t) dt$$

$$= \sum_{n=1}^{\infty}(1-\rho)\rho^n \int_0^1 \frac{(\mu t)^{n-1}}{(n-1)!} e^{-\mu t} . \mu dt = (1-\rho)\rho \int_0^1 \mu e^{-\mu t} \sum_{n=1}^{\infty} \frac{(\mu t \rho)^{n-1}}{(n-1)!} . dt$$

$$= (1 - \rho)\, \rho \int_0^1 \mu e^{-\mu t(1-\rho)} . dt$$

Hence, the waiting time of a customer who has to wait is given by

$$\psi_{(w)} = \frac{d}{dt}\left[\psi_w(t)\right] = \rho(\rho - 1).\, \mu e^{-\mu t(1-\rho)} = \lambda\left(1 - \frac{\lambda}{\mu}\right) e^{-(\mu-\lambda)t}$$

Characteristic of waiting time distribution

(*i*) Average waiting time of a customer (in the queue) is given by

$$E(w) = \int_0^\infty t.\psi(w)\,dt = \int_0^\infty t.\rho\mu(1-\rho)\,e^{-\mu t(1-\rho)}\,dt$$

$$= \rho\int_0^\infty \frac{xe^{-x}}{\mu(1-\rho)}dx, \text{ for } \mu(1-\rho)\,t = x$$

$$= \frac{\rho}{\mu(1-\rho)} = \frac{\lambda}{\mu(\mu-\lambda)}$$

(*ii*) Average waiting time of an arrival who has to wait is given by

$$E(w\backslash w > 0) = E(w)/P(w > 0) = \frac{\lambda}{\mu(\mu-\lambda)} / \frac{\lambda}{\mu} = \frac{1}{(\mu-\lambda)}$$

[Here $P(w > 0) = 1–P(w = 0) = 1–(1 - \rho) = p$]

(*iii*) For the busy period distribution, let the random variable v denote the total time that a customer has to spend in the system including service. Then the probability density of its cumulative density function is given by

$$\psi(w\backslash w > 0) = \psi(w)/P(w > 0) = \left\{\lambda\left(1-\frac{\lambda}{\mu}\right)e^{(\mu-\lambda)t}\right\} / \left(\frac{\lambda}{\mu}\right)$$

$$= (\mu - \lambda)\, e^{(\mu-\lambda)t+}, \ t > 0$$

(*iv*) Average waiting time that a customer spends in the system including service is given by

$$E(v) = \int_0^\infty t.\psi\,(w/w>0)dt = \int_0^\infty t(\mu-\lambda)e^{-(\mu-\lambda)t}\,dt$$

$$= \frac{1}{(\mu-\lambda)}\int_0^\infty xe^{-x}\,dx, \text{ , for } (\mu - \lambda)\,t = x$$

$$= \frac{1}{(\mu-\lambda)}$$

Relations between average queue length and average waiting time – Little's formula.

We have derived above the following important characteristics of M/M/1 queueing system:

$$E(n) = \frac{\lambda}{(\mu-\lambda)},$$

$$E(m) = \frac{\lambda^2}{\mu(\mu-\lambda)},$$

$$E(w) = \frac{\lambda}{\mu(\mu-\lambda)} \text{ and } E(v) = \frac{1}{(\mu-\lambda)}$$

Using these expressions, we observe that

$E(n) = \lambda E(v)$, $E(m) = \lambda E(w)$ and $E(v) = E(w) + 1/\mu$.

8.10 SOLVED PROBLEMS

Problem 1: A T.V. repairman finds that the time spent on his jobs has an exponential distribution with mean 30 minutes. If he repairs sets in the order in which they came in and if the arrival of sets is approximately Poisson with an average rate of 10 per 8 hour day, what is repairman's expected idle time each day? How many jobs are ahead of the average set just brought in?

Solution:

We are given,

$$\lambda = 10 \text{ sets per day, and } \mu = 16 \text{ sets per day.}$$

$$\therefore \quad \rho = \lambda/\mu = 10/16 = 0.625$$

The probability for the repairman to be idle is

$$P_0 = 1 - \rho = 1 - 0.625 = 0.375$$

(*i*) Expected idle time per day = 8 × 0.375 = 3 hours.

(*ii*) Expected (or average number of T.V. sets in the system)

$$E(n) = \frac{\rho}{(1-\rho)} = \frac{0.625}{(1-0.625)} = 2 \text{ (approximately T.V. sets)}$$

Problem 2: In a railway marshalling yard goods trains arrive at a rate of 30 trains per day. Assuming that the inter-arrival time follows an exponential distribution and the service time (the time taken to hump a train) distribution is also exponential with an average 36 minutes. Calculate the following :

(*i*) The average number of trains in the queue.

(*ii*) The probability that the queue size exceeds 10.

If the input of trains increases to an average 33 per day, what will be change in (*i*) and (*ii*)? Establish the formula you use in your calculations.

Solution:

Here

$$\lambda = \frac{30}{60 \times 24} = \frac{1}{48} \text{ trains/minute; } \mu = \frac{1}{36} \text{ trains/minute, and } \rho = \frac{\lambda}{\mu} = \frac{36}{48} = 0.75$$

(*i*) $L_s = \dfrac{\rho}{(1-\rho)} = \dfrac{0.75}{(1-0.75)} = 3$ trains

(*ii*) Probability [queue size ≥ 10] = $\rho^{10} = (0.75)^{10} = 0.06$

When the input increases to 33 trains per day $\lambda = 1/43$, $\mu = 1/36$

Therefore, $\rho = \lambda/\mu = 36/43 = 0.84$. Hence,

(*i*) $L_s = 0.84/0.16 = 5$ trains

(*ii*) Probability(queue size ≥ 10) = $(0.84)^{10} = 0.2$ (approx.)

Problem 3: Consider an example from a maintenance shop. The inter-arrival times at toolcrib are exponential with an average time of 10 minutes. The length of the service time (amount of time taken by the toolcrib operator to meet the needs of the maintenance man) is assumed to be exponentially distributed, with mean 6 minutes. Find

(*i*) The probability that a person arriving at the booth will have to wait.

(*ii*) Average length for the queue hat forms and the average time that an operator spends in the Q-system.

(*iii*) The manager of the shop will install a second booth when an arrival would have to wait 10 minutes or more for the service. By how much must the rate of arrival be increased in order to justify a second booth.

(*iv*) The probability that an arrival will have to wait for more than 12 minutes for service and to obtain his tools.

(*v*) Estimate the fraction of the day that toolcrib operator will be idle.

(*vi*) The probability that there will be six or more operators waiting for the service.

Solution:

Here $\lambda = 60/10 = 6$ per hour, $\mu = 60/6 = 10$ per hour

(*i*) A person will have to wait if the service facility is not idle.

Probability that the service facility is idle = probability of no customer in the system (P_0)

$\therefore$ Probability of waiting $= 1 - P_0 = 1 - (1 - \rho) = \rho = \lambda/\mu = 6/10 = 0.6$

(*ii*)
$$L_q = \frac{\rho^2}{(1-\rho)} = \frac{(0.6)^2}{(1-0.6)} = 0.9$$

$$L_s = L_q + \lambda/\mu = 0.9 + 0.6 = 1.5$$

$$\therefore \quad W_s = L_s/\lambda = 1.5/6 = 1/4 \text{ hours}$$

(*iii*) $W_q = L_q/\mu = 0.9/6$ hrs. $= 9$ minutes.

Let λ be the arrival rate when a second booth is justified, *i.e.*, $W_q \geq 10$ minutes.

$$W_q = \frac{l}{\mu(\mu-\lambda)} = \frac{10}{60}$$

$$6\lambda = 10(10-\lambda)$$

$$\text{or} \quad 16\lambda = 100 \text{ or } \lambda = 6.25$$

Hence if the arrival rate exceeds 6.25 per hour, the seconds booth will be justified.

(*iv*) Probability of waiting for 12 minutes or more is given by

$$\text{Probability} \quad (W \geq 12) = \int_{12}^{\infty} \rho(\mu-\lambda)e^{-(\mu-\lambda)w}\,dw = -\rho\left[e^{-(\mu-\lambda)w}\right]_{12/60}^{\infty}$$

$$= \rho e^{-(\mu-\lambda)12/60} = 0.6\, e^{-(10-6).12/60} = 0.6\, e^{-4/5} = 0.27$$

(*v*) $P_0 = 1 - \rho = 0.4$, 40% of the time of toolcrib operator is idle.

(*vi*) Probability of six or more operators waiting for the service $= \rho^6 = (0.6)^6$.

Problem 4: The mean rate of arrival of planes at an airport during the peak period is 20 hour, but the actual number of arrivals in any hour follows a Poisson distribution with the respective averages. When thereis congestion, the planes are forced to fly over the field in the stack awaiting the landing of other planes that arrived earlier.

(*i*) How many planes would be flying over the field in the stack on an average in good weather and in bad weather?

(*ii*) How long a plane would be in the stack and in the process of landing in good and in bad weather?

(*iii*) How much stack and landing time to allow so that priority to land out of order would have to be requested only one time in twenty?

Solution: Here

$$\lambda = 20 \text{ planes /hour}$$

And

$$\mu = \begin{cases} 60 \text{ planes/hour in good weather} \\ 30 \text{ planes/hour in bad weather} \end{cases}$$

(*i*)

$$L_q = \frac{\lambda^2}{\mu(\mu-\lambda)}$$

$$= \begin{cases} 20^2/60\,(60-20) = 1/6 \text{ in good weather} \\ 20^2/30(30-20) = 4/3 \text{ in bad weather} \end{cases}$$

(*ii*)

$$W_s = \frac{1}{(\mu-\lambda)}$$

$$= \begin{cases} 1/(60-20) = 1/40 \text{ hrs. in good weather} \\ 1/(30-20) = 1/10 \text{ hrs. in bad weather} \end{cases}$$

(*iii*) The waiting time '*w*' taken by the plane in landing and in the stack is given by

$$\int_0^w (\mu-\lambda)e^{-(\mu-\lambda)w}\,dw = 0.95$$

or $$(\mu-\lambda)\left[\frac{e^{-(\mu-\lambda)w}}{\lambda-\mu}\right]_0^w = 0.95$$

or $$1-e^{-(\mu-\lambda)w} = 0.95$$

or $$e^{-(\mu-\lambda)w} = 0.05 \qquad ...(1)$$

Now the value of w can be determined in both the cases.

Case 1: When weather is good:

Substituting $\lambda = 20$, $\mu = 60$ in equation (1), $e^{-40w} = 0.05$.

Taking the logarithm of both sides to the base e (instead of 10),

$-40w = \log_e(0.05) = -2.9957$ [Since $\log_e e = 1$, $\therefore \log_e = 1$, $\therefore \log_e(0.05) = -2.9957$]

or $$w = \frac{2.9957}{40} = 0.075 \text{ hour} = 4.5 \text{ minutes}$$

Case 2: When weather is bad :

Substitution $\lambda = 20$, $\mu = 30$ in equation (1)

$$e^{-10w} = 0.05$$

solving this equation as in case 1 above, we get $w = 0.3$ hour =18 minutes.

Problem 5: The rate of arrival of cu: tomers at a public telephone booth follows Poisson distribution, with an average time of 10 minutes between one customer and the next. The duration of a phone call is assumed to follow exponential distribution, with mean time of 3 minutes.

(*i*) What is the probability that a person arriving at the booth will have to wait?

(*ii*) What is the average length of the non-empty queues that from time to time?

(*iii*) The Mahanagar Telephone Nigam Ltd. will install a second booth when it is convinced that the customers would expect waiting for at least 3 minutes for their turn to make a call. By how much time should the flow of customers increase in order to justify a second booth?

(*iv*) Estimate the fraction of a day that the phone will be in use.

Solution:

Here we are given:

$$\lambda = \frac{1}{10} \times 60 \text{ or 6 per hour and } \mu = \frac{1}{3} \times 60 \text{ or 20 per hour}$$

(*i*) Probability that a person arriving at the booth will be justified if the arrival rate is greater than the waiting time.

Now, if λ^1 denotes the increased arrival rate, expected waiting time:

$$W_q^1 = \frac{\lambda^1}{\mu(\mu-\lambda^1)} \Rightarrow \frac{3}{60} = \frac{\lambda^1}{20\left(20-\lambda^1\right)} \text{ or } \lambda^1 = 10$$

$$P_0 = \text{prob. of no customer in the system} = 1 - \frac{\lambda}{\mu} = 0.5$$

Thus 50% of time an arrival will not have to wait.

(*ii*) Average queue length $= \dfrac{\lambda^2}{\mu(\mu-\lambda)} = \dfrac{5 \times 5}{10\,(10-5)} = 0.5$

(*iii*) Average time spent by a customer $= \dfrac{1}{(\mu-\lambda)} = \dfrac{1}{5}$ hour or 12 minutes

(*iv*) The management will deploy the person exclusively for Xeroxing when the average time spent by a customer exceeds 15 minutes. We wish to calculate the arrival rate that will lead to such situation. Let this arrival rate be λ^1. Then

$$\frac{1}{\left(\mu-\lambda^1\right)} > \frac{15}{16} \text{ or } \frac{1}{\left(10-\lambda^1\right)} > \frac{1}{4} \quad i.e.\ \lambda^1 > 6$$

Hence, if the arrival rate of customers is greater than 6 customers per hour, the average time spent by a customer will exceeds 15 minutes.

Problem 6: Telephone users arrive at a booth following a Poisson distribution with an average time of 5 minutes between one arrival and the next. The time taken for a telephone call is on an average 3 minutes and it follows an exponential distribution. What is the probability that the booth is busy? How many more booths should be established to reduce the waiting time to less than or equal to half of the present waiting time?

Solution:

Here we are given:

Arrival rate, $\lambda = 12$ per hour

Service rate, $\mu = 20$ per hour

$$\text{Probability that booth is busy} = 1 - P_0 = \frac{\lambda}{\mu} = \frac{12}{20} = 0.60$$

Average waiting time in queue :

$$W_q = \frac{\lambda}{\mu(\mu - \lambda)} = \frac{12}{20(20 - 12)} = \frac{3}{40} \text{ hour}$$

Average waiting time in system:

$$W_s = \frac{1}{(\mu - \lambda)} = \frac{1}{(20 - 12)} = \frac{1}{8} \text{ hour}$$

In case waiting time is required to be reduced to half,

$$W_s^1 = \frac{1}{(\mu - \lambda^1)} \Rightarrow \frac{1}{16} = \frac{1}{(\mu^1 - 12)}$$

or $\mu^1 = 28$ per hour .

Problem 7: The XYZ company's quality control deptt. is managed by a single clerk, who takes on an average 5 minutes in checking parts of each of the machine coming for inspection. The machines arrive once in every 8 minutes on the average. One hour of the machine is valued at Rs. 15 and a clerk's time is valued at Rs. 4 per hour. What are the average hourly queueing system costs associated with the quality control department?

Solution: Mean arrival rate $\lambda = \frac{1}{4}$ per min. $= \frac{60}{8}$ per hour

Mean service rate, $\mu = \frac{1}{5}$ per min. = 12 per hour

Average time spent by a machine in the system $= \frac{1}{(\mu - \lambda)} = \frac{2}{9}$ hour

Average queueing cost per machine is $= \frac{15 \times 2}{9} = \text{Rs. } \frac{10}{3}$

Average queueing arrival of $\frac{60}{8}$ machine per hour costs $\frac{10}{3} \times \frac{60}{8}$ or Rs. 25

Average hourly queueing cost = Rs. 25

Average hourly cost for the clerk = Rs. 4

Hence total cost = Rs. 29 per hour.

Problem 8: A road transport company has one reservation clerk on duty at a time. She handles information of bus schedules and makes reservations. Customers arrive at a rate of 8 per hour and the clerk can service 12 customers on an average per hour. After stating your assumptions, answer the following:

(*i*) What is the average number of customers waiting for the service of the clerk?

(*ii*) What is the average time a customer has to wait for the service of the clerk?

(*iii*) The management is contemplating to install a computer system to handle the information and reservations. This is expected to reduce the service time from 5 to 3 minutes. The additional cost of having the new system works out to be 12 paise per minute spent waiting before being served, should the company install the computer system? Assume 8 hour working day.

Solution:

(*i*) Here we are given :

Customer arrival rate = λ = 8 per hour and

Average number of customers waiting for the service of the clerk (in the system):

$$L_s = \frac{\lambda}{\mu - \lambda} = \frac{8}{12 - 8} = 2 \text{ customers.}$$

The average number of customers waiting for the service of the clerk (in the queue):

$$L_q = \frac{\lambda^2}{\mu(\mu - \lambda)} = \frac{8 \times 8}{12(12 - 8)} \text{ or } 1.33 \text{ customers.}$$

(*ii*) The average waiting time of a customer (in the system) before getting service:

$$W_s = \frac{1}{\mu - \lambda} = \frac{1}{12 - 8} \text{ hour or 15 minutes.}$$

(*iii*) The average waiting time of a customer (in the queue) before getting service:

$$W_s = \frac{\lambda}{\mu(\mu - \lambda)} = \frac{8}{12(12 - 8)} = \frac{1}{6} \text{ hour or 10 minutes.}$$

Here calculate the difference between the goodwill cost of customers with one system and the goodwill cost of customers with an additional computer system. This difference will be compared with the additional cost (of Rs. 50 per day) of installing another computer system.

Now an arrival waits for W_q hours before being served and there are λ arrivals per hour. Thus expected waiting time for all customers in an 8 hour day with one system

$= 8\lambda \,.\, W_q = 8 \times 8 \times \frac{1}{6}$ hrs. or $\frac{64}{6} \times 60$ minutes *i.e.* 640 minutes.

The goodwill cost per day with one system = Rs. 640 × $\frac{12}{100}$ = Rs.76.80

The expected waiting time of a customer before getting service when there is an additional computer system is

$$W_q^1 = \frac{8}{20(12 - 8)} = \frac{8}{20 \times 12} \text{ or } \frac{1}{30} \text{ hr.}$$

Thus expected waiting time of customers in an 8-hour day with an additional computer system is 8λ.

$$W_q^1 = 8 \times 8 \times \frac{1}{30} \text{ hrs.} = 128 \text{ minutes.}$$

The total goodwill cost with an additional computer system = Rs. 128 × $\frac{12}{100}$ = Rs. 15.36

Hence reduction in goodwill cost with the installation of a computer system

= Rs. 76.80 – Rs.15.36 = Rs. 61.44.

where as the additional cost of a computer system is Rs. 50 per day, Rs. 61.44 is the reduction in goodwill cost when additional computer system is installed, hence there will be net saving of Rs. 11.44 per day. It is therefore, worth while to install a computer.

8.11 MODEL II {(M/M/1): (∞/SIRO)}

This model is essentially the same as Model I except that the service discipline follows the SIRO-rule (service in random order) instead of FIFO-rule. As the derivation of P_n for Model I does not depend on any specific queue discipline, it may be concluded that for the SIRO-rule case, we must have

$$P_n = (1 - \rho)\, \rho^n,\ n \geq 0.$$

Consequently, whether the queue discipline follows the SIRO-rule or FIFO-rule the average number of customers in the system, E(n), will remain the same. In fact E(n) will remain the same for any queue discipline provided, of course, P_n remains unchanged. Thus, $E(v) = \frac{1}{\lambda} E(n)$ under the SIRO-rule is the same as under the FIFO-rule.

This result can be extended to any queue discipline as long as P_n remains unchanged. Specifically, the result applied to the three most common disciplines, namely, FIFO, LIFO and SIRO. The three queue discipline differ only in the distribution of waiting time where the probabilities of long and short waiting times change depending upon the discipline used. Thus we can use the symbol GD (general discipline) to represent the disciplines FIFO, LIFO and SIRO, when the waiting time distribution is not required.

8.12 MODEL III {(M/M/1): (N/FIFO)}

This model differs from that of Model I in the sense that the maximum number of customers in the system is limited to N. Therefore, the difference equations of Model I are valid for this model as long as n < N.

The additional difference equation for n = N is

$$P_N(t + \Delta t) = P_N(t)[1-\mu\Delta t] + P_{N+1}(t).[\lambda\,\Delta t][1-\mu\Delta t] + 0(\Delta t).$$

This gives, after simplification, the differential difference equation

$$\frac{d}{dt} P_N(t) = -\mu\, P_N(t) + \lambda\, P_{N-1}(t)$$

From which the resultant steady-state difference equations for this model, therefore, can be written as

$$\mu P_1 = \lambda P_0$$

$$\mu P_{n+1} = (\lambda + \mu)P_n - \lambda P_{n-1} \qquad 1 \le n \le N-1$$

and

$$\mu P_N = \lambda P_{N-1}$$

Using the iterative procedure (as in Model I), the first two difference equations give

$$P_n = (\lambda/\mu)^n P_0 \quad n \le N$$

Also we see that for this value of P_n, the third (last) difference equation holds for $n = N$

Therefore we have

$$P_n = (\lambda/\mu)^n P_0 = \rho^n P_0 \qquad n \le N$$

For obtaining the value of P_0, we make use of the boundary conditions $\sum_{n=0}^{N} P_n = 1$

Therefore

$$1 = P_0 \sum_{n=0}^{N} \rho_n = \begin{cases} P_0 \dfrac{1-\rho^{N+1}}{1-\rho}, & (\rho \ne 1) \\ P_0 (N+1) & (\rho = 1) \end{cases}$$

Thus

$$P_0 = \begin{cases} \dfrac{1-\rho}{1-\rho^{N+1}} & (\rho \ne 1) \\ \dfrac{1}{N+1} & (\rho = 1) \end{cases}$$

Hence

$$P_n = \begin{cases} \dfrac{(1-\rho)\rho^n}{1-\rho^{N+1}}, & (\rho \ne 1) \\ \dfrac{1}{N+1} \end{cases}$$

Characteristics of Model III

(*i*) Average number of customers in the system is given by

$$E(n) = \sum_{n=0}^{N} nP_n = P_0 \sum_{n=0}^{N} n\rho_n = P_0\,\rho \sum_{n=0}^{N} \frac{d}{d\rho} \rho_n$$

or

$$E(n) = P_0\,\rho \frac{d}{dt} \sum_{n=0}^{N} n\rho^n\, P_0\,\rho \frac{d}{d\rho}\left[\frac{1-\rho^{N+1}}{1-\rho}\right]$$

$$= P_0 \frac{\rho\left[1-(N+1)\rho^N + N\rho^{N+1}\right]}{(1-\rho)^2} = \frac{\rho\left[1-(N+1)\rho^N + N\rho^{N+1}\right]}{(1-\rho)\left(1-\rho^{N+1}\right)}$$

(Since $P_0 = \frac{1-\rho}{1-\rho^{N+1}}; \rho \neq 1$)

(*ii*) Average queue length is given by

$$E(m) = \sum_{n=1}^{N}(n-1)P_n = E(n) - \sum_{n=1}^{N} P_n = E(n)-(1-P_0)$$

$$= E(n) - \frac{\rho\left(1-\rho^N\right)}{1-\rho^{N+1}} \quad \text{(Since } P_0 = \frac{1-\rho}{1-\rho^{N+1}}, (\rho \neq 1))$$

$$= \frac{\rho^2\left[1-N\rho^{N-1} + (N-1)\rho^N\right]}{(1-\rho)\left(1-\rho^{N+1}\right)}$$

(*iii*) The average waiting time in the system can be obtained by using Little's formula, that is $E(v) = \{E(n)\}/\lambda^1$, where λ^1 is the mean rate of customers entering the system and is equal to $\lambda(1 - P_N)$. The average waiting time in the queue can be obtained by using the relations

$$E(w) = E(v)-1/\mu \text{ or } E(w) = \{E(m)\}/\lambda^1.$$

8.13 SOLVED PROBLEMS

Problem 1: At a railway station only one train is handled at a time. The railway yard is sufficient only for two trains to wait while other is given signal to leave the station. Trains arrive at the station at an average rate of 6 per hour and the railway station can handle them on an average of 12 per hour. Assuming Poisson arrivals and exponential service distribution, find the steady-state probabilities for the various number of trains in the system. Also find the average waiting time of a new train coming into the yard.

Solution:

Here $\lambda = 6$ and $\mu = 12$ so that $\rho = 6/12 = 1/2 = 0.5$

The maximum queue length is 2, *i.e.*, the maximum number of trains in the system is 3 (= N).

The probability that there is no train in the system (both waiting and in service) is given by

$$P_0 = \frac{1-\rho}{1-\rho^{N+1}} = \frac{1-0.5}{1-0.5^{3+1}} = 0.53$$

Now, since $P_n = P_0\,\rho^n$, therefore

$$P_1 = (0.53)(0.5) = 0.27,$$

$$P_2 = (0.53)(0.5)^2 = 0.13$$

And $$P_3 = (0.53)(0.5)^3 = 0.07$$

Hence, we get

$$E(n) = 1(0.27) + 2(0.13) + 3(0.07) = 0.74$$

Thus, the average number of trains in the system is 0.74 and each train takes on an average 1/12 (= 0.08) hours for getting service. As the arrival of new train expects to find an average of 0.74 trains in the system before it.

$E(w)$ = (0.74)(0.08) hours = 0.0592 hours or 3.5 minutes.

Problem 2: Assume that the goods trains are coming in a yard at the rate of 30 trains per day and suppose that the inter-arrival times follow an exponential distribution. The service time for each train is assumed to be exponential with an average of 36 minutes. If the yard can admit 9 trains at a time (there being 10 lines, one of which is reserved for shunting purposes), calculate the probability that the yard is empty and find the average queue length.

Solution: We have

$$\lambda = \frac{30}{60 \times 24} = \frac{1}{48} \text{ and } \mu = \frac{1}{16} \text{ trains per minute.}$$

$$\therefore \qquad \rho = \lambda/\mu = 36/48 = 0.75$$

The probability that the yard is empty is given by

$$P_0 = \frac{1-\rho}{1-\rho^{N+1}} = \frac{1-0.75}{1-0.75^{10}} \qquad \text{(Since N = 9)}$$

$$= 0.25/0.90 = 0.28$$

Average queue length is given by

$$E(m) = \frac{\rho^2\left[1 - N\rho^{N-1} + (N-1)\rho^N\right]}{(1-\rho)\left(1-\rho^{N+1}\right)} = \frac{(0.75)^2\left[1-9(0.75)^{9-1}+(9-1)(0.75)^9\right]}{(1-0.75)\left(1-(0.75)^{9+1}\right)}$$

$$= (2.22)\frac{(1-0.33)}{(1-0.005)} = (2.22)(0.70) = 1.55.$$

8.14 EXERCISE

1. On an average 96 patients per 24-hour day require the service of an emergency clinic. Also on average, a patient requires 10 minutes of active attentions. Assume that the facility can handle only one emergency at a time. Suppose that it costs the clinic Rs.100 per patient treated to obtain an average servicing time of 10 minutes, and that each minute of decrease in this average time would cost Rs. 10 per patient treated. How much would have to be budgeted by the clinic to decrease the average size of the queue from $1\frac{1}{3}$ patients to ½ patient?

2. A refinery distributes its products by trucks, loaded at the loading dock; both company trucks and independent distributor's trucks are loaded. The independent firms complained that sometimes they must wait in line and thus lose money paying for a truck and driver, that is only waiting. They have asked the refinery either to put in a second loading dock or to discount prices equivalent to the waiting time. Extra loading

dock cost Rs.100 per day whereas the waiting time for the independent firms cost Rs. 25 per hour. The following data have been accumulated. Average arrival rate of all trucks is 2 per hour and average service rate is 3 per hour. Thirty percent of all trucks are independent. Assuming that these rates are random according to the Poisson distributions, determine:

(*a*) The probability that a truck has to wait

(*b*) The waiting time of a truck that waits and

(*c*) The expected cost of waiting time of independent trucks per day.

Is it advantageous to decide in favour of a second loading dock to ward off the complaints?

3. An airlines organization has one reservation clerk on duty in its local branch at any given time. The clerk handles information regarding passenger reservation and flight timings. Assume that the number of customers arriving during any given period is Poisson distributed with an arrival rate of eight per hour and that the reservation clerk can serve a customer in six minutes on an average, with an exponentially distributed service time.

(*i*) What is the probability that the system is busy?

(*ii*) What is the average time a customer spends in the system?

(*iii*) What is the average length of the queue and what is the number of customers in the system?

4. If for a period of 2 hours in a day (8-10 A.M) planes arrive at the aerodrome for every 20 minutes but the service time continues to remain 32 minutes then calculate for this period:

(*i*) The probability that the aerodrome is empty

(*ii*) Average queue length on the assumption that the time capacity of the aerodrome is limited to 6 planes.

5. Customers arrive at a one-window drive according to a Poisson distribution with mean of 10 minutes and service time per customer is exponential with mean of 6 minutes. The space in front of the window can accommodate only three vehicles including the serviced one. Other vehicles have to wait outside this space. Calculate

(*i*) Probability that an arriving customer can drive directly to the space in front of the window.

(*ii*) Probability that an arriving customer will have to wait outside the directed space.

(*iii*) How long an arriving customer is expected to wait before getting the service?

6. A company distributes its products by trucks loaded at its only loading station. Both, company's truck and contractor's trucks, are used for this purpose. It was found out that on an average every five minutes, one truck arrived and the average loading time was three minutes. 50% of the trucks belong to the contractor. Find out

(*i*) The probability that a truck has to wait

(*ii*) The waiting time of truck that waits and

(*iii*) The expected waiting time of contractor's trucks per day, assuming a 24 hours shift.

7. A warehouse has only one loading dock manned by a three person crew. Trucks arrive at the loading dock at an average rate of 4 trucks per hour and the arrival rate is Poisson distributed. The loading of a truck takes 10 minutes on an average and can be assumed to be exponentially distributed. The operating cost of a truck is Rs.20 per hour and the members of the loading crew are paid @ Rs.6 each per hour. Would you advise the truck owner to add another crew of three persons?

8. In the production shop of a company, the breakdown of the machines is found to be Poisson distributed with an average rate of 3 machines per hour. Breakdown time at one machine costs Rs.40 per hour to the company. There are two choices before the company for hiring the repairmen. One of the repairmen is slow but cheap the other fast but expensive. The slow- cheap repairman demands Rs.20 per hour and will repair the broken down machines exponentially at the rate of 4 per hour. The fast-expensive repairman demands Rs. 30 per hour and will repair machines exponentially at an average rate of 6 per hour, which repairman should be hired?

9. In a factory, the machine breakdown on an average rate is 10 machines per hour. The idle time cost of a machine is estimated to be Rs. 20 per hour. The factory works 8 hours a day. The factory manager is considering 2 mechanics for repairing the machines. The first mechanic A takes about 5 minutes on an average to repair a machine and demands wages Rs. 10 per hour. The second mechanic B takes about 4 minutes in repairing a machine and demands wages at the rate of Rs. 15 per hour. Assuming that the rate of machine breakdown is Poisson-distributed and the repair rate is exponentially distributed, which of the two mechanics should be engaged?

10. A bank has two tellers working on savings accounts. The first teller handles withdrawals only. The second teller handles deposits only. It has been found that the service time distribution for both deposits and withdrawals is exponential with mean service time 3 minutes per customer. Depositors are found to arrive in Poisson fashion throughout the day with mean arrival rate of 16 per hour. Withdrawers also arrive in Poisson fashion with mean arrival rate of 14 per hour. What would be the effect on the average waiting time for depositors and withdrawers if each teller could handle both withdrawals and deposits? What could be the effect if this could be accomplished by increasing the mean service time to 3.5 minutes?

APPENDICES

1. Areas under the Standard Normal Curve from 0 to Z (Normal Tables)

z	0	1	2	3	4	5	6	7	8	9
0.0	.0000	.0040	.0080	.0120	.0160	.0199	.0239	.0279	.0319	.0359
0.1	.0398	.0438	.0478	.0517	.0557	.0596	.0636	.0675	.0714	.0754
0.2	.0793	.0832	.0871	.0910	.0948	.0987	.1026	.1064	.1103	.1141
0.3	.1179	.1217	.1256	.1293	.1331	.1368	.1406	.1443	.1480	.1517
0.4	.1554	.1591	.1628	.1664	.1700	.1736	.1772	.1808	.1844	.1879
0.5	.1916	.1950	.1985	.2019	.2054	.2088	.2123	.2157	.2190	.2224
0.6	.2258	.2291	.2324	.2357	.2389	.2422	.2454	.2486	.2518	.2649
0.7	.2580	.2612	.2642	.2673	.2704	.2734	.2764	.2794	.2823	.2852
0.8	.2881	.2910	.2939	.2967	.2996	.3023	.3051	.3078	.3106	.3133
0.9	.3159	.3186	.3212	.3238	.3264	.3289	.3315	.3340	.3365	.3389
1.0	.3413	.3438	.3461	.3485	.3508	.3531	.3554	.3577	.3599	.3621
1.1	.3643	.3665	.3686	.3708	.3729	.3749	.3770	.3790	.3810	.3830
1.2	.3849	.3869	.3888	.3907	.3925	.3944	.3962	.3980	.3997	.4015
1.3	.4032	.4049	.4066	.4082	.4099	.4115	.4131	.4147	.4162	.4177
1.4	.4192	.4207	.4222	.4236	.4251	.4265	.4279	.4292	.4306	.4319
1.5	.4332	.4345	.4357	.4370	.4382	.4394	.4406	.4418	.4429	.4441
1.6	.4452	.4463	.4474	.4484	.4495	.4505	.4515	.4525	.4535	.4545
1.7	.4654	.4564	.4573	.4582	.4591	.4599	.4608	.4616	.4625	.4633
1.8	.4641	.4649	.4656	.4664	.4671	.4678	.4686	.4693	.4699	.4706
1.9	.4713	.4719	.4726	.4732	.4738	.4744	.4750	.4756	.4761	.4767
2.0	.4772	.4778	.4783	.4788	.4793	.4798	.4803	.4808	.4812	.4817
2.1	.4821	.4826	.4830	.4834	.4838	.4842	.4846	.4850	.4854	.4857
2.2	.4861	.4864	.4868	.4871	.4875	.4878	.4881	.4884	.4887	.4890
2.3	.4893	.4896	.4898	.4901	.4904	.4906	.4909	.4911	.4913	.4916
2.4	.4918	.4920	.4922	.4925	.4927	.4929	.4931	.4932	.4934	.4936
2.5	.4938	.4940	.4941	.4943	.4945	.4946	.4948	.4949	.4951	.4952
2.6	.4953	.4955	.4956	.4957	.4959	.4960	.4961	.4962	.4963	.4964
2.7	.4965	.4966	.4967	.4968	.4969	.4970	.4971	.4972	.4973	.4974
2.8	.4974	.4975	.4976	.4977	.4979	.4978	.4979	.4979	.4980	.4981
2.9	.4981	.4982	.4982	.4983	.4984	.4984	.4985	.4985	.4986	.4986
3.0	.4987	.4987	.4987	.4988	.4988	.4989	.4989	.4989	.4990	.4990
3.1	.4990	.4991	.4991	.4991	.4992	.4992	.4992	.4992	.4993	.4993
3.2	.4993	.4993	.4994	.4994	.4994	.4994	.4994	.4995	.4995	.4995
3.3	.4995	.4995	.4995	.4996	.4996	.4996	.4996	.4996	.4996	.4997
3.4	.4997	.4997	.4997	.4997	.4997	.4997	.4997	.4997	.4997	.4998
3.5	.4998	.4998	.4998	.4998	.4998	.4998	.4998	.4998	.4998	.4998
3.6	.4998	.4998	.4999	.4999	.4999	.4999	.4999	.4999	.4999	.4999
3.7	.4999	.4999	.4999	.4999	.4999	.4999	.4999	.4999	.4999	.4999
3.8	.4999	.4999	.4999	.4999	.4999	.4999	.4999	.4999	.4999	.4999
3.9	.5000	.5000	.5000	.5000	.5000	.5000	.5000	.5000	.5000	.5000

2. t_α- Critical values of *t*-distribution

v	α						
	0.40	0.30	0.20	0.15	0.10	0.05	0.025
1	0.325	0.727	1.376	1.963	3.078	6.314	12.706
2	0.289	0.617	1.061	1.386	1.886	20920	4.303
3	0.277	0.584	0.978	1.250	1.638	2.353	3.182
4	0.271	0.569	0.941	1.190	1.533	2.132	2.776
5	0.267	0.559	0.920	1.156	1.476	2.015	2.571
6	0.265	0.553	0.906	1.134	1.440	1.943	2.447
7	0.263	0.549	0.896	1.119	1.415	1.895	2.365
8	0.262	0.546	0.889	1.108	1.397	1.860	2.306
9	0.261	0.543	0.883	1.100	1.383	1.833	2.262
10	0.260	0.542	0.879	1.093	1.372	1.812	2.228
11	0.260	0.540	0.876	1.088	1.363	1.796	2.201
12	0.259	0.539	0.873	1.083	1.356	1.782	2.179
13	0.259	0.537	0.870	1.079	1.350	1.771	2.160
14	0.258	0.537	0.868	1.076	1.345	1.761	2.145
15	0.258	0.536	0.866	1.074	1.341	1.753	2.131
16	0.258	0.535	0.865	1.071	1.337	1.746	2.120
17	0.257	0.534	0.863	1.069	1.333	1.740	2.110
18	0.257	0.534	0.862	1.067	1.330	1.734	2.101
19	0.257	0.533	0.861	1.066	1.328	1.729	2.093
20	0.257	0.533	0.860	1.064	1.325	1.725	2.086
21	0.257	0.532	0.859	1.063	1.323	1.721	2.080
22	0.256	0.532	0.858	1.061	1.321	1.717	2.074
23	0.256	0.532	0.858	1.060	1.319	1.714	2.069
24	0.256	0.531	0.857	1.059	1.318	1.711	2.064
25	0.256	0.531	0.856	1.058	1.316	1.708	2.060
26	0.256	0.531	0.856	1.058	1.315	1.706	2.056
27	0.256	0.531	0.855	1.057	1.314	1.703	2.052
28	0.256	0.530	0.855	1.056	1.313	1.701	2.048
29	0.256	0.530	0.854	1.055	1.311	1.699	2.045
30	0.256	0.530	0.854	1.055	1.310	1.697	2.042
40	0.255	0.529	0.851	1.050	1.303	1.684	2.021
60	0.254	0.527	0.848	1.045	1.296	1.671	2.000
120	0.254	0.526	0.845	1.041	1.289	1.658	1.980
∞	0.253	0.524	0.842	1.036	1.282	1.645	1.960

3. t_α-Critical values of t-distribution

v	α						
	0.02	0.015	0.01	0.0075	0.005	0.0025	0.0005
1	15.895	21.205	31.821	42.434	63.657	127.322	636.590
2	4.849	5.643	6.965	8.073	9.925	14.089	31.598
3	3.482	3.896	4.541	5.047	5.841	7.453	12.924
4	2.999	3.298	3.747	4.088	4.604	5.598	8.610
5	2.757	3.003	3.365	3.634	4.032	4.773	6.869
6	2.612	2.829	3.143	3.372	3.707	4.317	5.959
7	2.517	2.715	2.998	3.203	3.499	4.029	5.408
8	2.449	2.634	2.896	3.085	3.355	3.833	5.041
9	2.398	2.574	2.821	2.998	3.250	3.690	4.781
10	2.359	2.527	2.764	2.932	3.169	3.581	4.587
11	2.328	2.491	2.718	2.879	3.106	3.497	4.437
12	2.303	2.461	2.681	2.836	3.055	3.428	4.318
13	2.282	2.436	2.650	2.801	3.012	3.372	4.221
14	2.264	2.415	2.624	2.771	2.977	3.326	4.140
15	2.249	2.397	2.602	2.746	2.947	3.286	4.073
16	2.235	2.382	2.583	2.724	2.921	3.252	4.051
17	2.224	2.368	2.567	2.706	2.898	3.222	3.965
18	2.214	2.356	2.552	2.689	2.878	3.197	3.922
19	2.205	2.346	2.539	2.674	2.861	3.174	3.883
20	2.197	2.336	2.528	2.661	2.845	3.153	3.849
21	2.189	2.328	2.518	2.649	2.831	3.135	3.819
22	2.183	2.320	2.508	2.639	2.819	3.119	3.792
23	2.177	2.313	2.500	2.629	2.807	3.104	3.768
24	2.172	2.307	2.492	2.620	2.797	3.091	3.745
25	2.167	2.301	2.485	2.612	2.787	3.078	3.725
26	2.162	2.296	2.479	2.605	2.779	3.067	3.707
27	2.158	2.291	2.473	2.598	2.771	3.057	3.690
28	2.154	2.286	2.467	2.592	2.763	3.047	3.674
29	2.150	2.282	2.462	2.586	2.756	3.038	3.659
30	2.147	2.278	2.457	2.581	2.750	3.030	3.646
40	2.125	2.250	2.423	2.542	2.704	2.971	3.551
60	2.099	2.223	2.390	2.504	2.660	2.915	3.460
120	2.076	2.196	2.358	2.468	2.617	2.860	3.373
∞	2.054	2.170	2.326	2.432	2.576	2.807	3.291

4. **χ_{α}^{2}- Critical values of the chi-squared distribution**

v	α								
	0.995	0.99	0.98	0.975	0.95	0.90	0.80	0.75	0.75	0.50
1	0.04393	0.03157	0.03628	0.03982	0.00393	0.0158	0.0642	0.102	0.148	0.455
2	0.0100	0.0201	0.0404	0.103	0.103	0.211	0.446	0.575	0.713	1.386
3	0.0717	0.115	0.185	0.352	0.352	0.584	1.005	1.213	1.424	2.366
4	0.207	0.297	0.429	0.711	0.711	1.064	1.649	1.923	2.195	3.357
5	0.412	0.554	0.752	1.145	1.145	1.610	2.343	2.675	3.000	4.351
6	0.676	0.872	1.134	1.635	1.635	2.204	3.070	3.455	3.828	5.348
7	0.989	1.239	1.564	2.167	2.167	2.833	3.822	4.255	4.671	6.346
8	1.344	1.646	2.032	2.733	2.733	3.490	4.594	5.071	5.527	7.344
9	1.735	2.088	2.532	3.325	3.325	4.168	5.380	5.899	6.393	8.343
10	2.156	2.558	3.059	3.940	3.940	4.865	6.179	6.737	7.267	9.342
11	2.603	3.053	3.609	4.575	4.575	5.578	6.989	7.584	8.148	10.341
12	3.074	3.571	4.178	5.226	5.226	6.304	7.807	8.438	9.034	11.340
13	3.565	4.107	4.765	5.892	5.892	7.042	8.634	9.299	9.926	12.340
14	4.075	4.660	5.368	6.571	6.571	7.790	9.467	10.165	10.821	13.339
15	4.601	5.229	5.985	7.261	7.261	8.547	10.307	11.036	11.721	14.339
16	5.142	5.812	6.614	7.962	7.962	9.312	11.152	11.912	12.624	15.338
17	5.697	6.408	7.255	8.672	8.672	10.085	12.002	12.792	13.531	16.338
18	6.265	7.015	7.906	9.390	9.390	10.865	12.857	13.675	14.440	17.338
19	6.844	7.633	8.567	10.117	10.117	11.651	13.716	14.562	15.352	18.338
20	7.434	8.260	9.237	10.851	10.851	12.443	14.578	14.452	16.266	19.337
21	8.034	8.897	9.915	11.591	11.591	13.240	15.445	16.344	17.182	20.337
22	8.643	9.542	10.600	12.338	12.338	14.041	16.314	17.240	18.101	21.337
23	9.260	10.196	11.293	13.091	13.091	14.848	17.187	18.137	19.021	22.337
24	9.886	10.856	11.992	13.848	13.848	15.659	18.062	19.037	19.943	23.337
25	10.520	11.524	12.697	14.611	14.611	16.473	18.940	19.939	20.867	24.337
26	11.160	12.198	13.409	15.379	15.379	17.292	19.820	20.843	21.792	25.336
27	11.808	12.879	14.125	16.151	16.151	18.114	20.703	21.749	22.719	26.336
28	12.461	13.565	14.847	16.928	16.928	18.939	21.588	22.657	23.647	27.336
29	13.121	14.256	15.574	17.708	17.708	19.768	22.475	23.567	24.577	28.336
30	13.787	14.953	16.306	18.493	18.493	20.599	23.364	24.478	25.508	29.336

χ_α^2- **Critical values of the chi-squared distribution**

v	α									
	0.30	0.25	0.20	0.10	0.05	0.025	0.02	0.01	0.005	0.001
1	1.074	1.323	1.642	2.706	3.841	5.024	5.412	6.635	7.879	10.827
2	2.408	2.773	3.219	4.605	5.991	7.378	7.824	9.210	10.597	13.815
3	3.665	4.108	4.642	6.251	7.815	9.348	9.837	11.345	12.838	16.268
4	4.878	5.385	5.989	7.779	9.488	11.143	11.668	13.277	14.860	18.465
5	6.064	6.626	7.289	9.236	11.070	12.832	13.388	15.086	16.750	20.517
6	7.231	7.841	8.558	10.645	12.592	14.449	15.033	16.812	18.548	22.457
7	8.383	9.037	9.803	12.017	14.067	16.013	16.622	18.475	20.278	24.322
8	9.524	10.219	11.030	13.362	15.507	17.535	18.168	20.090	21.955	26.125
9	10.656	11.389	12.242	14.684	16.919	19.023	19.679	21.666	23.589	27.877
10	11.781	12.549	13.442	15.987	18.307	20.483	21.161	23.209	25.188	29.588
11	12.899	13.701	14.631	17.275	19.675	21.920	22.618	24.725	26.757	31.264
12	14.011	14.845	15.812	18.545	21.026	23.337	24.054	26.217	28.300	32.909
13	15.119	15.984	16.985	19.812	22.362	24.736	25.472	27.688	29.819	34.528
14	16.222	17.117	18.151	21.064	23.685	26.119	26.873	29.141	31.319	36.123
15	17.322	18.245	19.311	22.307	24.996	27.488	28.259	30.578	32.801	37.697
16	18.418	19.369	20.465	23.542	26.296	28.845	29.633	32.000	34.267	39.252
17	19.511	20.489	21.615	24.769	27.587	30.191	30.995	33.409	35.718	40.790
18	20.601	21.605	22.760	25.989	28.869	31.526	32.346	34.805	37.156	42.312
19	21.689	22.718	23.900	27.204	30.144	32.852	33.687	36.191	38.582	43.820
20	22.775	23.828	25.038	28.412	31.410	34.170	35.020	37.566	39.997	45.315
21	23.858	24.935	26.171	29.615	32.671	35.479	36.343	38.932	41.401	46.797
22	24.939	26.039	27.301	30.813	33.924	36.781	37.659	40.289	42.796	48.268
23	26.018	27.141	28.429	32.007	35.172	38.076	38.968	41.638	44.181	49.728
24	27.096	28.241	29.553	33.196	36.415	39.364	40.270	42.980	45.558	51.179
25	28.172	29.339	30.675	34.382	37.652	40.646	41.566	44.314	46.928	52.620
26	29.246	30.434	31.795	35.563	38.885	41.923	42.856	45.642	48.290	54.052
27	30.319	31.528	32.912	36.741	40.113	43.194	44.140	46.963	49.645	55.476
28	31.391	32.620	34.027	37.916	41.337	44.461	45.419	48.278	50.993	56.893
29	32.461	33.711	35.139	39.087	42.557	45.722	46.693	49.588	52.336	58.302
30	33.530	34.800	36.250	40.256	43.773	46.979	47.962	50.892	53.672	59.703

Values of $F_{0.05}(v_1, v_2)$										
v_2	v_1									
	10	12	15	20	24	30	40	60	120	∞
1	241.9	243.9	245.9	248.0	249.1	250.1	251.1	252.2	253.3	254.3
2	19.40	19.41	19.43	19.45	19.45	19.46	19.47	19.48	19.49	19.50
3	8.79	8.74	8.70	8.66	8.64	8.62	8.59	8.57	8.55	8.53
4	5.96	5.91	5.86	5.80	5.77	5.75	5.72	5.69	5.66	5.63
5	4.74	4.68	4.62	4.56	4.53	4.50	4.46	4.43	4.40	4.36
6	4.06	4.00	3.94	3.87	3.84	3.81	3.77	3.74	3.70	3.67
7	3.64	3.57	3.51	3.44	3.41	3.38	3.34	3.30	3.27	3.23
8	3.35	3.28	3.22	3.15	3.12	3.08	3.04	3.01	2.97	2.93
9	3.14	3.07	3.01	2.94	2.90	2.86	2.83	2.79	2.75	2.71
10	2.98	2.91	2.85	2.77	2.74	2.70	2.66	2.62	2.58	2.54
11	2.85	2.79	2.72	2.65	2.61	2.57	2.53	2.49	2.45	2.40
12	2.75	2.69	2.62	2.54	2.51	2.47	2.43	2.38	2.34	2.30
13	2.67	2.60	2.53	2.46	2.42	2.38	2.34	2.30	2.25	2.21
14	2.60	2.53	2.46	2.39	2.35	2.31	2.27	2.22	2.18	2.13
15	2.54	2.48	2.40	2.33	2.29	2.25	2.20	2.16	2.11	2.07
16	2.49	2.42	2.35	2.28	2.24	2.19	2.15	2.11	2.06	2.01
17	2.45	2.38	2.31	2.23	2.19	2.15	2.10	2.06	2.01	1.96
18	2.41	2.34	2.27	2.19	2.15	2.11	2.06	2.02	1.97	1.92
19	2.38	2.31	2.23	2.16	2.11	2.07	2.03	1.98	1.93	1.88
20	2.35	2.28	2.20	2.12	2.08	2.04	1.99	1.95	1.90	1.84
21	2.32	2.25	2.18	2.10	2.05	2.01	1.96	1.92	1.87	1.81
22	2.30	2.23	2.15	2.07	2.03	1.98	1.94	1.89	1.84	1.78
23	2.27	2.20	2.13	2.05	2.01	1.96	1.91	1.86	1.81	1.76
24	2.25	2.18	2.11	2.03	1.98	1.94	1.89	1.84	1.79	1.73
25	2.24	2.16	2.09	2.01	1.96	1.92	1.87	1.82	1.77	1.71
26	2.22	2.15	2.07	1.99	1.95	1.90	1.85	1.80	1.75	1.69
27	2.20	2.13	2.06	1.97	1.93	1.88	1.84	1.79	1.73	1.67
28	2.19	2.12	2.04	1.96	1.91	1.87	1.82	1.77	1.71	1.65
29	2.18	2.10	2.03	1.94	1.90	1.85	1.81	1.75	1.70	1.64
30	2.16	2.09	2.01	1.93	1.89	1.84	1.79	1.75	1.68	1.62
40	2.08	2.00	1.92	1.84	1.79	1.74	1.69	1.64	1.58	1.51
60	1.99	1.92	1.84	1.75	1.70	1.65	1.59	1.53	1.47	1.39
120	1.91	1.83	1.75	1.66	1.61	1.55	1.50	1.43	1.35	1.25
∞	1.83	1.75	1.67	1.57	1.52	1.46	1.39	1.32	1.22	1.00

Values of $F_{0.01}(v_1, v_2)$

v_2	v_1								
	1	2	3	4	5	6	7	8	9
1	405.2	4999.5	5403	5625	5764	5859	5928	5981	6022
2	98.50	99.00	99.17	99.25	99.30	99.33	99.36	99.37	99.39
3	34.12	30.82	29.46	28.71	28.24	27.91	27.67	27.67	27.35
4	21.20	18.00	16.69	15.98	15.52	15.21	14.98	14.80	14.66
5	16.26	13.27	12.06	11.39	10.97	10.67	10.46	10.29	10.16
6	13.75	10.92	9.78	9.15	8.75	8.47	8.26	8.10	7.98
7	12.25	9.55	8.45	7.85	7.46	7.19	6.99	6.84	6.72
8	11.26	8.65	7.59	7.01	6.63	6.37	6.18	6.03	5.91
9	10.56	8.02	6.99	6.42	6.06	5.80	5.61	5.47	5.35
10	10.04	7.56	6.55	5.99	5.64	5.39	5.20	5.06	4.94
11	9.65	7.21	6.22	5.67	5.32	5.07	4.89	4.74	4.63
12	9.33	6.93	5.95	5.41	5.06	4.82	4.64	4.50	4.39
13	9.07	6.70	5.74	5.21	4.86	4.62	4.44	4.30	4.19
14	8.86	6.51	5.56	5.04	4.69	4.46	4.28	4.14	4.03
15	8.68	6.36	5.42	4.89	4.56	4.32	4.14	4.00	3.89
16	8.53	6.23	5.29	4.77	4.44	4.20	4.03	3.89	3.78
17	8.40	6.11	5.18	4.67	4.34	4.10	3.93	3.79	3.68
18	8.29	6.01	5.09	4.58	4.25	4.01	3.84	3.71	3.60
19	8.18	5.93	5.01	4.50	4.17	3.94	3.77	3.63	3.52
20	8.10	5.85	4.94	4.43	4.10	3.87	3.70	3.56	3.46
21	8.02	5.78	4.87	4.37	4.04	3.81	3.64	3.51	3.40
22	7.95	5.72	4.82	4.31	3.99	3.76	3.59	3.45	3.35
23	7.88	5.66	4.76	4.26	3.94	3.71	3.54	3.41	3.30
24	7.82	5.61	4.72	4.22	3.90	3.67	3.50	3.36	3.26
25	7.77	5.57	4.68	4.18	3.85	3.63	3.46	3.32	3.22
26	7.72	5.53	4.64	4.14	3.82	3.59	3.42	3.29	3.18
27	7.68	5.49	4.60	4.11	3.78	3.56	3.39	3.26	3.15
28	7.64	5.45	4.57	4.07	3.75	3.53	3.36	3.23	3.12
29	7.60	5.42	4.54	4.04	3.73	3.50	3.33	3.20	3.09
30	7.56	5.39	4.51	4.02	3.70	3.47	3.30	3.17	3.07
40	7.31	5.18	4.31	3.83	3.51	3.29	3.12	2.99	2.89
60	7.08	4.98	4.13	3.65	3.34	3.12	2.95	2.82	2.72
120	6.85	4.79	3.95	3.48	3.17	2.96	2.79	2.66	2.56
∞	6.63	4.61	3.78	3.32	3.02	2.80	2.80	2.51	2.41

Values of $F_{0.01}(v_1, v_2)$										
v_2	v_1									
	10	12	15	20	24	30	40	60	120	∞
1	6056	6106	6157	6209	6235	6261	6287	6313	6339	6366
2	99.40	99.42	99.43	99.46	99.46	99.47	99.47	99.48	99.49	99.50
3	27.23	27.05	26.87	26.60	26.60	26.50	26.41	26.32	26.22	26.13
4	14.55	14.37	14.20	13.93	13.93	13.84	13.75	13.65	13.56	13.46
5	10.05	9.89	9.72	9.47	9.47	9.38	9.29	9.20	9.11	9.02
6	7.87	7.72	7.56	7.31	7.31	7.23	7.14	7.06	6.97	6.88
7	6.62	6.47	6.31	6.07	6.07	5.99	5.91	5.82	5.74	5.65
8	5.81	5.67	5.52	5.28	5.28	5.20	5.12	5.03	4.95	4.86
9	5.26	5.11	4.96	4.73	4.73	4.65	4.57	4.48	4.40	4.31
10	4.85	4.71	4.56	4.33	4.33	4.25	4.17	4.08	4.00	3.91
11	4.54	4.40	4.25	4.02	4.02	3.94	3.86	3.78	3.69	3.60
12	4.30	4.16	4.01	3.78	3.78	3.70	3.62	3.54	3.45	3.36
13	4.10	3.96	3.82	3.59	3.59	3.51	3.43	3.34	3.25	3.17
14	3.94	3.80	3.66	3.43	3.43	3.35	3.27	3.18	3.09	3.00
15	3.80	3.67	3.52	3.29	3.29	3.21	3.13	3.05	2.96	2.87
16	3.69	3.55	3.41	3.18	3.18	3.10	3.02	2.93	2.84	2.75
17	3.59	3.46	3.31	3.08	3.08	3.00	2.92	2.83	2.75	2.65
18	3.51	3.37	3.23	3.00	3.00	2.92	2.84	2.75	2.66	2.57
19	3.43	3.30	3.15	2.92	2.92	2.84	2.76	2.67	2.58	2.49
20	3.37	3.23	3.09	2.86	2.86	2.78	2.69	2.61	2.52	2.42
21	3.31	3.17	3.03	2.80	2.80	2.72	2.64	2.55	2.46	2.36
22	3.26	3.12	2.98	2.75	2.75	2.67	2.58	2.50	2.40	2.31
23	3.21	3.07	2.93	2.70	2.70	2.62	2.54	2.45	2.35	2.26
24	3.17	3.03	2.89	2.66	2.66	2.58	2.49	2.40	2.31	2.21
25	3.13	2.99	2.85	2.62	2.62	2.54	2.45	2.36	2.27	2.17
26	3.09	2.96	2.81	2.58	2.58	2.50	2.42	2.33	2.23	2.13
27	3.06	2.93	2.78	2.55	2.55	2.47	2.38	2.29	2.20	2.10
28	3.03	2.90	2.75	2.52	2.52	2.44	2.35	2.26	2.17	2.06
29	3.00	2.87	2.73	2.49	2.49	2.41	2.33	2.23	2.14	2.03
30	2.98	2.84	2.70	2.47	2.47	2.39	2.30	2.21	2.11	2.01
40	2.80	2.66	2.52	2.29	2.29	2.20	2.11	2.02	1.92	1.80
60	2.63	2.50	2.35	2.12	2.12	2.03	1.94	1.84	1.73	1.60
120	2.47	2.34	2.19	1.95	1.95	1.86	1.76	1.66	1.53	1.38
∞	2.32	2.18	2.04	1.79	1.79	1.70	1.59	1.47	1.32	1.00